# 1987 FLUID POWER STANDARDS

## VOLUME F
## CONTROL PRODUCTS

**Shirley C. Seal, Editor**

**National Fluid Power Association, Inc.
Milwaukee, Wisconsin**

Copyright 1987 by the

NATIONAL FLUID POWER ASSOCIATION, INC.

Printed in the USA

All technical reports, citations, references and related data including standards and practices approved and/or recommended are advisory only. Use thereof by anyone for any purpose is entirely voluntary and in any event without risk of any nature to the National Fluid Power Association, Inc., its officers, directors or authors of such work. There is no agreement by or between anyone to adhere to any NFPA Recommended Standard, policy or practice, and related matters. In formulating and approving technical reports, the Technical Board, its councils and committees and/or the National Fluid Power Association, Inc. will not investigate or consider citations, references or patents which may or may not apply to such subject matter since prospective users of such reports and data alone are responsible for establishing necessary safeguards in connection with utilization of such matters, including technical data, proprietary rights or patentable materials.

Recommended standards and/or policies and procedures are subject to periodic review and may be changed without notice. Recommended standards, after publication, may be revised or withdrawn at any time and current information on all approved recommended standards may be received by calling or writing the National Fluid Power Association, Inc.

An approved NFPA Recommended Standard implies a consensus of those substantially concerned with its scope and provisions and is intended as a guide to aid the manufacturer, the consumer and the general public. The publication of the NFPA Recommended Standard does not in any respect preclude anyone, whether they have participated in the development of or approved the recommended standard or not, from manufacturing, marketing purchasing or using products, processes or procedures not conforming to the recommended standard.

Participation by federal agency representative(s) or person(s) affiliated with industry is not to be interpreted as government or industry endorsement of this standard and/or policy and procedure.

This publication may not be reproduced in whole or in part without the written permission of the National Fluid Power Association, Inc.

```
Sci Ref
TJ
840
.N55
1987
v.F
```

ISBN 0-942220-82-X (Set)
ISBN 0-942220-88-9 (Volume F)

**NATIONAL FLUID POWER ASSOCIATION, INC., 1987**

# Contents

NFPA — Providing Services for the U.S. Fluid Power Industry ... v

Excerpts extracted from American National Standard Fluid power systems and products - Glossary, ANSI/B93.2-1986 ... 1

Glossary of Terms for Compressed Air Dryers, NFPA/T3.27.1-1972 (R1981) ... 31

Procedure for Self-Certification by Fluid Power Manufacturers, NFPA/T1.21.1-1978 (R1983) ... 43

Method for Verifying the Fatigue and Static Pressure Ratings of the Pressure Containing Envelope of a Metal Fluid Power Component, NFPA/T2.6.1-1974 (R1982) ... 51

Pneumatic Valve Pressure Rating Supplement No. 2 to NFPA Recommended Standard for Verifying the Fatigue and Static Pressure Ratings of the Pressure Containing Envelope of a Metal Fluid Power Component, NFPA/T2.6.1M S2 (T3.21.4M)-1977 (R1982) ... 81

Pressure Switch Pressure Rating Supplement No. 3 to NFPA Recommended Standard for Verifying the Fatigue and Static Pressure Ratings of the Pressure Containing Envelope of a Metal Fluid Power Component, NFPA/T2.6.1M S3 (T3.29.2M)-1976 (R1982) ... 99

Air Line Filter, Regulator and/or Lubricator Pressure Rating Supplement No. 4 to NFPA Recommended Standard for Verifying the Fatigue and Static Pressure Ratings of the Pressure Containing Envelope of a Metal Fluid Power Component, NFPA/T2.6.1M S4 (T3.12.10M)-1976 (R1981) ... 111

Hydraulic Valve Pressure Rating Supplement No. 9 to NFPA Recommended Standard for Verifying the Fatigue and Static Pressure Ratings of the Pressure Containing Envelope of a Metal Fluid Power Component, NFPA/T2.6.1M S9 (T3.5.26)-1977 (R1982) ... 121

American National Standard Hydraulic Fluid Power - Valves - Mounting interfaces ANSI/B93.7M-1986 ... 139

American National Standard Symbols for Marking Electrical Leads and Ports on Fluid Power Valves, ANSI/B93.9M-1969 (R1981) ... 185

American National Standard Series of Mounting Interfaces for 4567 Maximum psi (315 bar) Four-Port Hydraulic Fluid Power Directional Valves, ANSI/B93.40M-1976 (R1982) ... 199

American National Standard Hydraulic Fluid Power - Directional Control Valve - Method for Determining the Metering Characteristics, ANSI/B93.66M-1983 ... 209

American National Standard Hydraulic Fluid Power - Valves - Pressure Differential-flow Characteristic - Method of Measuring and Reporting, ANSI/B93.49-1980 ... 221

American National Standard Hydraulic Fluid Power - Solenoid-piloted Industrial Valves - Interface Dimensions for Electrical Connectors, ANSI/B93.55M-1981 ... 227

Hydraulic fluid power - Cylinder actuator mounted valves - Standard dimensions for mounting surfaces NFPA/T3.5.33M-1985 ... 235

American National Standard Hydraulic Fluid Power - Code for Identification of Valve Mounting Surfaces, ANSI/B93.65M-1983 ... 245

Graphic Symbols for Fluidic Devices and Circuits NFPA/T3.7.2M-1968 (R1980) ... 251

American National Standard Methods for Presenting Basic Performance Data for Fluidic Devices, ANSI/B93.14M-1971 (R1979) ... 267

American National Standard Requirements for Presentation of Catalog Data, Fluid Compatibility, Cleaning Media, Markings and Dimensional Identification Codes and Pressure Drop Characteristics for Fluid Power Air Line Filters, ANSI/B93.39M-1978 (R1986) ... 301

American National Standard Pneumatic Fluid Power - Pressure Regulators - Industrial Type, ANSI/B93.13M-1981 ... 317

American National Standard Interfaces for 4-Way General Purpose Industrial Pneumatic Directional Control Valves, ANSI/B93.33M-1974 (R1981) ... 333

Defining Interface Surfaces for each Pneumatic Valve Interface in NFPA Recommended Standard T3.21.1-1973, NFPA/T3.21.7M-1976 (R1981) ... 345

Definition of Port Communication for the Fluid Power Pneumatic Valve Interface to NFPA Recommended Standard T3.21.1 with the Valve in Position in Response to a Remote Pilot Signal or Electrical Energization, NFPA/T3.21.9M-1976 (R1981) ... 353

American National Standard Pneumatic Fluid Power - Five-Port Directional Control Valves - Mounting Surfaces - Optional Electrical Connector - Dimensions and Requirements, ANSI/B93.67M-1983 ... 363

# Contents continued

American National Standard Pneumatic Fluid Power - Compressed Air Dryers - Methods for Rating and Testing, ANSI/B93.45M-1982 . . . . . . . . . . . . . . . . . . . . . 371

American National Standard Method of Diagramming for Moving Parts Fluid Controls, ANSI/B93.38-1976 (R1981) . . . . . . . . . . . . . . . 381

A Bibliography of Hydraulic Valve Standards and Test Procedures, NFPA/T3.5.27-1976 (R1982) . . . . 445

A Bibliography of Fluid Power Pneumatic FRL Standards, NFPA/T3.12.9-1977 . . . . . . . . . . . . . . 459

A Bibliography of Fluid Power Pneumatic Valve Standards, NFPA/T3.21.5-1978 . . . . . . . . . . . . . . 465

A Bibliography of Compressed Air Dryers Standards, NFPA/T3.27.4-1979 . . . . . . . . . . . . . . 471

A Bibliography of Fluid Logic Devices NFPA/T3.28.11-1982 . . . . . . . . . . . . . . . . . . . . . 479

# Foreword

## FLUID POWER STANDARDS, Seventh Edition

Fluid Power Standards is comprised of ten separate volumes available as a set or separately. The titles of the volumes are:

| | |
|---|---|
| Volume A | Communication including graphic symbols and metric units |
| Volume B | Pressure Rating |
| Volume C | Pumps, Motors, Power Units and Reservoirs |
| Volume D | Filtration and Contamination |
| Volume E | Conductors and Associated Products |
| Volume F | Control Products |
| Volume G | Cylinders and Accumulators |
| Volume H | Fluids, Lubricants and Sealing Devices |
| Volume I | Testing |
| Volume J | Bibliographies |

## ORDERING INFORMATION

The volumes of Fluid Power Standards can be ordered by contacting the National Fluid Power Associations Publications Department, 3333 N. Mayfair Rd., Milwaukee, Wisconsin 53222 USA Telephone (414) 778-3344 Telex 26-898.

## WHAT IS CONTAINED IN FLUID POWER STANDARDS

NFPA's ten volume set, Fluid Power Standards contains both NFPA and ANSI (American National Standards Institute) Fluid Power standards.

## HOW FLUID POWER STANDARDS ARE DEVELOPED

The National Fluid Power Association coordinates development of Fluid Power standards on industry, national and international levels. The association holds the Secretariat for the Fluid Power Committee work of the American National Standards Institute (ANSI) and the International Organization for Standardization (ISO).

### On the Industry Level (NFPA)

NFPA standards originate in Product Sections and Technology Committees, and are submitted to the NFPA Technical Board and Board of Directors for approval. Product Sections are generally composed of individuals involved in the design, manufacture, performance and application of specific Fluid Power Products. Technology Committees are comprised of experts in a single broad area of Fluid Power technology, applying to many products.

More than 350 persons currently contribute to the standards writing work of NFPA's technical committees and are helping to develop approximately 140 projects as proposed NFPA Recommended Standards, Information Reports and Recommended Practices. Qualified engineers from NFPA member companies are eligible for participation in NFPA technical committees.

### On the National Level (ANSI)

After a standard has been approved by the NFPA, it may be submitted to Standards Committee B93, Fluid Power Systems and Products, accredited by the American National Standards Institute. NFPA serves as secretariat for this committee, which ballots the standard for approval as an American National Standard. To date, more than 60 NFPA Recommended Standards have advanced to become American National Standards.

### On the International Level (ISO)

Under authority granted by the International Organization for Standardization and the American National Standards Institute, NFPA administers the Secretariat of ISO Technical Committee 131 (TC 131) Fluid power systems. More than 35 nations participate in standards writing through TC 131. The U. S. Fluid Power Industry is represented through USA TAG to ISO/TC 131, a committee also administered by the NFPA.

On an average, the NFPA coordinates 44 deliberating sessions per year in various parts of the industrialized world.

## Participating in Standards Development

Participation in the NFPA, ANSI/B93 and ISO/TC 131 committees is open to qualified engineers. For details on how to become involved, contact the Director of Technical Services at National Fluid Power Association, 3333 North Mayfair Rd., Milwaukee, Wisconsin 53222 USA Telephone (414) 778-3344 Telex 26-898.

## Five Year Review of Fluid Power Standards

An NFPA Recommended Standard is subject to revision at anytime by the appropriate technical committee. They are reviewed every five years and if not revised, either reaffirmed or withdrawn. Comments are invited either for revision of any standard or for additional standards. Comments should be addressed to the Director of Technical Services at NFPA Headquarters. Comments will receive careful consideration at a meeting of the appropriate technical committee. Commentators are welcome to attend the meeting.

## Obsolete Fluid Power Standards

The 1987 edition of these Fluid Power Standards volumes makes the 1984 edition obsolete. For practical purposes, it is not wise to use obsolete volumes. However, for teaching purposes, the outdated volumes might be useful. Standards contained in each volume have an NFPA or ANSI reference number that includes the year the standard was first approved and the year it was reaffirmed.

## Disclaimer

See complete disclaimer on Copyright page (reverse of Title Page).

## NATIONAL FLUID POWER ASSOCIATION

The National Fluid Power Association is the trade association for manufacturers of hydraulic and pneumatic products and systems.

Founded in 1953, NFPA has 190 corporate members across the U.S. NFPA members produce more than 75 percent of all U.S. Fluid Power production.

NFPA was founded to pursue activities that advance the performance and application of Fluid Power products, better materials, designs and standards, organize committees to further the art and science of Fluid Power, support activities to advance knowledge and understanding of Fluid Power, and represent the members to the Federal Government, industry and user organizations.

Association activities in the areas of technical standards and services, marketing research and statistics, public affairs, industry promotion and management are directed by volunteer Boards and supported by full-time professional headquarters staff.

NFPA provides administrative services to these organizations in the Fluid Power Industry:

- Fluid Power Committee B93, the committee for national fluid power standards, accredited by American National Standards Institute,
- ISO/TC 131 Fluid power systems, the committee for international fluid power standards, of the International Organization for Standardization,
- USA Technical Advisory Group (USA TAG) to ISO/TC 131, the national committee for international Fluid Power standards,
- Fluid Power Appeals Boards,
- Fluid Power Coordinating Council,
- Fluid Power Educational Foundation,
- International Fluid Power Exposition
- International Fluid Power Exposition Applications Conference.

For information on membership in NFPA, or participation of any of the Fluid Power technical committees, contact the National Fluid Power Association, 3333 North Mayfair Rd., Milwaukee, Wisconsin 53222 USA (414) 778-3344 Telex 26-898.

- NOTES -

ANSI/B93.2-1986

---

**AN INDUSTRY STANDARD FOR FLUID POWER**


Excerpts Extracted from ANSI/B93.2

American National Standard

Fluid power systems and products -

Glossary

---

published by
# NATIONAL FLUID POWER ASSOCIATION, INC.
3333 N. Mayfair Road  /  Milwaukee, WI 53222  /  414-778-3344  /  TLX 26898

## GROUP 01 — PRIMARY TERMS

**01.01.080 Fluid Logic**

A branch of fluid power associated with digital signal sensing and information processing, using components with or without moving parts.

**01.01.100 Fluid Power**

Energy transmitted and controlled through use of a pressurized fluid.

**01.01.200 Fluid Power System**

A system that transmits and controls power through use of a pressurized fluid within an enclosed circuit.

**01.01.300 Fluidics**

Engineering science pertaining to the use of fluid dynamic phenomena to sense, control process information, and/or actuate.

**01.01.400 Hydraulics**

Engineering science pertaining to liquid pressure and flow.

**01.01.500 Hydrodynamics**

The engineering science which governs the movement of liquids and the forces opposing that movement.

**01.01.600 Hydrokinetics**

Engineering science pertaining to the energy of liquid flow and pressure.

**01.01.700 Hydropneumatics**

Pertaining to the combination of hydraulic and pneumatic fluid power.

**01.01.800 Hydrostatics**

Engineering science pertaining to the energy of liquids at rest.

**01.01.801 Hydrostatic Transmission**

Combination of one or more hydraulic pumps and motors forming a unit.

**01.01.850 Moving Parts Logic**

The technology of achieving logic control by means of fluid devices having moving parts.

**01.01.900 Pneumatics**

Engineering science pertaining to gaseous pressure and flow.

**01.01.910 Time**

An interval comprising a limited but continuous action.

**01.01.911 Time, Actuated**

The interval in which the component is under the influence of the actuating forces.

**01.01.912 Time, Fall**

The interval taken in a device for a quantity to change from a specified high level down to a specified lower level.

**01.01.913 Time, Operation**

The interval of an event measured from "start signal" to final "at rest".

**01.01.914 Time, Released**

The duration of time in which the component is not under the influence of the actuating forces.

**01.01.915 Time, Relative Duty**

(Expressed as a percentage)

$$\frac{t\,(\text{actuated})}{t\,(\text{actuated}) + t\,(\text{released})} \times 100$$

**01.01.916 Time, Response**

The elapsed time between the initiation of an action and the resulting reaction (measured under specified conditions).

**01.01.918 Time, Rise**

The interval in a device for a quantity to change from a specified low level up to a specified high level.

**01.01.919 Time, Start Up**

The interval needed to reach a steady state operating condition in the system from "start up".

**01.01.930 Torque**

Turning effort.

**01.01.931 Torque, Derived**

Torque corresponding to the derived hydraulic power.

**01.01.932 Torque, Effective**

Actual torque transmitted by the shaft under specified conditions.

**01.01.933 Torque, Geometric**

Torque corresponding to the geometric hydraulic power.

# GROUP 02 - FLUID POWER LAWS AND RELATED TERMS

**02.01.100 Bernoulli's Law**

If no work is done on or by a flowing frictionless liquid its energy due to pressure and velocity remains constant at all points along the streamline.

**02.01.200 Boyle's Law**

The absolute pressure of a fixed mass of gas varies inversely as the volume, provided the temperature remains constant.

**02.01.300 Charles' Law**

The volume of a fixed mass of gas varies directly with absolute temperature, provided the pressure remains constant.

**02.01.400 Continuity Equation**

The mass rate of fluid flow into a fixed space is equal to the mass flow rate out. Hence, the mass flow rate of fluid past all cross sections of a conduit is equal.

**02.01.500 Darcy's Formula**

A formula used to determine the pressure drop due to flow friction through a conduit.

$$h_f = \frac{fLv^2}{2Dg}$$

- $h_f$ = Head loss, feet (metre)
- $f$ = Friction factor (see 02.01.600)
- $L$ = Length of conduit, feet (metre)
- $v$ = Mean velocity of flow, ft/sec. (metre/sec.)
- $D$ = Internal diameter of conduit, feet (metre)
- $g$ = Acceleration due to gravity, 32.2 ft./sec.$^2$ (9.81 metre/sec.$^2$)

**02.01.600 Hagen Poiseuille Law**

The friction factor of Darcy's Formula is a ratio of 64 to the Reynolds Numbers when flow is laminar.

$$f = \frac{64}{N_r}$$

- $f$ = Friction factor
- $N_r$ = Reynolds Number

**02.01.700 Pascal's Law**

A pressure applied to a confined fluid at rest is transmitted with equal intensity throughout the fluid.

**02.01.800 Reynolds Number**

A numerical ratio of the dynamic forces of mass flow to the shear stress due to viscosity. Flow usually changes from laminar to turbulent between Reynolds Number 2,000 and 4,000.

$$N_r = \frac{\rho v D}{u} = \frac{vD}{\nu}$$

- $N_r$ = Reynolds Number
- $\rho$ (Rho) = Fluid density, pounds(mass)/ft.$^3$ (Kg/m$^3$)
- $v$ = Mean velocity of flow, ft./sec. (metre/sec.)
- $D$ = Internal diameter of conduit, feet (metre)
- $u$ (Mu) = Absolute viscosity, pounds/ft.sec. (Kg/cm$^5$)
- $\nu$ (Nu) = Kinematic viscosity, ft.$^2$/sec. (mm$^2$/sec.)

**02.01.900 Toricelli's Theorem**

The liquid velocity at an outlet discharging into the free atmosphere is proportional to the square root of the head.

$$v = \sqrt{2gh}$$

- $v$ = Velocity (mean), ft./sec. (m/sec.)
- $g$ = Acceleration, 32.2 ft./sec.$^2$ (9.81 m/sec.$^2$)
- $h$ = Pressure head, ft(m)

## GROUP 03 — FLOW TERMS

**03.01.200  Cavitation**

A localized gaseous condition within a liquid stream which occurs where the pressure is reduced to the vapor pressure.

**03.01.210  Coanda Effect**

The phenomenon, named after its discoverer, of the attachment of a free flowing turbulent jet to an adjacent, possibly curved, wall.

**03.01.409  Flow**

Movement of fluid generated by pressure differences.

**03.01.410  Flow, Laminar (Streamline)**

A flow situation in which fluid moves in parallel lamina or layers.

**03.01.420  Flow, Metered**

Flow at a controlled rate.

**03.01.430  Flow, Steady State**

A flow situation wherein conditions such as pressure, temperature, and velocity at any point in time do not change.

**03.01.450  Flow, Turbulent**

A flow situation in which the fluid particles move in a random fluctuating manner.

**03.01.460  Flow, Upsteady**

A flow situation wherein conditions such as pressure, temperature and velocity at points in the fluid change.

**03.01.600  Shock Wave**

A pressure wave front which moves at a sonic velocity.

**03.01.800  Surge**

A transient rise of pressure or flow.

## GROUP 04 — MENSURATION TERMS

**04.01.005  Air, Free**

Air at ambient temperature, pressure, relative humidity, and density.

**04.01.010  Air, Standard**

Air at a temperature of 68°F, a pressure of 14.70 pounds per square inch absolute, and a relative humidity of 36% (0.0750 pounds per cubic foot). In gas industries the temperature of "standard air" is usually given as 60°F.

**04.01.011  Amplification**

The ratio between the output signal variations and the control signal variations (for analogue devices only).

**04.01.012  Amplification, Flow**

Ratio between the output flow and the input (control) flow.

**04.01.013  Amplification, Power**

The ratio between the output power variation and the corresponding input (control) power variation (for analogue devices only).

**04.01.014  Amplification, Pressure**

Ratio between the outlet pressure and the inlet (control) pressure.

**04.01.020  Aniline Point**

The lowest temperature at which a liquid is completely miscible with an equal volume of freshly distilled aniline (ASTM Designation D611-64).

**04.01.025  Assurance Level**

The minimum percentage of pressure containing envelopes of a verified design that will sustain 10 million applications of its Rated Fatigue Pressure.

**04.01.028  Bistable**

A binary circuit or device which has two stable states and which in each state requires an appropriate impulse to cause a transition to the other state.

**04.01.030  Bulk Modulus**

The measure of resistance to compressibility of a fluid. It is the reciprocal of the compressibility.

**04.01.035  Capacity, Effective**

Actual volume displaced under specified conditions.

**04.01.036  Capacity, Geometric**

Volume displaced, calculated geometrically without reference to tolerances, clearances or deformation.

## 04.01.040 Compressibility

The change in volume of a unit volume of a fluid when subjected to a unit change in pressure.

## 04.01.041 Confidence Level

The degree to which a manufacturer can be assured that the desired assurance level is attained.

## 04.01.043 Displacement, Volumetric

Volume absorbed or displaced per stroke of a cylinder (See 81.01.012) or per cycle of a pump or motor (43.01.005).

## 04.01.046 Efficiency

Ratio of output to the corresponding input.

## 04.01.047 Expectancy, Life

The predicted working period during which a component or system will maintain a specified level of performance under specified conditions. Sometimes expressed in statistical terms as a probability.

## 04.01.048 Fire Point

The temperature to which a fluid must be heated to ignite and burn for five seconds (minimum) in the presence of air when a small flame is applied under controlled conditions.

## 04.01.050 Flash Point

The temperature to which a liquid must be heated under specified conditions of the test method to give off sufficient vapor to form a mixture with air that can be ignited momentarily by a flame.

## 04.01.051 Flow Degradation Ratio

The ratio of stabilized flow rate after a contaminant injection to the initial measured pump flow (Qr).

## 04.01.052 Flow Factor

Characterizes the conductance of a pneumatic or hydraulic device, flowline or connection.

## 04.01.060 Flow Rate

The volume, mass or weight of a fluid passing through any conductor per unit of time.

## 04.01.061 Flow Rate, Relief

The rate at which fluid can flow through the unloading device for each specific increase in controlled pressure above the original setting, measured under specified conditions.

## 04.01.063 Flow, Supply Port

The flow of fluid through the supply ports of the device or system.

## 04.01.070 Fluid Friction

Friction due to the viscosity of fluids.

## 04.01.072 Frequency Response

The changes, under steady-state conditions, in the output variable which are caused by a sinusoidal input variable.

## 04.01.170 Hammer, Liquid

Pressure and depression waves created by relatively rapid flow changes and transmitted through the system.

## 04.01.100 Head

The height of a column or body of fluid above a given point expressed in linear units. Head is often used to indicate gage pressure. Pressure is equal to the height times the density of the fluid.

## 04.01.110 Head, Friction

The head required to overcome the friction at the interior surface of a conductor and between fluid particles in motion. It varies with flow, size, type and condition of conductors and fittings, and the fluid characteristics.

## 04.01.115 Head, Pressure

The pressure due to the height of a column or body of fluid. (It is usually expressed in inches [mm]).

## 04.01.120 Head, Static

The height of a column or body of fluid above a given point.

## 04.01.130 Head, Static Discharge

The static head from the centerline of the pump to the free discharge surface.

## 04.01.140 Head, Static Suction

The head from the surface of the supply source to the centerline of the pump.

## 04.01.150 Head, Total Suction

The static head from the surface of the supply source to the free discharge surface.

### 04.01.160　Head, Velocity

The equivalent head through which the liquid would have to fall to attain a given velocity. Mathematically it is equal to the square of the velocity (in feet) divided by 64.4 feet per second squared.

$$h = \frac{v^2}{2g}$$

h = Head, feet (metre)
g = Acceleration due to gravity, 32.2 ft./sec.$^2$ (9.81 metre/sec.$^2$)
v = Mean velocity of flow, ft./sec. (metre/sec.)

### 04.01.200　Hydraulic Power

Power computed from flow rate and pressure differential (drop).

Hydraulic Power = .000583 $q_v p$ (0,002234 $q_v p$) expressed as horsepower where:
$q_v$ = Flow rate, gpm (L/min.)
p = Pressure, psi (bar)

Alternate formula = $\frac{q_v p}{1714}$ $\left(\frac{q_v p}{447.6}\right)$

### 04.01.210　Lift

The height of a column or body of fluid below a given point expressed in linear units. Lift is often used to indicate vacuum or pressure below atmosphere.

### 04.01.220　Lift, Static Suction

The lift from the centerline of the pump to the surface of the supply source. (See Head, Static Suction).

### 04.01.222　Linear Function

Describes a condition in which the relationship between two interdependent variables is constant.

### 04.01.223　Linear Region

The region of a given control characteristic over which the linearity remains within specified limits.

### 04.01.224　Linearity

The faithfulness with which an output signal of an electronic reproducing system reproduces an input signal.

### 04.01.230　Monostable

A binary circuit or device, which has one stable state and which requires an appropriate change of the input to cause a transition out of its stable state for a specified period of time. The specified period of time at which the circuit stays out of its stable state is independent of the duration of the appropriate change of the input signals.

### 04.01.250　Neutralization Number

A measure of the total acidity or basicity of an oil; this includes organic or inorganic acids or bases or a combination thereof (ASTM Designation D974-64).

### 04.01.260　Newt

A unit of kinematic viscosity in the English system. It is expressed in square inches per second (see 04.01.630 Stokes).

### 04.01.270　Poise

The standard unit of dynamic viscosity in the c.g.s. (centimeter-gram-second) system. It is the ratio of the shearing stress to the shear rate of fluid and is expressed in millipascal sec. (= 1 centipoise).

### 04.01.280　Pour Point

The lowest temperature at which a liquid will flow under specified conditions (ASTM Designation D97-66).

### 04.01.282　Power Consumption

The total power consumed by the device or system under specified conditions.

### 04.01.283　Power Supply, Fluid

Energy source which generates and maintains a flow of fluid under pressure.

### 04.01.290　Precipitation Number

The number of millilitres of precipitate formed when 10 mL of lubricating oil are mixed with 90 mL of ASTM precipitation naptha and centrifuged under prescribed conditions (ASTM Designation D91-61).

### 04.01.300　Pressure

Force per unit area, usually expressed in pounds per square inch (bar).

### 04.01.310　Pressure, Absolute

The pressure above zero absolute, i.e., the sum of atmospheric and gage pressure. In vacuum related work it is usually expressed in millimetres of mercury (mm Hg).

### 04.01.320　Pressure, Atmospheric

Pressure exerted by the atmosphere at any specific location. (Sea level pressure is approximately 14.7 pounds per square inch absolute, 1 bar = 14.5 psi).

### 04.01.330　Pressure, Back

The pressure encountered on the return side of a system.

**04.01.340        Pressure, Breakloose (Breakout)**

The minimum pressure which initiates movement.

**04.01.360        Pressure, Burst**

The pressure which causes failure of and consequential loss of fluid through the product envelope.

**04.01.370        Pressure, Charge**

The pressure at which replenishing fluid is forced into a fluid power system.

**04.01.375        Pressure, Control Range**

The permissible limits between which system pressure may be set.

**04.01.380        Pressure, Cracking**

The pressure at which a pressure operated valve begins to pass fluid.

**04.01.385        Pressure, Cyclic Test**

A pressure range applied in cyclic tests that are performed to verify a Rated Fatigue Pressure.

**04.01.390        Pressure, Differential (Pressure Drop)**

The difference in pressure between any two points of a system or a component.

**04.01.400        Pressure, Gage**

Pressure differential above or below ambient atmospheric pressure.

**04.01.412        Pressure, Induced**

Pressure generated by an externally applied force.

**04.01.414        Pressure, Inlet**

The pressure at the apparatus inlet port.

**04.01.415        Pressure, Intensified**

In a fluid power cylinder, the outlet pressure required to slow the piston rod extending under regulated pressure introduced at the cap end.

**04.01.417        Pressure, Maximum Inlet**

The maximum rated gage pressure applied to the inlet.

**04.01.419        Pressure, Nominal**

A pressure value assigned to a component or system for the purpose of convenient designation.

**04.01.420        Pressure, Operating**

The pressure at which a system is operated.

**04.01.425        Pressure, Outlet**

Pressure at the apparatus outlet port.

**04.01.427        Pressure, Overrange Rating**

The pressure to which a device can be subjected for extended time without change in operating characteristics, shift in set point, or damage to the device.

**04.01.430        Pressure, Override**

The difference between the cracking pressure of a valve and the pressure reached when the valve is passing its rated flow.

**04.01.435        Pressure, Peak**

The maximum pressure encountered in the operation of a component.

**04.01.440        Pressure, Pilot**

The pressure in the pilot circuit.

**04.01.450        Pressure, Precharge**

The pressure of compressed gas in an accumulator prior to the admission of a liquid.

**04.01.460        Pressure, Proof**

The non-destructive test pressure, in excess of the maximum rated operating pressure, which causes no permanent deformation, excessive external leakage, or other resulting malfuction.

**04.01.470        Pressure, Rated**

The qualified operating pressure which is recommended for a component or a system by the manufacturer.

**04.01.472        Pressure, Recovery**

The ratio of output pressure to the supply pressure.

**04.01.473        Pressure, Rated Fatigue**

A pressure that a pressure containing envelope is represented to sustain 10 million times without failure.

**04.01.474        Pressure, Regulation of**

Pertains to the control of pressure in a system.

**04.01.476 Pressure, Rated Static**

A pressure that a component pressure containing envelope is represented to sustain once, under test conditions without failure, after which the component must be discarded.

**04.01.477 Pressure, Residual**

The value of the output pressure in the "off" state of the device.

**04.01.480 Pressure, Shock**

The pressure existing in a wave moving at sonic velocity.

**04.01.481 Pressure, Shockwave**

A pressure pulse which moves at sonic speed in the liquid.

**04.01.490 Pressure, Static**

The pressure in a fluid at rest.

**04.01.495 Pressure, Static Test**

A pressure applied in a static test performed to verify a Rated Static Pressure.

**04.01.500 Pressure, Suction**

The absolute pressure of the fluid at the inlet of a pump.

**04.01.505 Pressure, Supply**

The pressure at the apparatus inlet port.

**04.01.510 Pressure, Surge**

The pressure resulting from surge conditions.

**04.01.520 Pressure, System**

The pressure which overcomes the total resistances in a system. It includes all losses as well as useful work.

**04.01.530 Pressure, Vapor**

The pressure, at a given fluid temperature, in which the liquid and gaseous phases are in equilibrium.

**04.01.540 Pressure, Working**

The pressure at which the apparatus is being operated in a given application.

**04.01.541 Pressure, Range, Working**

The tolerance (plus or minus) range of the working pressure.

**04.01.610 Reyn**

The standard unit of absolute viscosity in the English system. It is expressed in pound-seconds per square inch.

**04.01.611 Rotation**

The direction of rotation is always quoted as viewed looking at the shaft end. In dubious cases, provide a sketch.

**04.01.612 Rotation, Anti-Clockwise**

Rotation in the opposite sense to the clock.

**04.01.613 Rotation, Clockwise**

Forward rotation of the hands to the clock.

**04.01.614 Rotational Frequency**

Number of revolutions per unit of time.

**04.01.620 Specific Gravity, Liquid**

The ratio of the weight of a given volume of liquid to the weight of an equal volume of water.

**04.01.630 Stokes**

The standard unit kinematic viscosity in the c.g.s. (centimetre-gram-second) system. It is expressed in square centimetres per second; 1 centistokes equals .01 stokes.

**04.01.640 Surface Tension**

The surface force of a liquid in contact with a fluid by which it tends to assume a spherical form and to present the least possible surface. It is expressed in pounds per foot or dynes per centimetre.

**04.01.641 Temperature, Ambient**

The temperature of the environment in which the apparatus is working.

**04.01.642 Temperature, Equipment**

The temperature of the unit at a specified position and measured at a specified point.

**04.01.643 Temperature, Fluid**

The temperature of the pressure medium measured at a specified point.

**04.01.644 Temperature, Inlet**

Fluid temperature at the inlet port.

**04.01.645  Temperature, Outlet**

Fluid temperature at the outlet port.

**04.01.646  Temperature Range**

The permissible temperature range within which the apparatus or the fluid can operate satisfactorily.

**04.01.647  Torr**

A unit of pressure equal to 1/760 of an atmosphere and very nearly equal to 1mm Hg @ $0°C$.

**04.01.648  Unistable**

A binary circuit or device, which has one stable state and in which the output changes state for the duration of the appropriate change of the input signal.

**04.01.650  Vacuum**

Pressure less than ambient atmospheric pressure.

**04.01.657  Variability Factor**

A multiplier applied to the Rated Fatigue Pressure for calculating the Cyclic Test Pressure to account for the variability in fatigue strength of metals. It is also applied to the Rated Static Pressure for calculating the Static Test Pressure.

**04.01.700  Viscosity**

A measure of the internal friction or the resistance of a fluid to flow.

**04.01.710  Viscosity, Absolute**

The ratio of the shearing stress to the shear rate of a fluid. It is usually expressed in centipoise.

**04.01.720  Viscosity, Kinematic**

The absolute viscosity divided by the density of the fluid. It is usually expressed in centistokes.

**04.01.730  Viscosity, SAE Number**

The Society of Automotive Engineers' arbitrary numbers for classifying fluids according to their viscosities. The numbers in no way indicate the viscosity index of fluids.

**04.01.740  Viscosity, SUS**

Saybolt Universal Second (SUS), which is the time in seconds for 60 millilitres of oil to flow through a standard orifice at a given temperature (ASTM Designation D88-56).

**04.01.800  Viscosity Index**

A measure of the viscosity-temperature characteristics of a fluid as referred to that of two arbitrary reference fluids (ASTM Designation D2270-64).

**04.01.801  Viscosity Index Improver**

A chemical compound added to a fluid to modify its temperature/viscosity relationship.

**04.01.810  Vortex**

Spiral motion of a fluid resulting in a radial pressure gradient. The trajectories are curves which encircle a single line (axis).

## GROUP 05 — SYMBOLS

Section 01 — General
Section 02 — Symbol Types

### SECTION 01 — GENERAL

**05.01.100**      Symbol, Fluid Power

A representation of the characteristics of a fluid power component by means of lines on a flat surface.

### SECTION 02 — SYMBOL TYPES

**05.02.100**      Symbol, Combination

A symbol which combines graphical, cutaway and pictorial representations.

**05.02.200**      Symbol, Cutaway

A symbol showing principal internal parts, controls and actuating mechanisms, interconnecting lines, and functions of a component.

**05.02.300**      Symbol, Graphic (Schematic)

A simplified symbol which indicates essential characteristics applicable to all similar components.

**05.02.400**      Symbol, Pictorial

A symbol showing the actual shape of a component according to the manufacturer's description.

## GROUP 36 — LUBRICATORS

Section 01 — General

### SECTION 01 — GENERAL

**36.01.100**      Lubricator

A device which adds controlled or metered amounts of lubricant into a fluid power system.

## GROUP 58 — SUBPLATES

Section 01 — General
Section 02 — Types of Subplates

**58.01.100**      Subplate (Sub-Base)

Mounting to which a simple valve is fitted and which includes external ports for fluid connections.

**58.01.101**      Subplate (Sub-Base), Ganged Valve

Similar sub-bases of which two or more can be clamped together by tie bolts or other means. It can be arranged for the mating faces of the sub-bases to have matching ports, thus providing for a common supply and/or exhaust system. The sub-bases incorporate the various ports for connection of the external pipelines.

**58.01.102**      Subplate (Sub-Base), Multiple Valve

Mounting with appropriate ports matching those of two or more similar valves which are fitted to it and which include external ports for pipe connections.

## GROUP 70 — GENERAL CONTROLS

Section 01 — General
Section 02 — Types of General Controls
Section 03 — Restrictors
Section 04 — Logic Characteristics

## SECTION 01 — GENERAL

**70.01.100  Control**

A device used to regulate the function of a component or system.

**70.01.101  Controller**

A device which senses a change of fluid state and automatically makes adjustments to maintain the state of the fluid between predetermined limits, e.g., pressures, temperatures, etc.

**70.01.860  Fluid Memory, Off Return**

Fluid memory which receives a momentary signal and produces a change of state which continues to exist after the initiating signal has disappeared, providing an input is present at the supply port of the device. Upon loss of the supply pressure, the device reverts to its initial state.

**70.01.870  Fluid Memory, Retentive**

Fluid memory which receives a momentary signal and produces a change of state which continues to exist after the initiating signal has disappeared regardless of the presence or absence of supply pressure to the device. The device returns to its original state only upon receipt of a second reset control signal.

**70.01.900  Fluid Signal**

Fluid pressure or flow which can be detected or sensed.

**70.01.908  Fluid Signal, Maintained**

A fluid signal which exists indefinitely until caused to disappear by a secondary control action.

**70.01.910  Fluid Signal, Momentary**

A fluid signal which exists briefly and then disappears.

**70.01.914  Fluid Signal, Timed**

A fluid signal which exists for a definite period of time and then disappears.

## SECTION 02 — TYPES OF GENERAL CONTROLS

**70.02.105  Control, Automatic**

A control which actuates equipment in a predetermined manner.

**70.02.106  Control, Auxiliary**

A device, usually manual, fitted to a valve to provide an alternative method of control.

**70.02.110  Control, Combination**

A combination of more than one basic control.

**70.02.115  Control, Cylinder**

A control in which a fluid cylinder is the actuating device.

**70.02.117  Control, Detent**

A device which retains the moving part in position by means of artificially created resistance. Movement to a different position is achieved either by release of the detent, or by the application of sufficient force to overcome it.

**70.02.120  Control Electric**

A control actuated electrically.

**70.02.121  Control, Emergency**

A device, usually manual, fitted to a valve or circuit providing an alternative method of control in the case of failure of the normal method of control.

**70.02.122  Control, Feedback**

The means whereby the state of the controlled element is signaled.

**70.02.123  Control, Feedback, Mechanical**

Feedback using a mechanical transmission.

**70.02.124  Control, Feedback, Hydraulic**

Feedback using a hydraulic circuit.

**70.02.125  Control, Feedback, Electric**

Feedback using an electrical signal.

**70.02.126  Control, Feedback, Pneumatic**

Feedback using a pneumatic circuit.

**70.02.127  Control, Force Motor**

A type of electro-mechanical transducer having linear motion.

**70.02.130  Control, Hydraulic**

A control actuated by a liquid.

**70.02.133  Control, Impulse Generator**

A device so arranged that, if a continuous pneumatic signal is applied to the input port, a single pulse is produced at the output port.

ANSI/B93.2

**70.02.134     Control, Latch**

The moving parts are retained in a fixed position by means of a locking device which must be externally released.

**70.02.135     Control, Lever**

A pivoted arm which is hand operated by pushing or pulling.

**70.02.136     Control, Linkage**

Means of mechanical connection.

**70.02.137     Control, Liquid-Level**

A device which controls the liquid level by a float switch or other means.

**70.02.139     Control, Manual**

A control device which is manually operated.

**70.02.140     Control, Mechanical**

A control actuated by linkages, gears, screws, cams or other mechanical elements.

**70.02.141     Control, One-Way Trip**

A mechanism which will allow movement in one direction only of the actuating force.

**70.02.142     Control, Over Center**

The moving parts cannot be stopped in an intermediate position (dead center position).

**70.02.143     Control, Override**

An alternative method of control fitted to a valve, which takes precedence over the normal method of control.

**70.02.144     Control, Pedal**

A control device foot operated in one direction only.

**70.02.145     Control, Plunger**

A rod acting in the direct line of the application of force.

**70.02.146     Control, Pneumatic**

A control actuated by air or other gas pressure.

**70.02.147     Control, Pneumatic Programmer**

A calibrated device so arranged that, if a continuous signal is applied to the input port, one or more output signals will be produced. The duration of an interval between the outputs can be predetermined.

**70.02.148     Control, Pneumatic Programmer Cyclic**

Apparatus comprising a number of valves controlled by programming device with a repetitive action. The program may be either fixed or variable.

**70.02.149     Control, Pneumatic Time Delay**

A device so arranged that, if continuous pneumatic signal is applied to or removed from the input port, a signal will be produced at the output port after a predetermined time has elapsed. The time delay may be fixed or variable.

**70.02.150     Control, Pressure**

A control method operated by a change of fluid pressure in a pilot line.

**70.02.151     Control, Pressure Compensated**

A control in which a pressure signal operates a compensating device.

**70.02.152     Control, Pressure, Direct**

Control method in which the position of the moving parts is controlled directly by alteration of the control pressure.

**70.02.153     Control, Pressure, Indirect**

Control method in which the position of the moving parts is controlled by a change of the control pressure to a pivot device.

**70.02.154     Control, Pressure Pulse Generator**

A device so arranged that, if a continous pneumatic signal is applied to the input port, repetitive pulses are produced at the output port.

**70.02.155     Control, Pump**

A control applied to a positive displacement variable delivery pump to adjust the volumetric output or direction of flow.

**70.02.156     Control, Push-Pull Button**

A control device which is palm or finger operated by pushing or by pulling.

**70.02.157     Control, Roller**

A roller is attached to the operating mechanism to permit operation by means of a cam or slide acting at right angles to the mechanism.

#### 70.02.158 Control, Roller Lever

A lever with roller attached transmitting movement.

#### 70.02.159 Control, Roller Plunger

A plunger with roller attached permitting the transmission of movement at right angles to the mechanism.

#### 70.02.160 Control, Roller Rocker

A lever pivoted between rollers attached at each end, transmitting movement in both directions of the operating mechanism.

#### 70.02.161 Control, Rotating Shaft

A rotary mechanical control component.

#### 70.02.162 Control, Servo

A control actuated by a feedback system which compares the output with the reference signal and makes corrections to reduce the difference.

#### 70.02.163 Control, Solenoid, Single Acting (One - Way)

An electro-magnetic mechanism which has two positions, being operated against a bias to one extreme position by energizing the coil.

#### 70.02.164 Control, Solenoid, Double Acting (Two - Way)

An electro-magnetic mechanism which can take up two or three positions and is operated to either extreme position by energizing the appropriate coil.

#### 70.02.165 Control, Spring Return (Offset)

The moving parts of the unit are returned to the initial position by spring force after the actuating forces are removed.

#### 70.02.166 Control, Stepping Motor

An electric motor designed to provide displacement or speed variation in successive steps.

#### 70.02.167 Control, Torque Motor

A type of electro-mechanical transducer having rotary motion.

#### 70.02.168 Control, Tracer

A control operated by a system which follows the contours of master pattern.

#### 70.02.169 Control, Treadle

A control device foot operated in two directions.

#### 70.02.700 Logic Devices

The general category of components which perform logic functions; for example, AND, NAND, OR, and NOR. They can permit or inhibit signal transmission with certain combinations of control signals.

#### 70.02.710 Logical State

Signal levels in logic devices are characterized by two stable states, the logical 1 (one) state and the logical 0 (zero) state. The designation of the two states is chosen arbitrarily. Commonly the logical 1 state represents an "on" signal, and the 0 state represents an "off" signal.

#### 70.02.880 Sensing Device

A component which measures the state of a system variable such as fluid level, viscosity, temperature, pressure, or flow rate.

#### 70.02.900 Sensor

A device which detects a condition in a system and produces a signal.

### SECTION 03 — RESTRICTORS

#### 70.03.050 Diode, Fluid

A device with a passage for fluid flow with high, to infinitely high resistance in one direction and low resistance in the opposite direction.

#### 70.03.100 Restrictor

A device which reduces the cross-sectional flow area.

#### 70.03.130 Restrictor, Choke

A restrictor, the length of which is relatively large with respect to its cross-sectional area.

#### 70.03.160 Restrictor, Orifice

A restrictor, the length of which is relatively small with respect to its cross-sectional area. The orifice may be fixed or variable. Variable types of non-compensated, pressure compensated, or pressure and temperature compensated.

## SECTION 04 — LOGIC CHARACTERISTICS

**70.04.100    AND Device**

A control device which has its output in the logical 1 state if and only if all the control signals assume the logical 1 state.

**70.04.400    Flip Flop**

A digital component or circuit with two stable states and sufficient hysteresis so that it has "memory". Its state is changed with a control pulse; a continuous control signal is not necessary for it to remain in a given state.

**70.04.700    NAND Device**

A control device which has its output in the logical 0 state if and only if all the control signals assume the logical 1 (one) state.

**70.04.720    NOR Device**

A control device which has its output in the logical 1 state if and only if all the control signals assume the logical 0 state.

**70.04.740    NOT Device**

A control device which has its output in the logical 1 state if and only if the control signal assumes the logical 0 state. The NOT device is a single input NOR device.

**70.04.760    OR Device**

A control device which has its output in the logical 0 state if and only if all the control signals assume the logical 0 state.

---

## GROUP 71 — VALVES

Section 01 — General
Section 02 — Functional Types
Section 03 — Basic Designs
Section 04 — Positions
Section 05 — Flow Conditions
Section 06 — Actuators
Section 07 — Mountings
Section 08 — Features

---

## SECTION 01 — GENERAL

**71.01.100    Valve**

A device which controls fluid flow direction, pressure, or flow rate.

**71.01.500    Valve, Monoblock**

Unit comprising a number of similar valves in a common housing.

## SECTION 02 — FUNCTIONAL TYPES

**71.02.050    Valve, Air**

A valve for controlling air.

**71.02.055    Valve, Ball, Seat Action**

A valve design which utilizes a solid ball to obstruct the flow path.

**71.02.056    Valve, Ball, Shear Action**

A valve design which utilizes a ported ball that rotates on an axis normal to the flow path.

**71.02.060    Valve, Butterfly**

A straight-through shut-off in which the valve element consists of a flat disc rotating about a diametrical axis perpendicular to the flow of fluid.

**71.02.070    Valve, Cartridge**

A valve with working parts contained in a cylindrical body. The cylindrical body must be inserted into a housing for use. Ports through the body cooperate with ports in the containing housing.

**71.02.080    Valve, Diaphragm Type**

A valve in which the element is moved by forces acting on a diaphragm.

**71.02.100    Valve, Directional Control**

A valve whose primary function is to direct or prevent flow through selected passages.

**71.02.110    Valve, Directional Control, Check**

A directional control valve which permits flow of fluid in only one direction.

**71.02.111 Valve, Directional Control, Check, Spring Loaded**

A valve in which flow may occur in one direction only when fluid pressure overcomes spring pressure.

**71.02.112 Valve, Directional Control, Check, Pilot Operated**

A check valve in which the opening or closing is controlled by a pilot signal.

**71.02.113 Valve, Directional Control, Check, Cushioned**

A check valve in which the movement of the check device is damped, for use in systems with pulsating pressures.

**71.02.130 Valve, Directional Control, Four Way**

A directional control valve whose primary function is to pressurize and exhaust two ports.

**71.02.140 Valve, Directional Control, Selector (Diversion)**

A directional control valve whose primary function is to selectively interconnect two or more ports.

**71.02.160 Valve, Directional Control, Straight-way**

A two port directional control valve.

**71.02.170 Valve, Directional Control, Three Way**

A directional control valve whose primary function is to pressurize and exhaust a port.

**71.02.172 Valve, Directly Operated**

A valve in which the controlling forces acting on the element directly influence the movement of the control elements.

**71.02.174 Valve, Disc (Globe)**

A shut-off valve in which the flow at one point is at right angles to the normal direction of flow. The valve member is a flat disc which is lifted or seated to open or close the flow path.

**71.02.175 Valve, Disc (Swing)**

A shut-off valve design which utilizes a hinged disc to obstruct the flow path.

**71.02.190 Valve, Flow Combining, Pressure Compensated**

Pressure compensated valve which combines two input flow rates maintaining a pre-selected output.

**71.02.200 Valve, Flow Control (Flow Metering)**

A valve whose primary function is to control flow rate.

**71.02.201 Valve, Flow Control, Bypass**

A pressure compensated flow control valve which regulates the working flow diverting surplus fluid to reservoir or to a second service.

**71.02.202 Valve, Flow Control Adjustable Restrictor**

Valve in which the inlet and outlet ports are interconnected through a restricted passageway whose cross-sectional area can be varied within limits.

**71.02.203 Valve, Flow Control Fixed Restrictor**

A valve in which the inlet and outlet ports are interconnected through a restricted passageway whose cross-sectional area cannot be altered.

**71.02.210 Valve, Flow Control, Deceleration**

A flow control valve which gradually reduces flow rate to provide deceleration.

**71.02.216 Valve, Flow Control, One-Way Restrictor**

A valve which allows free flow in one direction and restricted flow in the other direction. Restricted flow path may be fixed or variable.

**71.02.220 Valve, Flow Control, Pressure Compensated**

A flow control valve which controls the rate of flow independent of system pressure.

**71.02.230 Valve, Flow Control, Pressure-Temperature Compensated**

A pressure compensated flow control valve which controls the rate of flow independent of fluid temperature.

**71.02.250 Valve, Flow Dividing**

A valve which divides the flow from a single source into two or more branches.

**71.02.280  Valve, Flow Dividing, Pressure Compensated**

A flow dividing valve which divides the flow at constant ratio regardless of the difference in the resistances of the branches.

**71.02.300  Valve, Gate**

A straight-through shut-off valve in which the valve element moves perpendicularly to the axis of the flow to control opening and closing.

**71.02.301  Valve, Gate, Spreader**

A gate valve which utilizes two companion discs which are positively seated by common spreaders to obstruct the flow path.

**71.02.302  Valve, Gate, Wedge**

A gate valve which utilizes a solid wedge shaped gate to obstruct the flow path.

**71.02.320  Valve, Globe**

(See: Disc Valve - 71.02.174)

**71.02.350  Valve, Hydraulic**

A valve for controlling liquid.

**71.02.380  Valve, Needle**

A flow control valve in which the adjustable control element is a tapered needle. Its usual purpose is the accurate control of the rate of volume of flow.

**71.02.400  Valve, Pilot**

A valve applied to operate another valve or control.

**71.02.401  Valve, Pilot Operated (Indirect)**

A valve in which a relatively small flow through an integral vent line relief (pilot) controls the movement of the main element.

**71.02.410  Valve, Pinch**

A straight-through shut-off valve in which the valve element consists of a flexible sleeve which is distorted to control the flow of the fluid.

**71.02.420  Valve, Piston**

A valve in which the element is moved by forces acting on a piston.

**71.02.430  Valve, Plug**

A shut-off valve in which ports are connected or sealed off by a rotating plug containing flow paths.

**71.02.431  Valve, Plug, Cylinder**

A valve in which the surface of contact between the plug and the valve body is cylindrical and requires a method of sealing.

**71.02.432  Valve, Plug, Shear Action**

A valve design which utilizes a ported plug that rotates on an axis normal to the flow path.

**71.02.433  Valve, Plug, Spherical**

A valve in which the surface of contact between the plug and the valve body is spherical and requires a method of sealing.

**71.02.434  Valve, Plug, Tapered**

A valve in which the surface of contact between the plug and the valve body is conical and provides the sealing surface.

**71.02.450  Valve, Pneumatic**

A valve for controlling compressed air.

**71.02.460  Valve, Poppet**

A valve in which the flow paths are opened or closed as the valve element (poppet) is lifted or seated.

**71.02.470  Valve, Power Control**

A valve which controls fluid power operating working devices.

**71.02.500  Valve, Prefill**

A valve which permits full flow from a tank to a cylinder during the advance portion of a cycle, permits the operating pressure to be applied to the cylinder during the working portion of the cycle, and permits free flow from the cylinder to the tank during the return portion of the cycle.

**71.02.550  Valve, Pressure Control**

A valve whose primary function is to control pressure.

**71.02.560**  Valve, Pressure Control, Counterbalance

A pressure control valve which maintains back pressure to prevent a load from falling.

**71.02.570**  Valve, Pressure Control, Decompression

A pressure control valve that controls the rate at which the contained energy of the compressed fluid is released.

**71.02.580**  Valve, Pressure Control, Load Dividing

A pressure control valve used to proportion pressure between two pumps in series.

**71.02.590**  Valve, Pressure Control, Pressure Reducing

A pressure control valve whose primary function is to limit outlet pressure.

**71.02.591**  Valve, Pressure Control, Reducing & Relieving

A valve which limits maximum pressure by exhausting fluid when the required pressure is reached.

**71.02.600**  Valve, Pressure Control, Relief

A pressure control valve whose primary function is to limit system pressure.

**71.02.620**  Valve, Pressure Control Relief, Safety

A relief valve whose primary function is to provide pressure limitation after malfunction.

**71.02.630**  Valve, Pressure Control, Unloading

A pressure control valve whose primary function is to permit a pump or compressor to operate at maximum load.

**71.02.635**  Valve, Pressure Proportioning

A pressure reducing valve in which the outlet pressure is maintained at a fixed ratio to the inlet pressure.

**71.02.640**  Valve, Pressure Sensing

A device similar to an electrical pressure switch, in which a signal to be sensed enters a control point, and actuates a mechanism which, at the proper pressure level, causes one or more flow passages to change condition. Removal of the signal allows the pressure sensing valve to reset.

**71.02.650**  Valve, Priority

A valve which directs flow to one operating circuit at a fixed rate and directs excess flow to another operating circuit.

**71.02.655**  Valve, Quick Exhaust

Valve in which, when air pressure falls at the inlet, the outlet is automatically opened to exhaust.

**71.02.660**  Valve, Relay

A logic device which receives control signals and changes flow conditions in one or more controlled flow passages.

**71.02.662**  Valve, Relay, Free Floating

A relay valve wherein the internal element moves freely without restraint and normally utilizes bias pressure at one control point.

**71.02.664**  Valve, Relay, One Shot

A relay valve wherein controlled flow passages immediately change conditions when a control point is pressurized by a maintained signal. After a period of time, the controlled flow passages return to their original conditions, even though the control point is pressurized. When the control signal is removed it resets for another operation.

**71.02.666**  Valve, Relay, Time Delay After Exhausting a Control Point

A relay valve with one control point which receives a maintained signal and causes immediate actuation of controlled flow passages. When the control signal is removed, a time delay occurs before controlled flow passages are reset.

**71.02.668**  Valve, Relay, Time Delay After Pressurizing a Control Point

A relay valve with one control point which receives a maintained signal and causes a time delay before the controlled flow passages are actuated. The device resets immediately upon exhausting the control point.

**71.02.670**  Valve, Relay, Time Delay

A relay valve which creates a time interval between the pressurizing of a control point and a change in the controlled flow passages.

**71.02.672**  Valve, Relay, Time Delay, Detented

A time delay relay valve having "A" and "B" control points arranged to accept and act on momentary signals. A momentary signal into the "A" control point starts actuation and a momentary signal into the "B" control points starts reset.

**71.02.674**      Valve, Relay, Time Delay, Detented, Delayed Action

A detented time delay relay valve in which a momentary signal into the "A" control point creates a time delay before controlled flow passages are actuated. The device resets immediately upon receipt of a signal in the "B" control point.

**71.02.676**      Valve, Relay, Time Delay, Detented, Delayed Reset

A detented time delay relay valve in which a momentary signal into the "A" control point produces immediate actuation of the controlled flow passages. A momentary signal to the "B" control points starts the reset action with a time interval before controlled flow passages reset.

**71.02.680**      Valve, Rotary Selector

A valve which utilizes rotary actuation to connect the inlet to any one of a number of outlets.

**71.02.690**      Valve, Separator Drain

A device whereby solid or liquid impurities which have collected in the installation can be removed. May be actuated automatically or manually.

**71.02.700**      Valve, Pressure Control, Sequence

When the inlet pressure exceeds the preset value, the valve opens to permit flow through the outlet port, (The effective setting, is not affected by the pressure on the outlet port.)

**71.02.750**      Valve, Shutoff

A valve which operates fully open or fully closed.

**71.02.751**      Valve, Shut Off, Automatic

A valve which closes automatically when the pressure drop across the valve, caused by increased flow, exceeds a predetermined amount.

**71.02.752**      Valve, Shut Off Sliding

A shut off valve whose flow paths are connected together or sealed off by means of a moveable sliding member. The movement may be axial, radial or both.

**71.02.753**      Valve, Shut Off, Sliding, Flat

A shut off valve in which the flow paths are connected together or sealed off by means of a movable flat faced valve member sliding on a flat seat.

**71.02.754**      Valve, Shut Off, Sliding, Spool

A shut off valve in which the flow paths are connected or sealed off by a cylindrical spool which slides within the matching bore of the valve body.

**71.02.800**      Valve, Shuttle

A connective valve which selects one of two or more circuits because flow or pressure changes between the circuits.

**71.02.801**      Valve, Shuttle, High Pressure

A valve in which the inlet at higher pressure is connected to the outlet, the other inlet is closed. The position is maintained under reverse flow.

**71.02.802**      Valve, Shuttle, Low Pressure

A valve in which inlet at lower pressure is connected to the outlet, the other inlet is closed. The position is maintained under reverse flow.

**71.02.810**      Valve, Slide

A valve in which the flow paths are connected or isolated by means of a flat movable sliding member. The movement may be axial, rotary, or both.

**71.02.811**      Valve, Slide, Linear

A valve in which the flow paths are connected or isolated by means of a flat faced valve member which slides on a flat seat.

**71.02.812**      Valve, Slide, Rotary

A sliding plate shear action valve design in which the motion of the plate is rotary.

**71.02.820**      Valve, Spool

A shear action valve design which utilizes a spool that slides through the flow path.

**71.02.850**      Valve, Surge Damping

A valve which reduces shock by limiting the rate of acceleration of fluid flow.

**71.02.900**      Valve, Time Delay

A valve in which the change of flow occurs only after a desired time interval has elapsed.

## SECTION 03 — BASIC DESIGNS

**71.03.200**      Flapper Action

A valve design in which output control pressure is regulated by a pivoted flapper in relation to one or two orifices.

**71.03.300**      Jet Action

A valve design in which flow effect is controlled by the relative position of a nozzle and a receiver.

**71.03.500**      Seating Action

A valve design in which flow is stopped by a seated obstruction in the flow path.

**71.03.700**      Shear Action

A valve design in which flow is modulated by an element which slides across the flow path.

## SECTION 04 — POSITIONS

**71.04.000**      Valve Position

The point at which flow directing elements provide a specific flow condition in a valve.

**71.04.002**      Valve Position, Actuated

One of the final positions of the valving element when under the influence of the actuating forces.

**71.04.100**      Valve Position, Center

The selective mid-position in a directional control valve.

**71.04.200**      Valve Position, Detent

A predetermined position maintained by a holding device acting on the flow-directing elements of a directional control valve.

**71.04.220**      Valve Position, Initial

The position of the valving element after main pressure is admitted and before the intended operating cycle begins under the influence of the actuating forces.

**71.04.222**      Valve Position, Intermediate (Transit)

Any position between the initial and the actuated position.

**71.04.300**      Valve Position, Normal

The valve position when signal or actuating force is not being applied.

**71.04.400**      Valve Position, Offset

An off-center position in a directional control valve.

**71.04.500**      Valve Position, Return

The initial valve position.

**71.04.600**      Valve, Four Position

A directional control valve having four positions to give four selections of flow conditions.

**71.04.700**      Valve, Three Position

A directional control valve having three positions to give three selections of flow conditions.

**71.04.701**      Valve, Three Position, Closed Center

All ports are closed in the initial position.

**71.04.702**      Valve, Three Position Open Center

All ports are connected in the initial position.

**71.04.800**      Valve, Two Position

A directional control valve having two positions to give two selections of flow conditions.

## SECTION 05 — FLOW CONDITIONS

**71.05.000**      Valve Flow Condition

A flow pattern in a directional control valve.

**71.05.100**      Valve Flow Condition, Closed

All ports are closed.

**71.05.200**      Valve Flow Condition, Float

Working ports are connected to exhaust or return.

**71.05.300**      Valve Flow Condition, Hold

Working ports are blocked to hold a powered device in a fixed position.

**71.05.400**      Valve Flow Condition, Open

All ports are open.

**71.05.500**      Valve Flow Condition, Regenerative

Working ports are connected to supply.

**71.05.600**      Valve Flow Condition, Tandem

Working ports are blocked and supply is connected to the return port.

## SECTION 06 — VALVE ACTUATORS

**71.06.000**      Valve Actuator

The valve part(s) through which force is applied to move or position flow-directing elements.

**71.06.200**      Valve Actuator, Manual

A valve actuator consisting of a hand lever, palm button, foot treadle, or other manual energizing devices.

**71.06.400**      Valve Actuator, Mechanical

A valve actuator consisting of a cam, lever, roller, screw, spring, stem, or other mechanical energizing devices.

**71.06.600**      Valve Actuator, Pilot

A valve actuator which utilizes pilot fluid.

**71.06.610**      Valve Actuator, Pilot, Barrier

A pilot valve actuator wherein the working fluid is isolated from the actuator.

**71.06.620**      Valve Actuator, Pilot, Differential Area

A pilot valve actuator wherein pilot fluid acts on unequal areas.

**71.06.630**      Valve Actuator, Pilot, Differential Pressure

A pilot actuator wherein pilot fluid acts at unequal pressure.

**71.06.640**      Valve Actuator, Pilot, External

A pilot valve actuator wherein fluid is received from an external source.

**71.06.650**      Valve Actuator, Pilot, Internal

A pilot valve actuator wherein pilot fluid is received from within the valve.

**71.06.660**      Valve Actuator, Pilot, Solenoid, Controlled

A pilot valve actuator wherein pilot fluid is controlled by the action of one or more solenoids.

**71.06.800**      Valve Actuator, Solenoid

A valve actuator which utilizes one or more solenoids.

## SECTION 07 — VALVE MOUNTINGS

**71.07.000**      Valve Mounting

The mounting characteristics of a valve.

**71.07.200**      Valve Mounting, Base

The valve is mounted on a plate which has top and side ports.

**71.07.201**      Valve Mounting, Base, Gang

Unit consisting of an assembly of a number of similar valves banked together, often with common supply and/or exhaust systems.

**71.07.400**      Valve Mounting, Line

The valve is mounted directly to system lines.

**71.07.600**      Valve Mounting, Manifold

The valve is mounted to a plate which provides multiple connection ports for two or more valves.

**71.07.800**      Valve Mounting, Sub-Plate

The valve is mounted to a plate which provides straight-through top and bottom ports.

## SECTION 08 — FEATURES

**71.08.200**      Flow Path (Gallery)

Passage through which fluid flows within a device.

## GROUP 72 — FLUIDIC DEVICES

Section 01 — General
Section 02 — Types of Fluidic Devices
Section 03 — Amplification
Section 04 — Gain
Section 05 — Interface Devices
Section 06 — Pneumatic Power for Fluidics
Section 07 — Fluidic Response Factors

### SECTION 01 — GENERAL

**72.01.100  Analog**

Of or pertaining to the general class of fluidic devices or circuits whose output varies as a continuous function of its input.

**72.01.105  Aspect Ratio, Nozzle**

The ratio of nozzle depth to nozzle width.

**72.01.250  Capacitance, Fluid**

Ratio of mass flow to rate of change of pressure drop.

**72.01.254  Characteristic, Switching**

Curve expressing output quantity as a function of control quantity.

**72.01.255  Collector (Receiver)**

Nozzle located downstream of a free flowing jet, normally used to catch the energy of the flowing medium of the jet.

**72.01.257  Conductance, Fluid**

Ratio between steady state mass flow and pressure drop (reciprocal value of fluid resistance).

**72.01.300  Digital**

Of or pertaining to the general class of fluidic devices or circuits whose output varies in discrete steps (i.e., pulses or "on-off" characteristics).

**72.01.400  Fan In Ratio**

The number of operating controls in a single fluidic device which individually and in combination will produce the same output.

**72.01.500  Fan Out Ratio**

The number of like devices to which operating controls are supplied by the output of the fluidic device.

**72.01.600  Fluidic**

Of or pertaining to devices, systems, assemblies, etc., utilizing fluidic components.

**72.01.650  Impedance, Capacitive, Fluid**

Imaginary ratio of pressure drop and transient mass flow in which pressure drop leads flow.

**72.01.651  Impedance, Fluid**

Complex ratio between pressure drop and transient mass flow.

**72.01.652  Impedance, Fluidic Input**

The impedance measured at an input port.

**72.01.653  Impedance, Fluidic Output**

The impedance measured at an output port.

**72.01.654  Impedance, Inductive, Fluid**

Imaginary ratio of pressure drop and transient mass flow in which pressure drop leads flow by phase.

**72.01.659  Inductance, Fluid**

Ratio of pressure drop and rate of change of mass flow.

**72.01.700  Interface**

A point or component where a transition is made between medium, power levels, modes of operation, etc.

**72.01.710  Jet**

Emission of a fluid from an orifice.

**72.01.711  Jet, Attached**

Jet which is attached to a wall by Coanda effect.

**72.01.712  Jet, Confined**

A jet influenced by its physical surrounding.

**72.01.713  Jet, Free**

A jet not influenced by its surroundings.

**72.01.714  Jet, Main Power**

A laminar or turbulent flow of fluid emitted from the supply channel or nozzle of a fluidic device.

**72.01.800  Load Line**

Curve expressing output pressure as a function of output flow. (The derivative of this curve is the expression of the output impedance.)

**72.01.801  Logic Threshold**

The minimum number of signals required at the inputs of a multi input device to change the output condition.

**72.01.900  Ratio, Series**

The number of identical devices mounted in series, which can be controlled by the output of a device.

**72.01.920 Ratio, Signal to Noise**

The ratio of the signal strength to that of the noise strength.

**72.01.930 Splitter**

That part of a fluid amplifier which separates alternative outputs.

**72.01.950 Volume, Fluidic Control**

The volume of the input chamber, including the pilot line.

## SECTION 02 — TYPES OF FLUIDIC DEVICES

**72.02.050 Fluidic Device, Active**

A device which requires a power supply independant of the value of input signals.

**72.02.100 Fluidic Amplifier**

A device which enables one or more fluid dynamic signals to control a source of power and thus is capable of delivering at its output an enlarged reproduction of the essential characteristics of the signal.

**72.02.110 Fluidic Amplifier, Closed**

A fluidic amplifier which has no vent port.

**72.02.111 Fluidic Amplifier, Flow**

A device which amplifies flow by use of a small valve which acts as a pilot for a larger one.

**72.02.115 Fluidic Amplifier, Digital**

An amplifier whose output varies in descrete steps related to the control signal.

**72.02.120 Fluidic Amplifier, Impact Modulator**

A fluidic amplifier in which the impact plane position of two opposed streams is controlled to alter the output.

**72.02.122 Fluidic Amplifier, Momentum**

An amplifier which functions on the interaction of momentum of the power and control jets.

**72.02.125 Fluidic Amplifier, Open**

A fluidic amplifier which has a vent port.

**72.02.130 Fluidic Amplifier, Stream Deflection**

A fluidic amplifier which utilizes one or more control streams to deflect a power stream, altering the output. It is usually analog.

**72.02.135 Fluidic Amplifier, Turbulence**

A fluidic amplifier in which the power jet is at a pressure such that it is in the transition region of laminar stability and can be caused to become turbulent by a secondary jet or by sound.

**72.02.140 Fluidic Amplifier, Vortex**

An amplifier which senses the pressure drop across a vortex, modulating the main flow.

**72.02.145 Fluidic Amplifier, Wall Attachment**

A fluidic amplifier in which the control of the attachment of a stream to a wall(s) alters the output. It is usually digital.

**72.02.700 Fluidic Device, Passive**

The general class of fluidic devices that operates on signal power alone.

**72.02.710 Fluidic Device, Interface**

A device which converts information between different types or levels of energy.

## SECTION 03 — AMPLIFICATION (GAIN)

**72.03.200 Gain, Flow**

The ratio of the change of output flow to the change of control flow at a given point.

**72.03.400 Gain, Power**

The ratio of the change of output power to the change of control power at a given point.

**72.03.600 Gain, Pressure**

The ratio of output pressure change to control pressure change at a given point.

## SECTION 06 — PNEUMATIC POWER FOR FLUIDICS

**72.06.050 Damping Parameter**

A measure of the time required as a function of the maximum pressure excursion of the power supply output to attain essentially steady state operation after an abrupt disturbance. Specifically, it is the transient recovery time divided by the maximum excursion.

**72.06.100 Drift**

The percentage above and below the operating pressure at a constant flow rate over a specified length of time.

72.06.120          **Flow, Minimum Control**

Flow through the control port at minimum control pressure.

72.06.121          **Flow/Pressure Characteristic**

The change of the specified controlled pressure due to change in the flow rate of the fluid, measured at specified pressure conditions.

72.06.140          **Interaction Region, Jet**

Chamber in which the power jet is affected by one or more control jets.

72.06.150          **Maximum Excursion**

The maximum pressure deviation from the operating pressure after an abrupt disturbance.

72.06.199          **Noise, Acoustic**

Spurious signals generated by external acoustic disturbances.

72.06.200          **Noise, Fluidic**

Random fluctuations of the signal level which may cause undesirable spurious signals in a circuit.

72.06.250          **Operating Band**

The range of pressures above and below the operating pressure within which it is desired to keep the supply output.

72.06.300          **Power Capacity**

The total volume of gas available at the operating pressure (applies to compressed gas storage supply source).

72.06.325          **Power, Output**

The power recoverable at the output port.

72.06.350          **Power Supply**

That component or group of components which supplies and processes the fluid for operating fluidic systems.

72.06.400          **Rated Flow**

The maximum flow that the power supply system is capable of maintaining at a specific operating pressure.

72.06.410          **Recovery, Flow Rate**

Ratio of no-load flow at the output to the supply flow.

72.06.420          **Recovery, Power**

A maximum ratio of power recovered at the output port to the supply power.

72.06.450          **Ripple**

A periodic variation of the pressure above and below the operating pressure. It is defined as a percentage of the operating pressure in terms of the maximum peak-to-peak value obtained at the point of rating.

72.06.500          **Start-Up Time**

The period of time needed to reach a steady state condition within the operating band starting from a long term off condition.

72.06.550          **Steady State Pressure Regulation**

A band indicating maximum and minimum pressure or a single curve with maximum deviation indicated in percent of operating pressure, all as a function of flow.

72.06.600          **Transient Recovery Time**

The period of time required for an abrupt change in the power supply output pressure to dampen out to within the operating band.

## SECTION 07 — FLUIDIC RESPONSE FACTORS

72.07.200          **Decay**

A falling pressure.

72.07.300          **Decay Rate**

The ratio of pressure decay to time.

72.07.600          **Output, Active**

Output power which, in all possible states of the device, is derived from supply power.

72.07.601          **Output, Passive**

Output the power of which in one or more states of the devices is derived solely from the input signals.

72.07.700          **Response Time**

Interval between the initiation of an operation and its completion.

72.07.800          **Rise Rate**

The ratio of pressure rise to time.

## GROUP 73 — SERVOVALVES

Section 01 — General
Section 02 — Types of Servovalves
Section 03 — Construction Features
Section 04 — Electrical Characteristics
Section 05 — Steady State Characteristics
Section 06 — Dynamic Characteristics

## SECTION 01 — GENERAL

**73.01.100  Servovalve**

A valve which modulates output as a function of an input command.

## SECTION 02 — TYPES OF SERVOVALVES

**73.02.200  Servovalve, Electrohydraulic**

A servovalve which is capable of continuously controlling hydraulic output as a function of an electrical input.

**73.02.300  Servovalve, Electrohydraulic, Flow Control**

An electrohydraulic servovalve whose primary function is control of output flow.

**73.02.400  Servovalve, Four-Way**

A multi-orifice flow control valve with supply, return and two control ports arranged so that the valve action in one direction opens supply to control port 1 and opens control port 2 to return. Reversed valve action opens supply to control port 2 and opens control port 1 to return.

**73.02.450  Servovalve, Mechanical Hydraulic**

A hydraulic servovalve in which the input command is mechanical.

**73.02.460  Servovalve, Pressure Control**

A hydraulic servovalve whose primary function is the control of output pressure.

**73.02.500  Servovalve, Three-Way**

A multi-orifice flow control valve with supply, return and one control port arranged so that valve action in one direction opens supply to control port and reversed valve action opens the control port to return.

**73.02.600  Servovalve, Two-Way**

A single orifice flow control valve with supply and one control port arranged so that action is in one direction only, from supply to control port.

## SECTION 03 — CONSTRUCTION FEATURES

**73.03.100  Force Motor**

A type of electromechanical transducer having linear motion used in the input stages of servovalves.

**73.03.200  Hydraulic Amplifier**

A fluid device which enables one or more inputs to control a source of fluid power and thus is capable of delivering at its output an enlarged reproduction of the essential characteristics of the input. Hydraulic amplifiers may utilize sliding spools, nozzle-flappers, jet pipes, etc.

**73.03.300  Output Stage**

The final stage of hydraulic amplification used in a servovalve.

**73.03.400  Stage**

A hydraulic amplifier used in a servovalve. Servovalves may be single stage, two stage, three stage, etc.

**73.03.500  Torque Motor**

A type of electromechanical transducer having rotary motion used in the input stages of servovalves.

## SECTION 04 — ELECTRICAL CHARACTERISTICS

**73.04.100  Dither**

A low amplitude, relatively high frequency periodic electrical signal, sometimes superimposed on the servovalve input to improve system resolution. Dither is expressed by the dither frequency (Hz) and the peak-to-peak dither current amplitude.

## SECTION 05 — STEADY STATE CHARACTERISTICS

**73.05.050  Servovalve Control Flow**

The flow through the servovalve control ports. Conventional test equipment normally measures no-load flow.

**73.05.060  Servovalve Control Flow, Loaded**

The flow through the servovalve control ports when there is load pressure drop.

**73.05.070  Servovalve Control Flow, No-Load**

The flow through the servovalve control ports when there is zero load pressure drop.

**73.05.100      Servovalve Flow Curve**

The graphical representation of control flow versus input current of a servovalve. This is usually a continuous plot of a complete cycle between plus and minus rated current valves (See Fig. 73.1).

**73.05.130      Servovalve Flow Curve, Normal**

The locus of the midpoints of the complete cycle flow curve, which is the zero hysteresis flow curve, of a servovalve. Usually valve hysteresis is sufficiently low, such that one side of the flow curve can be used for the normal flow curve. (See Fig. 73.1).

**73.05.150      Servovalve Flow Gain**

The slope of the control flow versus input current curve in any specific operating region, of a servovalve. Three operating regions are usually significant with flow-control servovalves: (1) the null region, (2) the region of normal flow control, and (3) the region where flow saturation effects may occur. Where this term is used without qualification, it is assumed to mean normal flow gain. (See Fig. 73.2)

**73.05.160      Servovalve Flow Gain, No-Load**

The normal flow gain of a servovalve with zero load pressure drop.

**73.05.170      Servovalve Flow Gain, Normal**

The slope of a straight line drawn from the zero flow point of the servovalve normal flow curve, throughout the range of rated flow current of one polarity, and drawn to minimize deviations of the normal flow curve from the straight line. Flow gain may vary with the polarity of the input, with the magnitude of load differential pressure and with changes in operating conditions. (See Fig. 73.2)

**73.05.200      Servovalve Flow Limit**

The condition wherein control flow no longer increases with increasing input current. Flow limitation may be deliberately introduced within the servovalve.

**73.05.250      Servovalve Flow Saturation Region**

The region where flow gain decreases with increasing input current, in a servovalve. (See Fig. 73.2)

**73.05.300      Servovalve Hysteresis**

The difference in the servovalve input currents required to produce the same output during a single cycle of valve input current when cycled at a rate below that at which dynamic effects are important. Hysteresis is normally specified as the maximum difference occuring in the flow curve throughout plus or minus rated current, and is expressed as percent of rated current. (See Fig. 73.1)

**73.05.350      Servovalve Internal Leakage**

The total internal servovalve flow from pressure to return with zero control flow. It is usually measured with control ports blocked. Leakage flow will vary with input pressure and input current.

**73.05.400      Servovalve Lap**

In a sliding spool servovalve the relative axial position relationship between the fixed and moveable flow metering edges with the spool at null. Lap is measured as the total separation at zero flow of straight line extensions of the nearly straight portions of the normal flow curve, drawn separately for each polarity, expressed as percent of rated current.

**73.05.405      Servovalve Null Leakage**

Total internal leakage from the valve in the null position.

**73.05.410      Servovalve, Overlap**

The lap condition which results in a decreased slope of the normal flow curve in the null region. (See Fig. 73.3)

**73.05.420      Servovalve, Underlap**

The lap condition which results in an increased slope of the normal flow curve in the null region. (See Fig. 73.4)

**73.05.430      Servovalve Zero Lap**

The lap condition in which there is no separation of the straight line extensions of the nomal flow curve. (See Fig. 73.5).

**73.05.450      Servovalve Flow Linearity**

The degree to which the normal flow curve conforms to the normal flow gain line with other operational variables held constant in a servovalve. Linearity is measured as the maximum deviation of the normal flow curve from the normal flow gain line, expressed as percent of rated current.

**73.05.500      Servovalve Null**

The condition in which the servovalve supplies zero control flow at zero load pressure drop.

**73.05.510      Servovalve Null Bias**

The input current required to bring the servovalve to null, excluding the effects of valve hysteresis, expressed as percent of rated current. (See Fig. 73.1).

**73.05.520      Servovalve Null Pressure**

The pressure existing at both control ports in a servovalve at null.

#### 73.05.530 Servovalve Null Region

The region in a servovalve about null wherein effects of lap in the output stage predominate.

#### 73.05.540 Servovalve Null Shift

A change in null bias in a servovalve, expressed as percent of rated current. Null shift may occur with changes in supply pressure minus the return pressure minus the load pressure drop.

#### 73.05.550 Servovalve Pressure Drop

The sum of the differential pressures across the control orifices of the output stage in a servovalve. Pressure drop will equal the supply pressure minus the return pressure minus the load pressure drop.

#### 73.05.570 Servovalve Load Pressure Drop

The differential pressure between the control ports of a servovalve.

#### 73.05.600 Servovalve Pressure Gain

The change in load pressure drop per unit input current with zero control flow (control ports blocked). Pressure gain is specified as the average slope of the curve of load pressure drop versus input current in the region between $\pm$ 40% of maximum load pressure drop (See Fig. 73.7).

#### 73.05.650 Servovalve Rated Flow

The specified control flow corresponding to rated current and specified servovalve pressure drop.

#### 73.05.660 Servovalve Resolution

The increment of input signal required to produce a change in valve output at a specified signal level, expressed as a percentage of rated signal. Resolution is normally specified as the minimum signal required to cause either an increase or a decrease of valve output. If these signals differ, the larger of the two should be quoted.

#### 73.05.700 Servovalve Symmetry

The degree of equality between the servovalve normal flow gain of one polarity and that of the reversed polarity. Symmetry is measured as the difference in normal flow gain of each polarity, expressed as percent of the greater.

#### 73.05.750 Servovalve Threshold

The increment of input current required to produce a change in servovalve output, expressed as percent of rated current. Threshold is normally specified as the current increment required to revert from a condition of increasing output to a condition of decreasing output.

### SECTION 06 — DYNAMIC CHARACTERISTICS

#### 73.06.200 Servovalve Amplitude Ratio

The ratio of the servovalve control flow amplitude to a sinusoidal input current-amplitude at a particular frequency divided by the same ratio at the same input-current amplitude at a specified low frequency. Amplitude ratio may be expressed in decibels where db = $20 \log_{10} AR$ (See Fig. 73.6).

#### 73.06.500 Servovalve Frequency Response

The complex ratio of servovalve control flow to input current as the current is varied sinusoidally over a range of frequencies. Frequency response is normally measured with constant input current amplitude and zero load pressure drop, expressed as amplitude ratio, and phase lag. Servovalve frequency response may vary with the input-current amplitude, temperature, supply pressure, and other operating conditions (See Fig. 73.6).

#### 73.06.800 Servovalve Phase Lag

The instantaneous time by which the servovalve sinusoidal flow follows the sinusoidal input current, measured at a specified frequency and expressed in degrees (See Fig. 73.6).

## SERVOVALVES
### FLOW CURVE

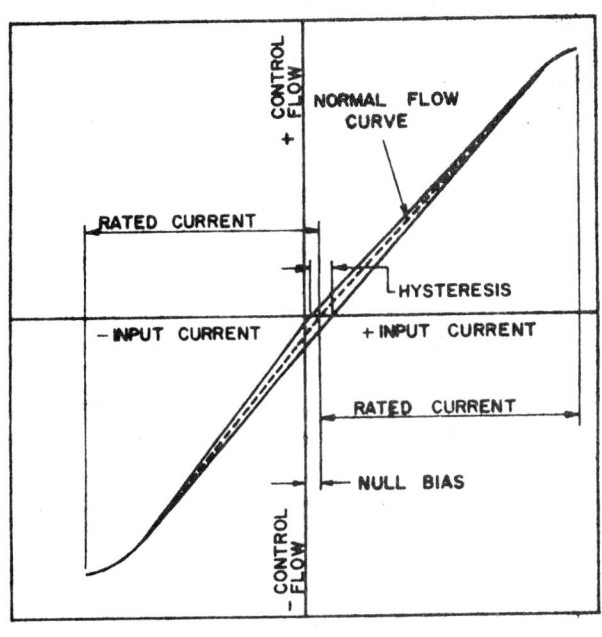

Fig. 73.1

## SERVOVALVES
### FLOW GAIN

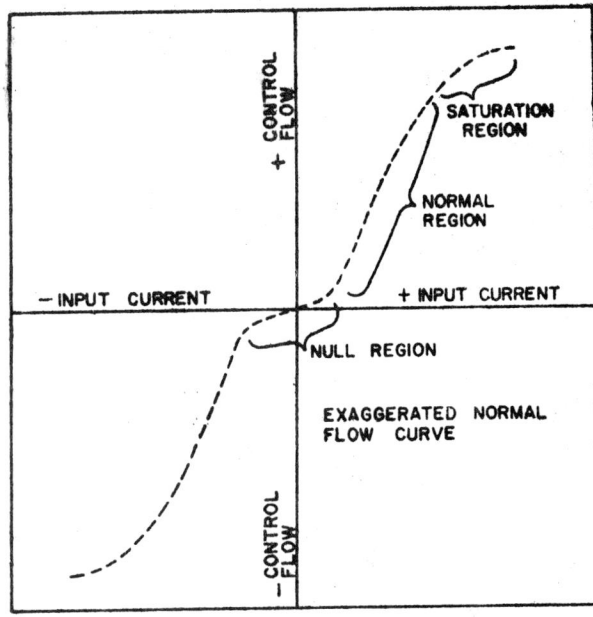

Fig. 73.2

## SERVOVALVES
### OVERLAP CONDITION

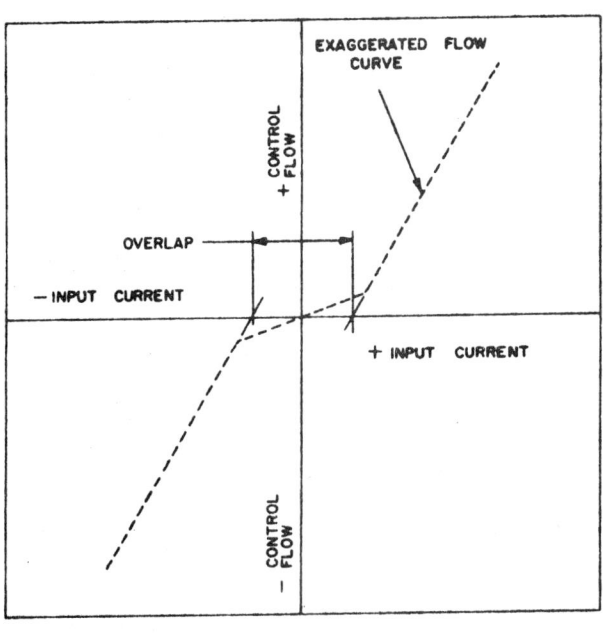

Fig. 73.3

## SERVOVALVES
### UNDERLAP CONDITION

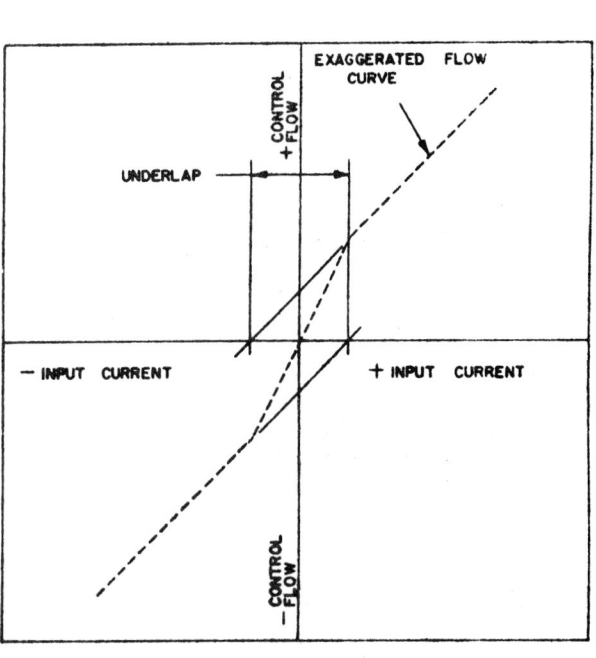

Fig. 73.4

## SERVOVALVES
### ZERO LAP CONDITION

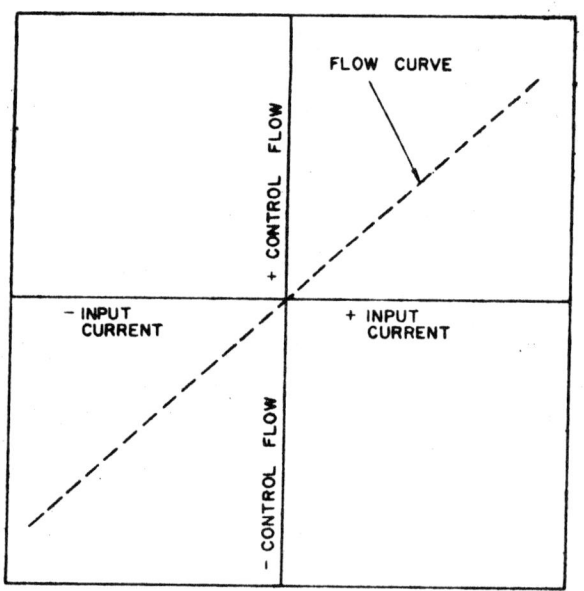

Fig. 73.5

## SERVOVALVES
### EXAMPLE OF SINUSOIDAL FREQUENCY RESPONSE CHARACTERISTICS

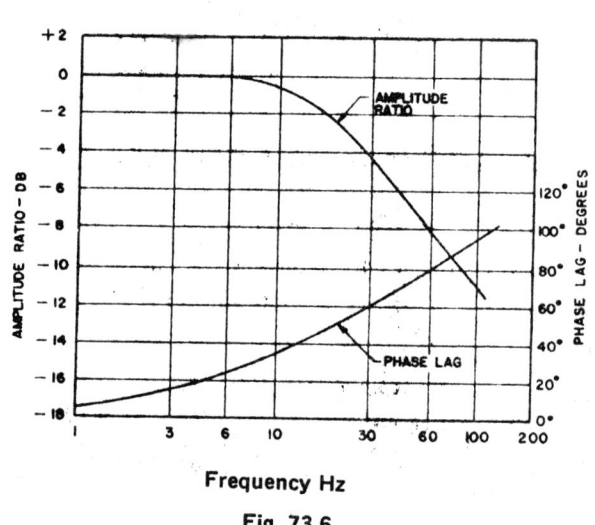

Frequency Hz

Fig. 73.6

## PRESSURE GAIN

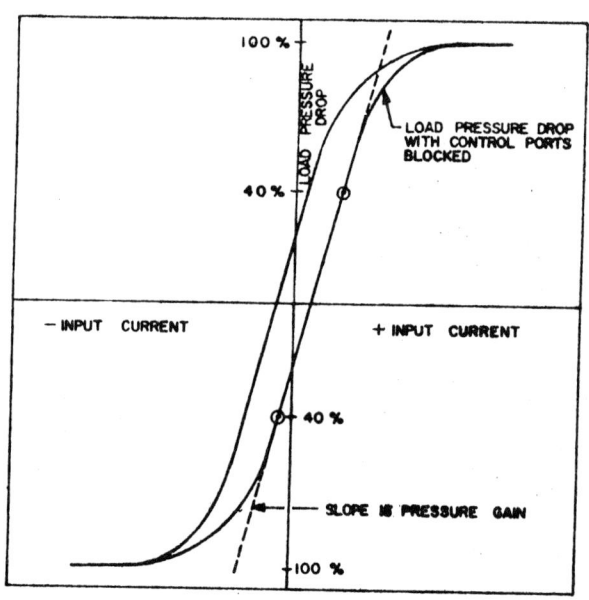

Fig. 73.7

# GROUP 74 — REGULATORS

Section 01 — General
Section 02 — Regulator Types
Section 03 — Operational Factors

## SECTION 01 — REGULATOR TYPES

**74.02.100**     Regulator, Air Line Pressure

A regulator which transforms a fluctuating air pressure supply to provide a constant lower pressure output.

**74.02.120**     Regulator, Constant Bleed Air Line Pressure

An Air Line Pressure Regulator that depends upon a bleed to atmosphere for proper operation.

**74.02.160**     Regulator, Pressure Relieving Air Line

An Air Line Pressure Regulator which automatically vents over-pressures applied to the regulated (secondary) pressure.

## SECTION 03 — OPERATIONAL FACTORS

**74.03.100**     Droop

The deviation between no flow secondary pressure and secondary pressure at a given flow.

**74.03.200**     Flow Characteristic Curve

The change in regulated (secondary) pressure occuring as a result of a change in the rate of air flow over the operating range of the regulator.

**74.03.300**     Maximum Inlet Pressure

The maximum rated gage pressure applied to the inlet port of the regulator.

**74.03.400**     Reduced Pressure Range

The adjustment range of the regulator.

**74.03.500**     Regulation Characteristic Curve

The change in regulated (secondary) pressure occuring as a result of a change in the supply (primary) pressure to a regulator.

**74.03.501**     Pressure, Regulation Steady State Supply

The region between the permitted limits of the supply pressure.

**74.03.600**     Relief Characteristic Curve

The change in the relief flow rate in a relieving type air line regulator which occurs as a result of an increase in regulated (secondary) pressure over set pressure.

# GROUP 75 — SWITCHES

Section 01 — General
Section 02 — Switch Type

## SECTION 02 — SWITCH TYPES

**75.02.400**     Switch, Float

An electric switch which is responsive to liquid level.

**75.02.500**     Switch, Flow

An electric switch operated by fluid flow.

**75.02.600**     Switch, Liquid Level

A device incorporating an electrical switch in which actuation of the contacts is effected at a predetermined level of the liquid.

**75.02.700**     Switch, Pressure

An electric switch operated by fluid pressure.

**75.02.750**     Switch, Pressure Differential

An electric switch operated by a difference in pressure.

# NOTES

NFPA Recommended Standard
T3.27.1-1972 (R1981)

**AN INDUSTRY STANDARD FOR FLUID POWER**

Glossary of Terms

for

Compressed Air Dryers

Approved as an NFPA Recommended Standard
27 January 1972

published by
# NATIONAL FLUID POWER ASSOCIATION, INC.
3333 N. Mayfair Road / Milwaukee, WI 53222 / 414-778-3344 / TLX 26898

# NFPA/T3.27.1

## FOREWORD

(This Foreword is not part of NFPA Recommended Standard Glossary of Terms for Compressed Air Dryers, NFPA/T3.27.1-1972)

Compressed air dryers are widely accepted as necessary equipment for reliable operation of many types of pneumatic systems. Unfortunately, the absence of standards for this important category of pneumatic components has resulted in confusion by users regarding the design, construction, and performance of dryers.

In 1968, NFPA recognized the need for a component section to oversee development of standards for compressed air dryers. All U.S. manufacturers of this type of equipment were invited to participate in a general conference held 15 July 1969 to determine if a consensus existed to justify the establishment of the proposed component section and project groups. As a result of that meeting, Section T3.27 was organized and charged with the responsibility to provide general and specific standards for compressed air dryers.

The initial project group T3.27.1, prepared this Glossary of Terms. Suggested lists of terms, with definitions, were submitted by individual members of the group. These suggestions were reviewed at a meeting in September 1969 and were then combined into a single draft of the glossary.

The draft was edited at later meetings and circulated for general industry review in May 1970. In order to minimize confusion, certain terms; term positions; and definitions that were superceded by T2.1-1970 (Glossary of Terms) were deleted or changed.

The Section reached consensus on the draft on 20 October 1970. The Ballot Draft was prepared and circulated on 4 June 1971. The document was approved in the balloting which ended 9 July 1971. Technical Board approval was granted on 22 September 1971.

The NFPA Board of Directors approved this document as NFPA Recommended Standard T3.27.1-1972 on 27 January 1972.

NFPA/T3.27.1

## MEMBERS OF THE PROJECT GROUP RESPONSIBLE FOR THE DEVELOPMENT OF THIS STANDARD

| | | |
|---|---|---|
| Seabrook, Derek | Project Chairman and Section Chairman | Flick-Reedy Corp. |
| Coffin, Tom | Section Vice Chairman | Pall Trinity Micro |
| Tomlinson, Ephraim | Section Secretary | Pall Trinity Micro |
| Kudlaty, Walter | Terminology Coordinating Committee Chairman | Marvel Engineering |
| Morgan, James I. | NFPA, ANSI and ISO Liaison | National Fluid Power Association |

| | |
|---|---|
| Anders, E. | Anders-Reufeisen, Inc. |
| Baughman, J. | Van Air, Inc. |
| Bloxham, A. | KelAir Division |
| Bowbin, J. | Flick-Reedy Corp. |
| Bowen, K. | Cargocaire Eng. Corp. |
| Brown, J. | Arrow Pneumatics |
| Bryan, J. | Hankison Corp. |
| Brzezinski, D. | Johnson Service Co. |
| Donikowski, R. | Van Air, Inc. |
| Fitch, S. | Bry-Air, Inc. |
| Funk, E. | Desomatic Products |
| Gifford, E. | E. Gifford Co. |
| Graham, H. | Deltech Engineering, Inc. |
| Harrison, D. | General Air Systems (Zurn) |
| Hartzer, M. | Scovill Fluid Power Div. |
| Henderson, J. | Henderson Engineering Co. |
| Huffman, L. | Deltech Engineering, Inc. |
| Kantro, R. | Miller Fluid Power Div. |
| Kuppe, J. | Arrow Pneumatics |
| Luciana, J. | ITT Bell & Gossett |
| Rapp, V. | General Air Drying (Zurn) |
| Shenk, A. | Van Air, Inc. |
| Shikasho, S. | ITT Bell & Gossett |
| Singh, G. | Pneumaflow, Inc. |
| Surprise, C. | Johnson Service Co. |
| Tucker, A. | Scovill Fluid Power Div. |
| VanderHorst, J. | Wilkerson Corp. |
| Watkins, L. | C. A. Norgren Co. |

## REFERENCES

1. American National Standard Glossary of Terms for Fluid Power, ANSI/B93.2-1971. (ISO/TC 131/SC 1 (USA-___)___)

NFPA/T3.27.1

# GLOSSARY OF TERMS FOR COMPRESSED AIR DRYERS

## INTRODUCTION

In pneumatic fluid power systems, power is transmitted and controlled thru a gas under pressure within an enclosed circuit. For reliable operation, many applications require installation of a compressed air dryer to remove water vapor and prevent condensation within the pneumatic system.

1. SCOPE

    To include a listing of terms, their definitions, and a numerical classification for technical terms used in the compressed air dryer segment of the fluid power industry.

2. PURPOSE

    2.1   To provide a unified glossary for the compressed air segment of the fluid power industry, educational programs and users of fluid power.

    2.2   To provide a convenient reference for technicians.

    2.3   To clarify terms and definitions for beginners.

    2.4   To simplify technical communications.

    2.5   To reduce interpretation errors.

    2.6   To encourage individuals to expand their working vocabulary.

    2.7   To promote a common understanding of compressed air dryer terms.

    2.8   To promote greater use of compressed air dryers.

3. TERMS AND DEFINITIONS

    For definition of other terms used herein, see Reference No. 1.

    (The terms set forth herein are presented in format for direct insertion in Section 37 of Reference No. 1 at its next revision.)

NFPA/T3.27.1

# GROUP 37 - COMPRESSED AIR DRYERS

Section 01 - General
Section 02 - Types of Compressed Air Dryers
Section 03 -
Section 04 - Operational Factors
Section 05 - Compressed Air Dryer Components

## SECTION 01 - GENERAL

Compressed Air Dryer                                       37.01.100

A device that lowers the dew point of compressed air.

NOTE. - "Dehumidifiers", "Dehydrators" and basic "Desiccators" are often erroneously considered as Compressed Air Dryers. Their primary function is to remove moisture from atmospheric pressure air or from solids and do not meet the strict definition of Compressed Air Dryers. Therefore, these terms are not included in the text.

## SECTION 02 - TYPES OF COMPRESSED AIR DRYERS

Compressed Air Dryer, Automatic                            37.02.100

A compressed air dryer in which all functions are automatically controlled.

Compressed Air Dryer, Demand Cycle                         37.02.200

A compressed air dryer which switches chambers in accordance with the drying load on the unit, rather than on a fixed cycle.

Compressed Air Dryer, Desiccant                            37.02.300

A compressed air dryer which lowers the dew point of compressed air by passing the air through a bed of desiccant.

Compressed Air Dryer, Deliquescent                         37.02.350

A desiccant compressed air dryer which utilizes absorbent desiccant and requires periodic addition of desiccant.

Compressed Air Dryer, Fixed Cycle                          37.02.400

A compressed air dryer which operates on a time-controlled cycle.

Compressed Air Dryer, Manual　　　　　　　　　　　　　　37.02.500

　　A compressed air dryer in which all functions are manually controlled.

Compressed Air Dryer, Refrigerated　　　　　　　　　　　37.02.600

　　A compressed air dryer that utilizes a mechanical refrigeration device to cool compressed air, condensing and separating moisture.

Compressed Air Dryer, Refrigerated, Cycling　　　　　　　37.02.620

　　A refrigerated compressed air dryer which automatically cycles on and off according to load.

Compressed Air Dryer, Refrigerated, Non-Cycling　　　　　37.02.660

　　A refrigerated compressed air dryer that runs continuously.

Compressed Air Dryer, Regenerative　　　　　　　　　　　37.02.700

　　A compressed air dryer which periodically processes the desiccant to restore its ability to absorb water vapor.

Compressed Air Dryer, Regenerative, Closed System　　　37.02.710

　　A regenerative compressed air dryer which does not exhaust purge air to the atmosphere.

Compressed Air Dryer, Regenerative, Dual Tower　　　　　37.02.720

　　A regenerative compressed air dryer which includes two desiccant chambers which are alternately cycled from drying to reactivation.

Compressed Air Dryer, Regenerative, Externally
　(Convection) Heated　　　　　　　　　　　　　　　　　37.02.730

　　A regenerative compressed air dryer which utilizes a heater(s) located externally to the desiccant bed.

Compressed Air Dryer, Regenerative, Internally Heated　　37.02.740

　　A regenerative compressed air dryer which utilizes a heater(s) located within the desiccant bed.

Compressed Air Dryer, Regenerative, Open System　　　　37.02.750

　　A regenerative compressed air dryer which exhausts purge air to the atmosphere.

Compressed Air Dryer, Semi-Automatic                          37.02.800

    A compressed air dryer in which some functions are automatically controlled while others are manually controlled.

## SECTION 03 -

(Reserved for future additions)

## SECTION 04 - OPERATIONAL FACTORS

Ambient Temperature Range                                      37.04.125

    The range of temperature of the air surrounding the dryer in which the equipment will perform as recommended.

Capacity, Compressed Air Dryer                                 37.04.190

    The amount of dry air delivered at recommended operating conditions.

Condensation                                                   37.04.200

    The process of changing a vapor into liquid condensate by the extraction of heat.

Contact Time                                                   37.04.250

    The time required for a molecule in a stream of air to pass completely through a desiccant bed based on superficial bed velocity.

Cycle, Adsorbent Dryer                                         37.04.280

    The time required for an adsorbent desiccant bed to pass through one drying period and one regeneration period.

Dew Point                                                      37.04.300

    The temperature at which vapors condense. For practical purposes, it must be referred to a stated pressure.

Dew Point, Atmospheric                                                  37.04.310

    The dew point of the air at atmospheric pressure.

Dew Point, Pressure                                                     37.04.340

    The dew point of the air at the actual operating pressure.

Dew Point Depression                                                    37.04.350

    The difference between inlet and outlet dew points of a compressed air dryer referred to the same operating conditions.

Evaporator Freeze-Up                                                    37.04.400

    Blocking of the air passages through the evaporator due to freezing of the condensate.

Fluidization Velocity                                                   37.04.450

    The rate of air flow upward through a desiccant bed, which if exceeded, will physically disturb the desiccant.

Period, Cooling                                                         37.04.560

    That portion of the regeneration cycle during which the desiccant is cooled.

Period, Drying (Sorption)                                               37.04.570

    That portion of the cycle during which the desiccant bed is onstream in drying service.

Period, Heating                                                         37.04.580

    That portion of the regeneration cycle during which the desiccant is heated.

Period, Regeneration (Desorption)                                       37.04.590

    That portion of the cycle during which the desiccant bed is removed from drying service and its efficiency restored.

Purge Flow                                                              37.04.600

    A flow of air to regenerate a desiccant.

NFPA/T3.27.1

Purge Flow, Co-Current     37.04.620

    Purge flow is in the same direction through the desiccant as drying flow.

Purge Flow, Countercurrent     37.04.630

    Purge flow direction through the desiccant is opposite to the direction of drying flow.

Regeneration (Reactivation)     37.04.650

    The process of restoring the capacity of adsorbent desiccant.

Regeneration, Atmospheric     37.04.660

    Reactivation of the desiccant bed at atmospheric pressure.

Regeneration, Heat     37.04.670

    Reactivation of desiccant by increasing its temperature.

Regeneration, Heatless     37.04.680

    Reactivation of desiccant without heat. It is usually done with dry air purge flow.

Regeneration, Pressure     37.04.690

    Reactivation of the desiccant bed at or near operating pressure.

Repressurization     37.04.700

    Return of the regenerated desiccant chamber from atmospheric pressure to operating pressure prior to chamber switchover.

Superficial Bed Velocity     37.04.750

    The rate of air flow through the cross-sectional area of the desiccant bed, without regard for the area consumed by the desiccant.

## SECTION 05 - COMPRESSED AIR DRYER COMPONENTS

Afterfilter     37.05.050

    A filter which follows the compressed air dryer usually for the protection of the downstream equipment from desiccant dust.

Automatic Drain 37.05.100

A device which automatically discharges condensate from the moisture separator.

Condensing Unit 37.05.200

A specific refrigerant machine combination for a given refrigerant consisting of one or more power driven compressors, air or water cooled condensers, etc.

Desiccant 37.05.250

Material that tends to remove moisture from compressed air.

Desiccant, Absorbent (Deliquescent) 37.05.255

A desiccant that dissolves into the moisture it removes from the compressed air and is slowly consumed in the process.

Desiccant, Adsorbent 37.05.260

A solid desiccant which is capable of removing moisture from compressed air by adherence of moisture to its surface.

Desiccant, Tabular Support 37.05.280

Inert material which supports a desiccant bed and diffuses air.

Evaporator 37.05.300

The heat exchanger where the refrigerant absorbs heat.

Evaporator Back-Pressure Valve 37.05.350

A valve which maintains the refrigerant pressure, and consequently its temperature, in the evaporator at a predetermined level.

Expansion Device 37.05.400

A device which controls expansion of high pressure liquid refrigerant to a lower pressure. It may be a capillary tube, thermostatic expansion valve, or automatic expansion valve.

High Side Components 37.05.440

Parts of a refrigeration system under condenser pressure or higher.

NFPA/T3.27.1

Hot Gas By-Pass Valve  37.05.450

    A modulating valve which by-passes hot refrigerant gas from the high pressure to the low pressure side of the system in order to reduce refrigeration capacity commensurate with reduction in load and to control evaporation temperature.

Low Side Components  37.05.490

    Parts of a refrigeration system at or below evaporation pressure.

Moisture Separator  37.05.500

    A device which removes liquids from an air stream.

Precooler Reheater  37.05.550

    A heat exchanger which lowers the temperature of the inlet air and raises the temperature of the exiting air.

Prefilter  37.05.600

    A filter which precedes the compressed air dryer usually for the protection of desiccant or heat transfer surfaces.

Refrigerant  37.05.650

    A substance which produces a cooling effect by its absorption of heat while expanding or vaporizing.

Refrigerant Gauge  37.05.700

    A pressure gauge calibrated to indicate the corresponding temperature of saturated refrigerant.

Refrigerant Receiver  37.05.750

    A liquid refrigerant storage tank in a refrigeration system.

Refrigeration Compressor  37.05.800

    The part of a refrigeration system which takes refrigerant at low pressure and compresses it to a smaller volume at a higher pressure.

Refrigeration Compressor, Hermetic  37.05.830

    A refrigeration compressor which is sealed in a housing with its driving motor.

NFPA/T3.27.1

Refrigeration Compressor, Open-Type  37.05.840

A refrigeration compressor which is driven by a physically-separated power source through mechanical power transmission equipment.

NFPA Recommended Standard
T1.21.1-1978 (R1983)

**AN INDUSTRY STANDARD FOR FLUID POWER**

Procedure for Self-Certification

by Fluid Power Manufacturers

Approved as an NFPA Recommended Standard
25 October 1978

published by
**NATIONAL FLUID POWER ASSOCIATION, INC.**

3333 N. Mayfair Road  /  Milwaukee, WI 53222  /  414-778-3344  /  TLX 26898

NFPA/T1.21.1

# FOREWORD

This Foreword is not part of NFPA Recommended Standard Procedures for Self-Certification by Fluid Power Manufacturers, NFPA/T1.21.1-1978.

In the early 1970's the Technical Board recognized the need to establish guidelines for self-certification. Work was undertaken, but in 1974 the Technical Board halted its self-certification work on recommendation of legal counsel and pending release of a FTC opinion regarding certification programs. The Technical Board, after consulting legal counsel, reactivated the ad-hoc committee on self-certification at its 7 April 1976 meeting.

The project was initiated and Draft No. 1 was written at the first meeting of the Self-Certification Ad-Hoc group meeting of 14 September 1976. The document was reviewed and Draft No. 2 resulted from the 12 January 1977 meeting. The Technical Board, at its 2 Feburary 1977 meeting approved the Title, Scope and Purpose. Draft No. 2 was reviewed at the 9 August 1977 meeting and the resulting changes were incorporated into a final working draft.

Headquarters prepared the document for General Review on 12 August 1977. The Project Group met on 26 October 1977 to discuss and answer the negative comments received through General Review. Chairman Kay responded to the commentors in writing during December 1977. One comment concerning self-certification to corporate standards remained unresolved until the 18 January 1978 meeting of the NFPA Technical Board. At that meeting, the Board agreed to include corporate standards on the basis that their use would enhance the overall usefulness of the self-certification procedure. The Board also granted approval to ballot. NFPA Headquarters prepared the Ballot Draft on 14 February 1978.

Three negative ballots were received on this document. At it's 21 September 1978 meeting, the NFPA Technical Board discussed the negative comments and concurred that the negative votes could not be resolved. The Technical Board voted to recommend to the Board of Directors that this document be approved as an NFPA Recommended Standard.

The NFPA Board of Directors granted final approval to NFPA/T1.21.1-1978 on 25 October 1978.

NFPA/T1.21.1

## PROJECT GROUP MEMBERS WHO DEVELOPED THIS STANDARD

| | | |
|---|---|---|
| Kay, Robert | Project Chairman | Sperry Vickers |
| Ratkay, Edward | Technical Auditor | Commercial Shearing, Inc. |
| Luecke, John R. | Director of National Technical Services | National Fluid Power Association* |

| | |
|---|---|
| Chenoweth, R. | Abex/Denison Div. |
| Johnson, J. | Milwaukee School of Engineering |
| Sallberg, D. | HUSCO/Pegasus |
| Stockwell, S. | J I Case Company |

## REFERENCES

1. American National Standard Glossary of Terms for Fluid Power, ANSI/B93.2-1971, and Supplement ANSI/B93.2A-1978. (ISO/DP 5598)

2. American National Standard, General Requirements for a Quality Program, ANSI/Z1.8-1971.

\* Company affiliation has changed since work with the Project Group.

# NFPA/T1.21.1

## PROCEDURE FOR SELF-CERTIFICATION

## BY FLUID POWER MANUFACTURERS

### INTRODUCTION

In fluid power systems, power is transmitted and controlled thru a fluid under pressure within an enclosed circuit. Self-certification provides a standard procedure for fluid power manufacturers who wish to certify that their products or services meet specific industry standards and/or specifications.

1. SCOPE

   To include:

   1.1   A standard self-certification procedure for use by fluid power manufacturers.

   1.2   Products and/or services supplied by fluid power manufacturers and certified to technical society, trade association, national, international, government, and publicly available corporate standards or specifications.

2. PURPOSE

   To provide a standardized procedure to be followed when a fluid power manufacturer certifies the product or service to technical society, trade association, national, international, or government and publicly available corporate standards or specification(s).

3. TERMS AND DEFINITIONS

   For definition of terms not defined below, see Reference No. 1.

   3.1   Certification. The procedure by which a product or service becomes certified.

   3.2   Certified. Attested by the manufacturer under the procedures of this recommended standard as satisfying the requirements of the referenced standard(s) and/or specification(s).

3.3 Inspection. The process of examining, measuring, testing, gaging, or otherwise verifying that the product or service conforms with applicable provisions of referenced standard(s) and/or specification(s).

3.4 Self-Certification. Certification to the user by the manufacturer, on his own authority, that a product or service is in compliance with referenced standard(s) and/or specification(s).

3.5 Quality Control. The overall system of activities of the manufacturer to assure that manufactured products or services do in fact comply with requirements of the referenced standard(s) and/or specification(s).

3.6 Specification. A statement of requirements.

3.7 Standard. A document (or an object for physical comparison) for use in defining product characteristics, products, or processes. It is prepared by a consensus of a properly constituted group of those substantially affected and having the qualifications to prepare the standard.

4. STANDARD(S) AND/OR SPECIFICATION(S) REFERENCED

 4.1 Select referenced standard(s) and/or specification(s) from the following:

  4.1.1 International, national, and foreign national organizations.

  4.1.2 Technical societies, trade associations, agencies, or other organizations of national scope and recognition.

  4.1.3 Government

  4.1.4 Corporations whose standard(s) and/or specification(s) are publicly available.

  NOTE: To the fullest extent possible, NFPA will assist the reader in finding sources for referenced standard(s) and/or specification(s).

NFPA/T1.21.1

    4.2    Use the referenced standard(s) and/or specification(s) in its entirety, unless full disclosure of limitations is made in the statement of self-certification.

5. QUALITY CONTROL

    5.1    Maintain a quality control system sufficient to ensure that product or service provides and consistently meets the criteria contained in the referenced standard(s) and/or specification(s).

    5.2    Maintain the quality control system in accordance with American National Standard "Specification of General Requirements for a Quality Program," ANSI/Z1.8-1971, (Reference No. 2) and any other quality control standards agreed to by the manufacturer and buyer.

    NOTE: ANSI/Z1.8-1971 includes requirements for:

        (A)    Quality Management
        (B)    Design Information
        (C)    Procurment
        (D)    Material Control
        (E)    Manufacture
        (F)    Acceptance
        (G)    Measuring Instruments
        (H)    Quality Information

6. VERIFICATION TESTS AND DOCUMENTATION

    6.1    Carry out tests and/or inspection in accordance with the referenced standard(s) and/or specification(s).

    6.2    Use managerial controls to assure that <u>samples</u> used for testing and/or inspection are representative of the product or service supplied.

    6.3    Maintain a record of results from samples used to verify compliance.

    6.4    Make results available to the buyer, upon legitimate and responsible requests.

NFPA/T1.21.1

7. STATEMENT OF SELF-CERTIFICATION

    7.1    Include the following in a formal statement of self-certification.

        7.1.1    Manufacturer's designation for the product or service certified.

        7.1.2    Statement of conformance with applicable referenced standard(s) and/or specification(s).

        7.1.3    Identification of the manufacturer as the certifier.

    7.2    Examples:

Word the statement of self-certification as follows:

        7.2.1    <u>Valve XYZ-0123</u> conforms
(Manufacturer's designation for product or service)

            to  <u>ANSI/B93.7-1968 (R1973)</u> as certified
(Referenced standard(s) and/or specification(s))

            by  <u>XYZ Valve Manufacturing, Inc.</u> in accordance
(Identification of Manufacturer)

with NFPA/T1.21.1-1978.

        7.2.2    <u>Pneumatic FRL Pressure Drop Test</u> conducted per
(Manufacturer's designation for product or service)

<u>NFPA/T3.12.6-1975</u> as certified
(Referenced standard(s) and/or specification(s))

            by  <u>FRL's Inc.</u> in accordance
(Identification of Manufacturer)

with NFPA/T1.21.1-1978

    7.3    Certification extends beyond the initial buyer <u>only</u> when such buyer has not made alterations, changes, or additions to the product or service as produced or received, except by express agreement with the manufacturer. Any such alterations, changes, or additions not expressly approved by the manufacturer invalidates the manufacturer's certification.

NFPA/T1.21.1

       7.4    Use of this Recommended Standard does not authorize the use of NFPA Register Trademark.

             NOTE: Only NFPA members are permitted to use the trademark on product literature, etc.

8. KEY WORDS

The following Key Words, useful in indexes and in information retrieval systems are suggested for this recommended standard:

self-certification, documentation

self-certification, statement

self-certification, verification

fluid power

NFPA Recommended Standard
T2.6.1-1974 (R1982)

## AN INDUSTRY STANDARD FOR FLUID POWER

**This Standard is now under review for possible revision see TSP**

Method for Verifying the Fatigue and Static Pressure Ratings of the Pressure Containing Envelope of a Metal Fluid Power Component

Approved as an NFPA Recommended Standard
18 February 1974

published by
**NATIONAL FLUID POWER ASSOCIATION, INC.**
3333 N. Mayfair Road  /  Milwaukee, WI 53222  /  414-778-3344  /  TLX 26898

# NFPA/T2.6.1

## FOREWORD

This Foreword is not part of NFPA Recommended Standard Method for Verifying the Fatigue and Static Pressure Ratings of the Pressure Containing Envelope of a Metal Fluid Power Component, NFPA/T2.6.1-1974.

Early in 1968 producers and users of fluid power components expressed a need for a uniform method for verifying a Pressure Rating for individual components. It was felt that the time had come for industry to establish a suitable set of rules in the form of a standard.

A search of the existing codes in thirty-one states indicated that those quoted were inapplicable to our industry since they covered products not related to hydraulics, pneumatics, fluidics and/or other fluid power areas. This search showed further that there were no known federal regulations covering this subject; and that regulations written by private agencies such as ASME, ASTM, etc., were not applicable. (NFPA will continue to reference material specifications judged applicable and prepared by such societies.)

As the result of the search, the committee developed a belief that there is no applicable reference point suitable for adoption or supplementation. The committee would, therefore, have to prepare its own guidelines and cite references as developed.

The Pressure Rating General Technical (Coordinating) Committee was organized in February 1968. It consisted of Members-at-large who are skilled in the theory of Pressure Rating. In addition, each component section of the NFPA Technical Board was asked to name a representative who would be a specialist for that particular component.

Thru a series of meetings and theoretical investigations, the committee was able to produce their first draft in December, 1970. After a series of five drafts and reviews, committee consensus was reached on 1 February 1972. The General Review Draft was prepared on 6 March 1972.

To aid in understanding the proposal, and to facilitate the General Review process, a general conference was held in Chicago on 23 March. More than 100 representatives of 50 member companies took advantage of the conference to learn the document's history, background and theory and to have their questions answered by those preparing the proposal.

All comments on record were resolved by the Committee at their 18 May 1972 meeting and at their 29 June 1972 meeting when those submitting comments were in attendance or represented. A Ballot Draft was prepared on 11 July 1972 taking into account all the technical refinements and editorial clarifications agreed upon in the process of resolving all comments. This preliminary Ballot Draft was reviewed and revised by the Committee officers and the Ballot Draft was prepared on 19 July 1972.

At their regular meeting on 27 September 1972, the NFPA Technical Board reviewed the proposed Ballot Draft and all the comments which had been submitted. The Board made minor editorial modifications to the Introduction, Section 21 and clause 17.2 and released the proposal for ballot by Member Companies. The Ballot Draft was prepared on 2 October 1972. Balloting closed on 6 November 1972.

Upon recommendation by the Technical Board, the Committee met with representatives of the Cylinder Section, who constituted the majority of the negative voters, to resolve their comments. Agreement was reached by providing for RFP Verification by Similarity. In a related action, the Technical Board Steering Committee proposed, and the Board of Directors and Technical Board approved, a modified sense of direction whereby document T2.6.1 should be immediately approved as an industry-wide philosophy and general intent with limited specific applicability. Document T2.6.1 was modified to facilitate its becoming useable within 12 months of its date of approval thru approximately 15 supplementary NFPA Recommended Standards, each setting forth specific details for testing and extending test data to short-run and one-of-a-kind situations thru verification by similarity.

The Committee met on 28 March 1973 and 6 June 1973 and modified this proposal to incorporate the foregoing and to resolve other comments. An accelerated rereview and reballot procedure was adopted in accordance with the Board of Directors' mandate for "immediate action." A rereview draft was completed on 11 June 1973 and was immediately circulated to all Member Companies, all members of the Pressure Rating Project Groups and members of the NFPA Board of Directors and Technical Board for a review which closed on 25 June 1973.

Eighteen pages of comments were received from 8 Member Companies. All comments were resolved thru meetings on 28 June and 26 July 1973; and thru mail action concluded on 31 August 1973. In addition to many editorial improvements and clarifications, a new fourth paragraph was added to the Introduction suggesting a method for estimating the pressure that an individual component can normally be subjected to based upon its relationship with the rated fatigue and static pressures. Figure 1 was also added. Another significant modification was the addition of clause 14.10, thereby permitting the selection of singular assurance and confidence levels in specific component standards.

A 27 September 1973 draft was circulated for a 10-day review by the committee and others involved in the resolution of comments. As a result of this 10-day review, Figure 1 was improved; the listing of Project T3.8.12 was merged with T3.15.8; and many editorial improvements were made to clarify proposal T2.6.1. Also, Reference A16 was added and improvements made to the Tutorial Reference.

Ballot Draft No. 2 was prepared on 24 October 1973. The balloting of this draft closed on 16 November 1973, and three of the five negative ballots were resolved on 20 November 1973.

The Technical Board recommended approval of this proposed standard to the Board of Directors on 28 November 1973. The Board of Directors approved T2.6.1 as an NFPA Recommended Standard on 18 February 1974.

MEMBERS OF THE PROJECT GROUP THAT DEVELOPED THIS STANDARD:

**Schroeder, Reed**
Committee Chairman
Schroeder Brothers

**Skaistis, Stan**
Committee Vice Chairman (May 1972 to present)
Sperry Vickers

**Barnes, Robert**
Committee Vice Chairman (February 1968 to May 1972)
AMF Cuno Division

**Morgan, James I.**
Secretariat
National Fluid Power Association

# NFPA/T2.6.1

**Members-at-large**

**Cordes, H.**
Pall Corporation

**Faust, D.**
C. A. Norgren

**Forster, W.**
WABCO Fluid Power Division

**Glidden, J.**
Hydreco

**Johnson, J.**
MSOE

**Stephens, T.**
Gresen Mfg. Co.

**Members-Section Representatives**

**Barthe, H.**
Schroeder Brothers
Filter & Separator

**Berninger, J.**
Parker Hannifin Corp.
Cylinder

**Bowbin, J.**
Miller Fluid Power
Air Dryer

**Dodson, R.**
Clayton Mark
Conductor

**Jacoby, H.**
Webster Electric
Hydraulic Valve

**Lamb, T.**
Parker Hannifin Corp.
Quick Disconnect

**Ratkay, E.**
Commercial Shearing
Pump & Motor

**Schwarz, A.**
ITE Imperial Corp.
Fittings

**Sessoms, W.**
Scovill Fluid Power
Pneumatic FRL

**Zajdler, A.**
Superior Hydraulics
Accumulators

**Other Participants**

**Allen, C.**
WABCO Fluid Power Division

**Bailey, R.**
C. A. Norgren Co.

**Brake, C.**
Scovill Fluid Power

**Czarnecki, G.**
DeLaval - IMO Division

**\*Flock, H.**
Garlock/OM

**Greenwood, M.**
ITE Imperial Corp.

**Kosmak, M.**
Marvel Engineering Co.

**Larson, B.**
Eaton Corp.

**May, R.**
Aeroquip

**Olson, J.**
Applied Power/Dynex

**Pauken, D.**
Double A Products

**Prevallet, D.**
Gresen Mfg. Co.

**Ritchie, R.**
Aeroquip

**Roth, R.**
Flodar Corp.

**Schultz, H.**
Garlock/OM

**Sethi, I.**
Abex Corp.

**Smith, H.**
U. S. Army - MERDC

**Trainor, T.**
Battelle Institute

**Woodworth, R.**
Moog, Inc.

## REFERENCES

1. **American National Standard Glossary of Terms for Fluid Power, ANSI/B93.2-1971, and Supplement thereto.** (ISO/TC 131/SC 1 [USA-2] 3)

2. **SI units and recommendations for the use of their multiples and of certain other units, ISO 1000-1973.**

## TUTORIAL REFERENCE
(Attached as Appendix)

1. **National Fluid Power Association Tutorial Reference, File T2.6.1, 24 October 1973.** (Includes listing of all background references.)

\*Company affiliation has changed since work with the Project Group.

NFPA/T2.6.1

# METHOD FOR VERIFYING THE FATIGUE AND STATIC PRESSURE RATINGS OF THE PRESSURE CONTAINING ENVELOPE OF A METAL FLUID POWER COMPONENT

## INTRODUCTION

In fluid power systems, power is transmitted and controlled thru a fluid (liquid or gas) under pressure within an enclosed circuit. A basic requirement of fluid power components is that they should be capable of adequately containing the pressurized fluid.

This recommended standard establishes a group of common requirements intended to provide an industry-wide philosophy and basic standard, providing a rationale for judging a component's ability as a pressure containing envelope. Although this recommended standard's specific applicability is limited, it does immediately establish a uniform base for subsequent, more specific NFPA Proposed Recommended Standards for individual fluid power components. The proposed documents* listed below will implement Recommended Standard NFPA/T2.6.1-1974.

| | |
|---|---|
| Accumulators | NFPA/T3.4.7-19xx |
| Hydraulic Valves | NFPA/T3.5.26-19xx |
| Cylinders | NFPA/T3.6.29-19xx |
| Cylinders | NFPA/T3.6.31-19xx |
| Pumps and Motors | NFPA/T3.9.22-19xx |
| Hydraulic Filters & Separators | NFPA/T3.10.5.1-19xx |
| Pneumatic Filters, Regulators and Lubricators | NFPA/T3.12.10-19xx |
| Tube and Fitting Assemblies | NFPA/T3.15.8-19xx |
| Hydraulic Reservoirs | NFPA/ T3.16.8-19xx |
| Quick Disconnect Couplings | NFPA/T3.20.8-19xx |
| Pneumatic Valves | NFPA/T3.21.4-19xx |
| Compressed Air Dryers | NFPA/ T3.27.5-19xx |
| Fluid Logic Devices | NFPA/ T3.28.10-19xx |
| Pressure Switches | NFPA/T3.29.2-19xx |

Ratings verified in accordance with this standard do not replace pressure ratings based on considerations such as performance, bearing capacity, leakage and heat rejection. Instead, these new ratings are intended to supplement those that are provided by present practice and may be numerically different from prior ratings.

* NOTE - The above listing is correct as of the date of publication, but is subject to change.

The pressure that an individual component can normally be subjected to has a relationship with the rated fatigue and static pressures. This relationship may be estimated and used as a basis of total life expectancy for the component in an individual application. Such an estimate must be applied by the user and factors such as shock, heat, misuse, etc., must be judged by the user in its application. The selection of a specific pressure and life expectancy for a component in such an individual application may be based upon the rated fatigue and static pressure as prescribed in Figure 1.

Although for test verification, RFP is defined as $10^7$ cycles (see clause 3.1), the generally accepted conservative design practice when using this type of S-N curve requires that the locus for the line separating the verified/unverified is based upon $10^6$ cycles.

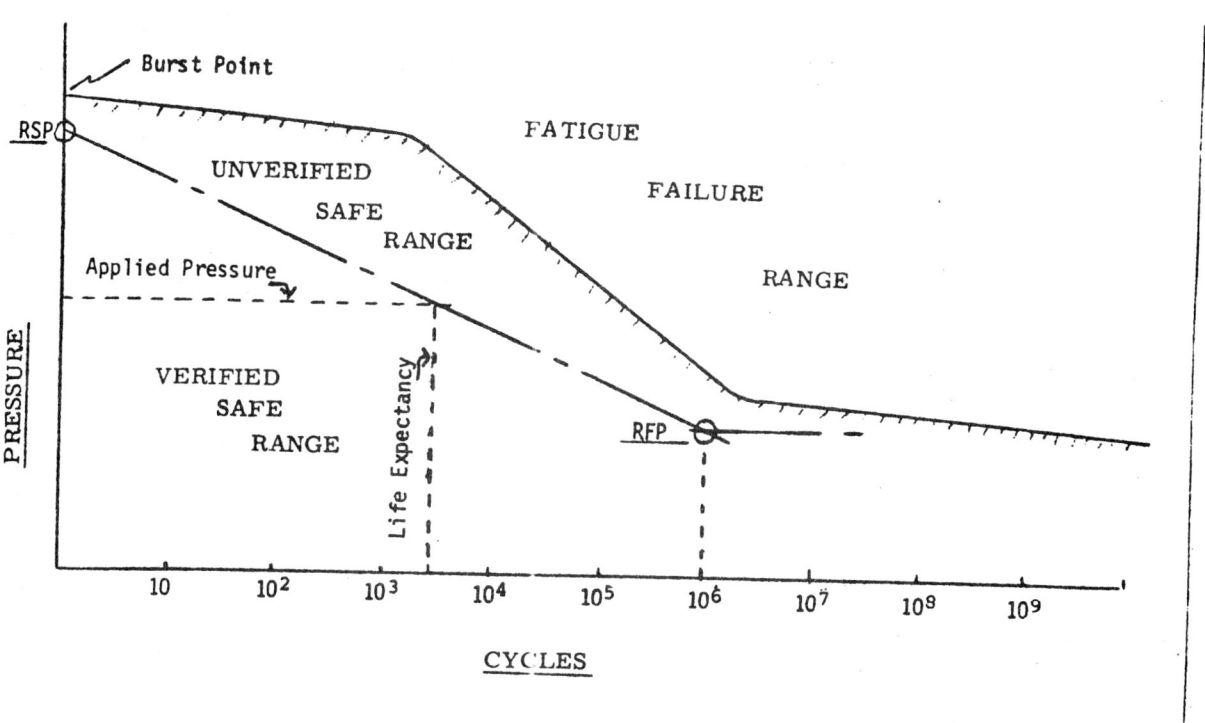

FIGURE 1 - Possible S-N curve method for estimating finite life rating

It should be noted that this document deals solely with verifying the fatigue and static pressure ratings of a component's pressure containing envelope. Separate from this verification procedure, manufacturers have the continuing responsibility to utilize managerial controls necessary to maintain similarity between the test and production pressure containing envelopes.

NFPA/T2.6.1

1. SCOPE

    1.1    To include methods for verifying ratings of the fatigue and static strengths of metal fluid power component pressure containing envelopes by test; or by analysis by geometric similarity (based on tests conducted in accordance with this recommended standard) that are:

        1.1.1    Subjected to pressure induced stresses.

        1.1.2    Used in the -20° to 200°F (-29° to 93°C) fluid temperature range.

        1.1.3    Not subject to loss of strength due to corrosion or other chemical action.

        1.1.4    Made of aluminum, magnesium, steel, iron, copper-based alloys and stainless steels. (Other metals will be included in subsequent revisions as soon as suitable data is available.)

    1.2    Hydraulic hoses and components constructed primarily of non-metals are not included in this recommended standard.

    1.3    This recommended standard does not establish a standard means for determining the magnitude of pressure transients in systems.

    NOTE - These magnitudes are usually determined by the system designer who then selects components with adequate strengths. However, it is generally recognized that there is very little authoritative data available to assist the system designer in making this determination. It is anticipated that the promulagation of this recommended standard will also stimulate the study of the realtionships between peak and operating pressures.

2. PURPOSE

To provide standard methods for verifying the fatigue and static pressure ratings of a metal fluid power component's pressurized boundaries with regard to sustaining cyclic and steady pressure loads that can be used as part of specific component standards.

NFPA/T2.6.1

3. TERMS AND DEFINITIONS

(For definition of other fluid power terms used, see Reference No. 1.)

3.1 Rated Fatigue Pressure. A pressure that a component pressure containing envelope is represented to sustain 10 million times without failure.

3.2 Rated Static Pressure. A pressure that a component pressure containing envelope is represented to sustain without failure.

3.3 Cyclic Test Pressure. A pressure range applied in cyclic tests that are performed to verify a Rated Fatigue Pressure.

3.4 Static Test Pressure. A pressure applied in static tests performed to verify a Rated Static Pressure.

3.5 Test Duration Factor. A multiplier applied to the Rated Fatigue Pressure for calculating the Cyclic Test Pressure which will permit shortening the qualification test from 10 million to one million cycles.

3.6 Variability Factor. A multiplier applied to the Rated Fatigue Pressure for calculating the Cyclic Test Pressure to account for the variability in fatigue strength of metals. It is also applied to the Rated Static Pressure for calculating the Static Test Pressure.

3.7 Assurance Level. The minimum percentage of pressure containing envelopes of a verified design that will sustain 10 million applications of its Rated Fatigue Pressure.

3.8 Confidence Level. The degree to which a manufacturer can be assured that the desired assurance level is attained.

3.9 Pressure Containing Envelope Failure. A structural fracture, crack or excessive seal leakage caused by deformation resulting from an RFP verification test; or a structural fracture, crack, or excessive seal leakage caused by deformation or permanent deformation which interfered with component functioning resulting from an RSP verification test.

NFPA/T2.6.1

The following terms, common in fluid power terminology, are defined in general usage dictionaries so therefore do not appear in Reference No. 1. Because these terms have more than one accepted meaning, we have set forth below the meaning applied to their use in this document.

3.10    Component. A complete product or assembly (e.g., a valve, pump, fitting).

3.11    Pressure Containing Envelope. That part of the component which contains the pressurized fluid.

3.12    Elements. The integral pieces which make up a part or component (e.g., cylinder end cap, filter element, etc.).

The following terms, not included in this document but found in related reference material, are defined within the meaning that the Committee applied when using the reference.

3.13    Peak Pressure. The maximum pressure actually experienced by a fluid power component in application. It is caused by any number of events, individually or in combination, such as (but not limited to), impact, inertia, shock, heat, etc.

3.14    Nominal Pressure. A pressure value assigned to a component or a system for the purpose of convenient designation. This may be an average of normal values without peaks, or a design figure for stress calculations.

4.    UNITS

4.1    The "Customary US" units are used.

4.2    Approximate conversions to the International System of Units (SI) are given per Reference No. 2. These appear after their "Customary US" counterpart in parentheses.

5.    LETTER SYMBOLS

|  |  |
|---|---|
| RFP | Rated Fatigue Pressure |
| RSP | Rated Static Pressure |
| CTP | Cyclic Test Pressure |
| STP | Static Test Pressure |
| $K_N$ | Test Duration Factor |
| $K_V$ | Variability Factor |

NFPA/T2.6.1

6. OUTLINE OF PROCEDURES FOR RFP VERIFICATION BY TEST

Select an RFP, then:

6.1 Select a $K_N$ factor per clause 12.2.1.

6.2 Select a $K_V$ factor per clause 12.2.2.

6.3 Calculate the CTP, per clause 12.2.

6.4 Use test equipment per clause 10.1.

6.5 Test the number of pressure containing envelopes upon which the factor $K_V$ was selected in clause 6.2.

6.6 Provide pressure pulses per clause 10.1.

6.7 Subject the pressure containing envelope(s) being tested per section 12 to the number of test cycles upon which the factor $K_N$ was selected in clause 6.1.

6.8 Verify the RFP per section 15.

6.9 Record data per section 17.

7. RFP VERIFICATION BY SIMILARITY

7.1 If it is desired to verify the Rated Fatigue Pressure of components by geometric similarity, base verification upon tests conducted in accordance with proposed recommended standards for individual fluid power components; and use the rules for similarity also included therein.

7.2 Assurance and confidence levels of verification made by geometric similarity are the values derived in the test that provides the basis of the similarity analysis.

8. OUTLINE OF PROCEDURES FOR RSP VERIFICATION BY TEST

Select an RSP, then:

8.1 Select a $K_V$ factor per clause 12.2.2.

8.2 Calculate the STP per clause 13.2.

8.3 Use test equipment per clause 10.2.

8.4 Test the number of pressure containing envelopes upon which the factor $K_V$ was selected in clause 8.1.

8.5 Conduct static tests per section 13.

8.6 Verify the RSP per section 16.

8.7 Record data per section 17.

9. PREPARATIONS FOR TESTING

9.1 Use liquid or gas to pressurize pressure containing envelopes being tested.

9.2 CAUTION: Take special precautions, while conducting any of the pressure tests, in the design of test rigs, and in training the operators to maximize operator safety.

9.3 Bleed the entrapped air from the circuit and from the pressure containing envelope being tested; and bleed gas charges from accumulators before tests are started when liquids are used.

9.4 Apply different ratings to separate portions of the pressure containing envelope as required by operating or component failure mode conditions.

9.5 Allow all drains and low pressure ports that are not part of the pressure containing envelopes to drain freely and keep them at atmospheric pressure.

9.6 Test pressure containing envelopes whose geometries, material properties and manufacturing methods are known and representative of normal production.

9.7 Do not perform manufacturer's proof tests prior to cyclic testing, unless such tests are made on all production pressure containing envelopes.

9.8 Perform cyclic and static tests on separate pressure containing envelopes, if possible.

9.9 Perform the cyclic tests first, when both RFP and RSP verification tests must be made on the same pressure containing envelopes.

NFPA/T2.6.1

9.10  Complete machining and processing of test components to the degree necessary for duplicating operating stress distributions and final strength in the pressure containing envelope.

9.11  It is permissible to make modifications to the test piece to facilitate cyclic or static tests providing that such modifications do not increase the pressure capabilities of the pressure containing envelope being tested.

10. TEST EQUIPMENT

   10.1  Cyclic Test

   10.1.1  Use a circuit on the test stand which will produce repeatable pressure pulses which cycle between approximately zero and a specified maximum pressure.

   NOTE - The specific design of the test stand and circuit is optional.

   10.1.2  Use any suitable non-corrosive test fluid.

   10.1.3  Mount the pressure measuring instrument directly into the pressure containing envelope being tested, or closely to it, thru a pressurized port that is not being used to supply the test fluid, whenever possible.

   10.1.4  Minimize restrictions between the instrument and the pressure containing envelope being tested, when measurements must be made in the pressure supply line.

   NOTE - It is recommended that it be verified that the pressure generated in the pressure containing envelope at the actual cycling rate be the intended value.

   10.1.5  Provide adequate personnel protection.

   10.2  Static Test

   10.2.1  Use a circuit on the test stand which provides a stable and controllable hydrostatic fluid pressure.

   NOTE - The specific design of the test stand is optional.

10.2.2 Provide adequate personnel protection.

10.2.3 Use any suitable non-corrosive test fluid for the test stand.

## 11. TEST CONDITIONS ACCURACY

Set up and maintain equipment accuracy so that data is accurate within (±) the following:

    CTP          3%
    STP          2%
    Temperature  5°F (3°C)

NOTE - Unless the frequency response of the measurement system or its components is high enough to faithfully reproduce the pressure waveform, the actual Cyclic Test Pressure will be higher than measured, thereby penalizing the component under test.

## 12. PROCEDURES FOR CYCLIC TEST

### 12.1 General Provisions

12.1.1 Place metal shot or loosely fitting metal pieces in the pressure containing envelope being tested to reduce the volume of the pressurized fluid when higher cycling rates are desired.

12.1.2 Subject all pressure containing envelopes to the same pressure loadings when testing multiple samples.

12.1.3 In each complete cycle, apply test pressure to all parts of the pressure containing envelope being tested, in combination or sequentially, in a manner that simulates operating modes occurring in service.

12.1.4 Replace gaskets, seals and other expendable items that fail due to wear during the test so long as pre-loads in stressed elements are not affected such that clause 9.11 is not violated.

12.1.5 Verify with straingages that the ratio of induced stresses to pressure under static loading is attained at the test cycling rate where pressures must penetrate between close fitting parts, very large components are tested, or where hysteresis in joints can significantly affect stresses.

## 12.2 Cyclic Test Pressure

The Cyclic Test Pressure (CTP) is a product of two test parameter factors ($K_N$ and $K_V$) and the Rated Fatigue Pressure (RFP), calculated as follows:

$$CTP = RFP\,(K_N \times K_V)$$

### 12.2.1 Test Duration Factor ($K_N$).

12.2.1.1 Select Test Duration Factor from Table 1.

12.2.1.2 Use factors specified for non-ferrous metals in testing all pressure containing envelopes with non-ferrous elements.

TABLE 1 - Test duration factor, $K_N$

| Test Cycles (minimum) | Ferrous | Non-Ferrous |
|---|---|---|
| 10 million | 1.0 | 1.0 |
| 1 million | 1.15 | 1.25 |

### 12.2.2 Variability Factor ($K_V$).

12.2.2.1 Select Variability Factor from Table 2.

12.2.2.2 Use the Variability Factor for the materials contained in the pressure containing envelope being tested which have the highest value.

TABLE 2 - Variability factor, $K_V$

| Assurance Levels* | Number of Test Units | | |
|---|---|---|---|
| | 1 | 2 | 5 |
| Irons | | | |
| 99.9% | 1.90 | 1.42 | 1.22 |
| 99 | 1.55 | 1.30 | 1.15 |
| 90 | 1.30 | 1.15 | 1.07 |
| Aluminums, Magnesiums & Steels | | | |
| 99.9% | 1.55 | 1.30 | 1.15 |
| 99 | 1.45 | 1.25 | 1.12 |
| 90 | 1.30 | 1.15 | 1.06 |
| Copper Based Alloys | | | |
| 99.9% | 1.50 | 1.20 | 1.09 |
| 99 | 1.35 | 1.15 | 1.07 |
| 90 | 1.20 | 1.10 | 1.02 |
| Stainless Steels | | | |
| 99.9% | 1.30 | 1.17 | 1.09 |
| 99 | 1.25 | 1.15 | 1.07 |
| 90 | 1.17 | 1.10 | 1.02 |

\* Assurance levels are based on a 90 percent confidence level. Test twice the specified number of pressure containing envelopes to achieve a 99 percent confidence level.

12.3 Pressure Pulses

12.3.1 Provide a test pressure that decreases to below 0.05 CTP in each cycle.

12.3.2 Provide a maximum test pressure that exceeds the minimum test pressure (per clause 12.3.1) by an amount equal to the CTP.

12.3.3 Provide pressure pulses having any shape with maximum values that repeat cycle to cycle, as required by section 11, and do not have superimposed oscillations whose range exceeds 0.6 CTP.

12.3.4 Apply pressure pulses at any rate, up to 30 per second, that permits stable operation in which the pressure range repeats within ($\pm$) 2 percent.

## 13. PROCEDURES FOR STATIC TEST

### 13.1 General Provisions

13.1.1 Apply test pressure to all elements, individually or in combination, in a manner that simulates pressure containing envelope operating modes occurring in actual service.

13.1.2 Apply test pressure slowly and maintain for one minute.

### 13.2 Static Test Pressure

13.2.1 Calculate the Static Test Pressure as follows:

$$STP = K_V \times RSP$$

## 14. ADDITIONAL PROVISIONS

NOTE - There are other provisions, such as those listed below, that pertain to the verifying of presssre ratings for individual components. Each of these provisions will be taken into account as more detailed supplementary standards are prepared for individual components.

Give consideration to the following additional provisions:

14.1 Integrating the pressure rating test procedures into performance tests, where possible.

14.2 Specifying sequences or combinations to be used in applying pressures to specific components.

14.3 Permissible methods for restraining functional motions (essentially to attain higher cyclic test rates).

14.4 Positioning of internal parts to produce maximum stresses.

NFPA/T2.6.1

14.5 Parts of components that may be omitted for test purposes.

14.6 Component modifications that will be permitted to facilitate testing.

14.7 Additional criteria for judging failure.

14.8 Additional designated information.

14.9 Criteria and procedures required for verifying Rated Fatigue Pressures by geometric and material similarity to components having pressure ratings that were previously verified by testing in accordance with this recommended standard.

14.10 Specifying singular assurance and confidence levels for specific components or component usage.

15. CRITERIA FOR RFP VERIFICATION

NOTE - More specific details for this requirement will be set forth in the individual component standards.

15.1 Consider structural fracture a failure.

15.2 Consider any crack produced by metal fatigue due to pressure cycling, as verified by magnetic particle or fluorescent penetrant techniques after testing, a failure.

15.3 Consider excessive leakage at seals or sealing surfaces, that result from deformation of the metallic portion of the pressure containing envelope, a failure.

16. CRITERIA FOR RSP VERIFICATION

NOTE - More specific details for this requirement will be set forth in the individual component standards.

16.1 Consider structural fracture a failure.

16.2 Consider any crack produced by internal static pressure, as verified by magnetic particle or fluorescent penetrant techniques after testing, a failure.

NFPA/T2.6.1

16.3 Consider excessive leakage at seals or sealing surfaces, that result from deformation of the metallic portion of the pressure containing envelope, a failure.

16.4 Consider any permanent deformation, which interferes in any way with the proper functioning of the pressure containing envelope, a failure.

## 17. DATA PRESENTATION

17.1 Include the following minimum information in all component manufacturer's specifications, test reports, drawings, catalogs and sales literature referencing this recommended standard:

17.1.1 Rated Fatigue Pressure, RFP, and the assurance and confidence categories upon which the RFP is based.

17.1.2 Rated Static Pressure, RSP, and the assurance and confidence categories upon which the RSP is based.

17.2 Have available a record of all the following minimum test data for all component manufacturer's test reports referencing this recommended standard.

17.2.1 All physical values pertaining to the test.

17.2.2 All additional provisions or modifications pertaining to the test (see Section 14).

## 18. SUMMARY OF DESIGNATED INFORMATION

The following designated information is needed when applying this recommended standard to a particular application or use:

18.1 Desired Assurance Level

18.2 Desired Confidence Level

18.3 Desired RFP

18.4 Desired RSP

18.5 Expected ambient and fluid temperature range

NFPA/T2.6.1

19. JUSTIFICATION STATEMENT

This recommended standard verification procedure is based upon the combined expert experiences of those who have participated in its preparation and/or review.

Further justification is set forth in the Tutorial Reference, attached hereto as an Appendix.

Additional justification data will be obtained and compiled as the procedures set forth in this document become more widely used. Recommendations for the further improvement of this recommended standard are always welcome. Forward them directly to NFPA at the following address:

>   National Fluid Power Association
>   3333 North Mayfair Road
>   Milwaukee, Wisconsin 53222

20. TEST/PRODUCTION SIMILARITY

Utilize managerial controls necessary to maintain substantial similarity between test and production components or elements.

21. IDENTIFICATION STATEMENT

Use the following statement in test reports, catalogs and sales literature when electing to comply with this voluntary standard:

> "Method of verifying rated fatigue and rated static pressures of the pressure containing envelope conforms to NFPA Recommended Standard, NFPA/T2.6.1-1974, category _____*."

*NOTE - Category code numbers are as follows:

|  |  | Assurance Level | | |
|---|---|---|---|---|
|  |  | 90% | 99% | 99.9% |
| Confidence Level | 90% | 1/90 | 2/90 | 3/90 |
|  | 99% | 1/99 | 2/99 | 3/99 |

(For explanation of percent values, see clause 12.2.2.)

- 68 -

NFPA/T2.6.1

EXAMPLE - RFP (1/90) = _____ psi (bar, kPa)
This means that the component's Rated Fatigue Pressure is _____ psi (bar), based upon a 90% Assurance Level and 90% Confidence Level. (The use of the code 1/90 in this example does not suggest that these levels are preferred.)

# - NOTES -

# APPENDIX

NFPA/T2.6.1

TUTORIAL REFERENCE

FOR USE WITH

NFPA RECOMMENDED STANDARD

METHOD FOR VERIFYING THE FATIGUE AND STATIC

PRESSURE RATINGS OF THE PRESSURE CONTAINING ENVELOPE

OF A METAL FLUID POWER COMPONENT

---

NOTE. - This appendix is intended for reference purposes only. It is not a part of NFPA Recommended Standard Method for Verifying the Fatigue and Static Pressure Ratings of the Pressure Containing Envelope of a Metal Fluid Power Component, NFPA/T2.6.1-1974. Its contents do not necessarily represent an agreed-upon position by NFPA or its Pressure Rating Coordinating Committee. However, its material may be helpful towards a fuller understanding of the philosophy which led to the development of the basic standard.

---

Published by

NATIONAL FLUID POWER ASSOCIATION, INC.

MILWAUKEE, WISCONSIN 53222

NFPA/T2.6.1

## REFERENCES

A1. "The Statistical Aspects of Fatigue Strength" by S. J. Skaistis, September 14, 1971. (Unpublished)

A2. "Some Quantitative Aspects of Fatigue of Materials" by Harold N. Cummings, WADD Technical Report 60-42, July 1960.

A3. Fatigue Design Handbook by J. A. Graham, Editor, SAE.

A4. "Reliability Prediction - Mechanical Stress/Strength Interference" by Lipson, Sheth and Disney, Technical Report RADC-TR-66-710, March 1967, AD 813574.

A5. "Reliability Prediction - Mechanical Stress/Strength Interference (Nonferrous)" by Lipson, Sheth, Disney and Altun, Technical Report RADC-TR-68-403, February 1969, AD 856021.

A6. "Relating Probabilistic Methods to Reliability Considerations in Product Design" by Gerhard Reethof, ASME Paper 70 DE 70, May 1970.

A7. "Estimation of Fatigue Life with Particular Emphasis on Cumulative Damage" by Milton A. Miner, Chapter 12 of Metal Fatigue, Sines and Waisman Editors, McGraw-Hill Book Co., 1959.

A8. "The Fatigue of Metals and Structures" by Grover, Gordon and Jackson, NAVWEPS 00-25-534, revised June 1960.

A9. "NFPA Pressure Cycling Test Study - Evaluation of Endurance Test" by E. A. Ostreicher, August 17, 1970. (Unpublished)

A10. "Feasibility of Including $10^5$ Cycle Tests in the Pressure Rating Standard" by S. J. Skaistis, January 27, 1972. (Unpublished)

A11. "Evaluating Component Fatigue Performance Under Programmed Random, and Programmed Constant Amplitude Loading" by S. R. Swanson, SAE Paper 690050, January 1969.

A12. "The Effect of Test Pulse Oscillations" by S. J. Skaistis, November 20, 1970. (Unpublished)

A13. "Aluminum and Aluminum Alloys", ASM Metals Handbook, Vol. 1, 8th Edition, 1961.

A partial listing of other sources of information not referenced in this appendix:

A14. "Statistical Design and Analysis of Engineering Experiments" by Lipson and Sheth, McGraw-Hill Book Co., 1973.

A15. "Fatigue-Based Pressure Standards Introduced by Fluid Power Group" Product Engineering, June 1973.

A16. "Fatigue Testing of Hydraulic Equipment Components", by R. Stephens, Sperry Vickers.

# TUTORIAL REFERENCE FOR USE WITH
# NFPA RECOMMENDED STANDARD
# METHOD FOR VERIFYING THE FATIGUE AND STATIC
# PRESSURE RATINGS OF THE PRESSURE CONTAINING ENVELOPE
# OF A METAL FLUID POWER COMPONENT

## INTRODUCTION

The following is tutorial and is presented to aid in developing a greater understanding of rating specific fluid power components and in utilizing the ratings verified by recommended standard T2.6.1-1974.

These discussions are summaries of many of the technical studies that were made in preparing recommended standard T2.6.1-1974. Most of them deal with metal fatigue technology because it is believed that most structural failures in the Fluid Power Industry are due to metal fatigue. Static strength is believed to be less important.

1. THE STATISTICAL NATURE OF FATIGUE

    The fatigue strengths of components are subject to considerable variability. Because of this it could be expected that extensive testing would be necessary to verify a Rated Fatigue Pressure. However, it was found that by using statistical techniques it is possible to develop a standard method for verifying ratings with a single sample test and still attain a specified level of confidence (Reference No. A1).

    With this method, the allowance for the statistical nature of fatigue is made by increasing test pressures by a specified Variability Factor, $K_V$. This factor was determined by using a statistical analysis of fatigue strength data found in the referenced literature (References No. A2, A3, A4, and A5). The values selected were based on attaining a 90, 99, or 99.9 percent Assurance Level with a 90 percent Confidence Level. When desired, a 99 percent Confidence Level can be achieved by successfully testing twice the specified number of units.

NFPA/T2.6.1

Variability Factors are also given for multiple sample tests. These permit the use of lower test pressures. Tests with multiple samples using these factors are statistically equivalent to the single sample tests.

2. ASSURANCE LEVEL

As used in recommended standard T2.6.1-1974, the Assurance Level is the minimum percentage of pressure containing envelopes of a verified design that will sustain 10 million applications of its Rated Fatigue Pressure. In practice, a given design must have a fatigue strength greater than that implied by the Assurance Level to reduce the risk of its failing in its cyclic test. This conservatism is implicit in the use of a small number of samples to verify fatigue strengths. It can be decreased when the number of samples used to verify the rating is increased.

Even without this conservatism, a 90 percent Assurance Level does not imply a 10 percent service failure rate. The reason for this is that only a portion of the service pressure containing envelopes of a given design will be subjected to the equivalent of 10 million applications of the Rated Fatigue Pressure (References No. A3, A4, A6). In the case of a 90 percent Assurance Level, if only 5 percent of the pressure containing envelopes encounter the rated loading, the service failure rate will be less than 1/2 percent.

Service failure rates, even with the lowest Assurance Level, can be acceptably low where only a small percentage of the pressure containing envelopes will be subjected to the rated loading. The higher Assurance Levels are needed where a larger percentage of service pressure containing envelopes are used at the rated loading or where the consequences of service failure require that failure rates be lower.

3. RATED FATIGUE PRESSURE

Recommended standard T2.6.1-1974 in verifying the Rated Fatigue Pressure for a component's pressure containing envelope provides a datum point for judging whether or not a component has adequate strength for use in a specific application.

Fatigue failures are caused by the actual extreme values of stresses induced by changing pressures, regardless of whether or not the stresses remain at their extremes for hours or for a few micro-seconds. Therefore, the peak transition pressures, rather than the normal system or working pressures, determine the strength requirements of pressure containing envelopes in a given system.

Usually pressure transients are a characteristic of complete circuits rather than single components, so it must be left to the system designer to determine their magnitudes and to select components with adequate strengths. There is very little data available to assist the system designer in making this determination. It is anticipated that the promulgation of recommended standard T2.6.1-1974 will stimulate the study of the relationships between peak and working pressures.

In most cases the Rated Fatigue Pressure of components will have to equal the system pressure peaks to provide adequate strength. In cases where some of the materials in the pressure containing envelope, like aluminum, do not have true endurance limits and where more than 10 million cycles can occur, the Rated Fatigue Pressure must be greater than the pressure peaks to allow for the higher cycle life.

Pressure peaks can be allowed to exceed the Rated Fatigue Pressure where it can be proved that less than 1,000,000 of these peaks will occur. When such peaks are all of the same magnitude, the effect can be judged by using the "method for estimating finite life capability", given in the Introduction. This method can also be used as a basis to evaluate the cumulative fatigue damage effect of peaks of various magnitudes. (References A3 and A7)

4. REDUCED TEST CYCLE

Rated Fatigue Pressure is based on sustaining 10 million cycles. It can be verified with tests consisting of one million cycles, if the test pressure is increased by the factor $K_N$ (the ratio of the fatigue strengths at one and at ten million cycles). This factor was determined by using a large volume of data found in the referenced literature (Reference No. A8). Since most of this data was for reversed stressing, the modified Goodman relationship was used to convert this to the equivalent zero-to-max stressing usually occurring in pressurized structures (Reference No. A9).

The values used in recommended standard T2.6.1-1974 are those that equal or exceed 90 percent of the ratios calculated. This was done to eliminate data points that did not appear to be consistent. While this action tends to reduce the true Assurance Levels in a small percentage of cases, the use of the Test Duration Factor is still felt to be conservative.

Further decreases in the required number of test cycles were considered. It was found that this might affect the validity of the tests, therefore, shorter tests have not been utilized in recommended standard T2.6.1-1974 (See Reference No. A10). Such tests may still be used for production monitoring where they are first correlated with the tests prescribed in recommended standard T2.6.1-1974.

## 5. TEST RATE

Pressure changes travel at the speed of sound which approximate 50,000 IPS in oil and 13,000 IPS in air at atmospheric pressure. Stresses travel through metals at even higher speed.

However, exceptions to this rule can occur. These would include: (1) very large pressure containing envelopes where pressure may be measured far from the point where maximum stresses occur, (2) assemblies where appreciable joint slippage must occur before maximum stressing can be attained, and (3) pressure containing envelopes where large deflections can produce high internal inertial forces. Where it is suspected that such factors are having an effect, strain gages on the test pressure containing envelope should be used to verify whether or not the ratio of the measured stress to the measured pressure remains constant from a static condition to the actual test cycling rate.

The specified maximum cycling rate of 30 cycles per second is not based on stress propagation limits. Experience determined this to be the approximate practical limit of test hydraulic and measuring systems under very favorable conditions. Therefore, in most cases, limits for specific tests will be well below this rate.

Testing economics favor using the highest cycling rate permitted by the test circuit used. For this reason, good practice includes restraining all operational motions and displacing fluid in test pressure containing envelope cavities with metal to reduce the pressurized oil volume. These procedures reduce the fluid flow required for each pressure cycle so that a higher cycling rate can be used with a given test circuit flow capacity.

## 6. PULSE SHAPE

Fatigue failure is governed by the extreme stresses experienced by a pressure containing envelope and by the fact that the shape of the stress-time curve is usually not significant (Reference No. A11). However, in some cases, the presence of high amplitude oscillations superimposed on the test pulses can have an effect. This possibility has been minimized by using tests of one million cycles or more. Analyses show that in such tests oscillations must have amplitudes greater than 60 percent of the test pulse to contribute to fatigue failure (Reference No. A12).

NFPA/T2.6.1

The standard recognizes that the pressure in a test piece decays very slowly as it approaches zero. To make it possible to achieve practical cycling rates, it does not require that the pressure return to zero in each cycle (12.3.1). The effect of this deviation is minimized by requiring the pressure range to equal the Cyclic Test Pressure (12.3.2). When the minimum and maximum cyclic pressures are measured individually, the pressure range is subject to the total errors in the two measurements. Since the cyclic test pressure range is of prime importance, it is desirable to use instrumentation that measures it directly so that these measurements are only subject to one set of errors.

It is also recognized that the minimum and maximum pressures may not repeat exactly from cycle to cycle so a limit in the variability of the cycle to cycle pressure range is also specified (12.3.4). This tolerance should not be confused with the accuracy specified for measuring pressure range (Section 11). In the presence of variations, accuracy refers to how closely the measured average of the pressure range agrees with the actual value of this average. In general, it is the resultant of instrumentation and reading errors only and does not include the effect of cyclic variation.

7. TEMPERATURE

The tests specified in recommended standard T2.6.1-1974 are to be run at room temperature and the results are stated to be applicable to operating temperatures up to 200°F (93°C). This upper limit was selected because some aluminum alloys only retain their strength to this temperature (Reference No. A13). For many other materials, this limit is conservative and it did not seem practical to provide the temperature-strength characteristics of all materials. For applications above 200°F (93°C) special tests at the application temperature are considered necessary.

8. STATIC TESTS

Burst tests have often been used to evaluate the strength of fluid power components because they are simple to perform. The interpretation sometimes given to such tests is that the fatigue strength of a component is one-fourth or one-fifth its burst strength. No substantiation for this relationship was found in the referenced literature. Therefore, cyclic, rather than burst tests, are employed in recommended standard T2.6.1-1974 to verify fatigue strength.

NFPA/T2.6.1

The static test included in recommended standard T2.6.1-1974 is only for the purpose of verifying the Rated Static Pressure. Generally, this pressure will be higher than the Rated Fatigue Pressure. In addition to being a pressure limit for static applications, it may also provide a datum for estimating the effect of a limited number of load cycles as indicated in the "method for estimating finite life capability" given in the Introduction.

In most cases, the duration of the static load is not important as failure will occur almost as fast as pressure is applied. A duration of one minute was specified to be conservative and allow for possible cases where appreciable yielding or buckling must take place before failure becomes apparent. It is still possible that some components will require longer times for failure to be completed. Since it is in the best interests of the producer to detect such cases, longer durations are advised.

Static strength generally has less variability than fatigue strength. However, sufficient data for calculating allowances for this variability were not found. To be conservative, it is assumed that this variability is the same as that for fatigue strength.

When the static test pressure is applied, it will often cause localized yielding at points of higher stress concentration such as in the roots of threads or in sharp corners. So little metal is affected that such yielding cannot be detected by inspection methods, but it results in residual compressive stresses in the highest stressed areas, after the pressure is removed. These stresses subtract from the stresses that will be produced in subsequent loadings so the fatigue strength is improved. Care must be taken, therefore, that units used in cyclic tests are not previously strengthened by static testing unless such prestressing is also performed regularly in the production of these units.

# NOTES

NFPA Recommended Standard
T2.6.1M S2 (T3.21.4M)-1977 (R1982)

## AN INDUSTRY STANDARD FOR FLUID POWER

Pneumatic Valve Pressure Rating

Supplement No. 2 to

NFPA Recommended Standard for Verifying the

Fatigue and Static Pressure Ratings of the

Pressure Containing Envelope of a Metal Fluid Power Component

Approved as an NFPA Recommended Standard
1 December 1977

published by
**NATIONAL FLUID POWER ASSOCIATION, INC.**

3333 N. Mayfair Road   /   Milwaukee, WI 53222   /   414-778-3344   /   TLX 26898

## NOTES

NFPA/T2.6.1 S2

## FOREWORD

This Foreword is not part of Pneumatic Valve Pressure Rating Supplement No. 2 to NFPA Recommended Standard for Verifying the Fatigue and Static Pressure Ratings of the Pressure Containing Envelope of a Metal Fluid Power Component, NFPA/T2.6.1 S2 (T3.21.4)-1977.

Early in 1974, the approval of NFPA/T2.6.1-1974 established a group of common requirements intended to provide an industry-wide philosophy and basic standard, providing a rationale for judging a component's ability as a pressure containing envelope. Although the specific applicability of NFPA/T2.6.1-1974 is limited, it immediately established a uniform base for subsequent, more specific proposed NFPA recommended standards for individual components.

Work was officially begun by members of the Pneumatic Valve Section, T3.21, with the appointment of the Project Group in December 1972. Specific activity, however, did not begin until the "kickoff" meeting of 20 November 1973, although several members had actively followed the development of the basic standard, NFPA/T2.6.1-1974.

One major point of discussion at the 20 November 1973 meeting concerned that of interpretation of the term "failure". It was believed that a key component, such as a control valve, be able to function properly after testing, since its performance involves other things than just the integrity of its overall pressure envelope.

Following the November meeting, Chairman K. Johnson (MSOE) prepared the first draft of the document. He circulated it among members of the Project Group, who in turn submitted written comments by 1 May 1974. On 6 May 1974, Chairman Johnson incorporated these comments into the second draft and submitted it to NFPA Headquarters. On 14 May the NFPA Technical Staff edited the second draft for discussion at the upcoming T3.21 Section meeting.

On 21 May 1974, the T3.21 Section approved the concept that elements normally considered as internal elements may become part of the pressure containing envelope under some operating conditions, e.g., the spool-passage wall combination in a spool valve particularly with the exhaust ports open to the atmosphere.

(continued)

NFPA/T2.6.1 S2

## FOREWORD (continued)

Chairman Johnson circulated the second draft to members of the Project Group for their review on 28 May 1974, and on 23 July 1974 he submitted the Final Working Draft to NFPA Headquarters. The Headquarters Technical Staff prepared the General Review Draft on 12 November 1974. Ten comments were received during General Review; they were answered by Chairman Johnson on 30 April 1975.

On 12 June 1975, the NFPA Pressure Rating Coordinating Committee, T2.6, met to discuss suggested revisions to the basic pressure rating document, NFPA/T2.6.1-1974. Several changes for rating by similarity were agreed upon. As a result of this agreement, Chairman Johnson revised his responses to those who had offered General Review comments on T3.21.4-19xx; the new responses were mailed on 8 August 1975.

Project Group T3.21.4 was reorganized at the 1 October 1975 Pneumatic Valve Section meeting. John Berninger (Parker Hannifin Corp.) was appointed to replace Chairman Johnson, who had resigned. The Project Group met on 18 March 1976 and modified the General Review Draft to incorporate both the applicable original comments as well as the latest input from the Pressure Rating Coordinating Committee. The document was again reviewed and modified at the 27 April 1976 Project Group meeting. On 28 April 1976, the Pneumatic Valve Section voted to recommend that the document be submitted for a Second General Review. The document was submitted to NFPA Headquarters on 18 May 1976, and the NFPA Technical Staff prepared the Second General Review Draft on 13 August 1976.

General Review comments were resolved during the Project Group Meeting held on 9 November 1976. The NFPA Technical Board granted approval to the ballot on 11 May 1977, and the NFPA Technical Staff prepared the Ballot Draft on 2 June 1977. Affirmative comments resulting from ballot were accepted by the Project Group, and incorporated into the document.

On 2 November 1977, the NFPA Technical Board voted to recommend to the Board of Directors that this document be approved as an NFPA Recommended Standard. The NFPA Board of Directors granted final approval to NFPA/T2.6.1 S2 (T3.21.4)-1977 on 1 December 1977.

NFPA/T2.6.1 S2

## PROJECT GROUP MEMBERS WHO DEVELOPED THIS STANDARD

| | | |
|---|---|---|
| Berninger, John | Project Chairman (1975 - present) | Parker Hannifin Corp. |
| Johnson, Kenneth | Project Chairman (1972 - 1975) | Milwaukee School of Engineering* |
| Bailey, Richard | Project Vice Chairman (1976 - present) | C. A. Norgren Co. |
| Grassl, Roman | Project Vice Chairman (1972 - 1976) | Galland Henning Nopak, Inc. |
| Lytle, John | Section Chairman (1975 - present) | Ross Operating Valve Co. |
| Brake, Clifford | Section Chairman (1970 - 1975) | Schrader Fluid Power Division |
| Sallberg, David | Technical Auditor | HUSCO Division |
| Luecke, John R. | Director of National Technical Services | National Fluid Power Association |

Allen, C.    WABCO Fluid Power Division*
Bowbin, J.   Miller Fluid Power Corp.

\* Company affiliation has changed since work with the Project Group.

## REFERENCES

1. American National Standard Glossary of Terms for Fluid Power, ANSI/B93.2-1971, and Supplements thereto. (ISO Draft Proposal 5598)

2. SI units and recommendations for the use of their multiples and of certain other units, ISO 1000-1973.

3. National Fluid Power Association Recommended Standard Method for Verifying the Fatigue and Static Pressure Ratings of the Pressure Containing Envelope of a Metal Fluid Power Component, NFPA/T2.6.1-1974.

4. National Fluid Power Association Recommended Standard for Defining Port Communication for Pneumatic Valve Interface to NFPA Recommended Standard T3.21.1-1973, With the Valve in Position in Response to a Remote Pilot Signal or Electrical Energization, NFPA/T3.21.9-1976.

NFPA/T2.6.1 S2

# PNEUMATIC VALVE PRESSURE RATING

SUPPLEMENT NO. 2 to

NFPA Recommended Standard for Verifying the

Fatigue and Static Pressure Ratings of the

Pressure Containing Envelope of a Metal Fluid Power Component

---

HEADQUARTERS NOTE - The Project Group which developed this Supplement intended it for use with only the basic pressure rating document, NFPA/T2.6.1-1974. This Supplement provides a list of additions, deletions and changes to the basic document which are necessary for the establishment of pressure ratings for metal pneumatic valves. All clauses which have been modified are so indicated in this Supplement. Read all other clauses as they appear in the basic pressure rating document, NFPA/T2.6.1-1974.

Since revisions of the basic pressure rating document will apply to this Supplement, the reader is cautioned to always use the most recent edition of the basic document. When reading this Supplement with the basic document, <u>substitute the phrase "pneumatic valve(s)" whenever the term "component(s)" appears in the basic document.</u>

---

## INTRODUCTION

CHANGE the first and second paragraphs to read:

In pneumatic fluid power systems, power is transmitted and controlled thru air under pressure. A pneumatic valve is a device which controls air flow direction, pressure or flow rate. A basic requirement of fluid power pneumatic valves is that they should be capable of adequately containing the pressurized fluid as well as properly controlling its flow.

The basic pressure rating document, NFPA/T2.6.1-1974, established a group of common requirements intended to provide an industry-wide philosophy and basic standard, providing a rationale for judging a component's ability as a pressure containing envelope. Although the

NFPA/T2.6.1 S2

specific applicability of NFPA/T2.6.1-1974 is limited, it immediately established a uniform base for subsequent, more specific NFPA Recommended Standards for individual fluid power components. This Recommended Standard implements NFPA/T2.6.1-1974 and specifically applies to pneumatic valves.

1. SCOPE

   ADD the following clause:

   1.4 Pressure Regulators are not included in this Recommended Standard but are included in NFPA/T2.6.1 S4 (T3.12.10-1976).

2. PURPOSE

   NO CHANGE from the basic document.

3. TERMS AND DEFINITIONS

   CHANGE clauses 3.11 up to and including 3.12 to read:

   3.11 **Pressure Containing Envelope. That part of the component (i.e., pneumatic valve) which contains the pressurized fluid.**

   3.12 Elements. The integral pieces which make up a part or component (i.e., pneumatic valve), (e.g., spool, stem, diaphragm, seat).

   The following terms, not included in this document but found in related reference material, are defined with the meaning that the Committee applied when using the reference:

   3.13 Peak Pressure. The maximum pressure actually experienced by a fluid power component (i.e., pneumatic valve) in an application. It is caused by any number of events, individually or in combination, such as (but not limited to) impact, inertia, shock, heat, etc.

   3.14 Nominal Pressure. A pressure value assigned to a component (i.e., pneumatic valve) or a system for the purpose of convenient designation. This may be an average of normal values without peaks, or a design figure for stress calculations.

NFPA/T2.6.1 S2

4. UNITS OF MEASUREMENT

   NO CHANGE from the basic document.

5. LETTER SYMBOLS

   NO CHANGE from the basic document.

6. OUTLINE OF PROCEDURES FOR RFP VERIFICATION BY TEST

   NO CHANGE from the basic document.

7. RFP AND RSP VERIFICATION BY SIMILARITY

   CHANGE the entire Section to read:

   7.1 Verification of the rated fatigue and rated static pressure ratings of a component's pressure containing envelope by similarity is allowed when the similar component differs from the tested component only in the size of the female port openings, provided that those ports are of the same type and smaller than in the tested component; or in the size of male port openings, provided that those ports are of the same type, are either smaller in O.D. (Outside Diameter) and not less in wall thickness than in the tested component, or larger in O.D. if the I.D. (Inside Diameter) remains the same as the tested component.

   7.2 Use an assurance level for rating by similarity that is identical to the component rating verified by test.

8. OUTLINE OF PROCEDURES FOR RSP VERIFICATION BY TEST

   NO CHANGE from the basic document.

9. PREPARATIONS FOR TESTING

   NO CHANGE from the basic document.

10. TEST EQUIPMENT

   CHANGE clause 10.1.2 to read:

   10.1.2   Use any test fluid which is compatible with the valve elements.

   CHANGE clause 10.1.3 to read:

   10.1.3   Mount the pressure sensing device directly into the pressure containing envelope, thru a pressurized port which is not being used to supply the test fluid.

   CHANGE clause 10.1.4 to read:

   10.1.4   When no such port (clause 10.1.3) is available, mount the pressure sensing device on the pressure supply line, minimizing restrictions between the device and the pressure containing envelope.

   CHANGE clause 10.2.3 to read:

   10.2.3   Use any fluid which is compatible with the valve elements.

11. TEST CONDITIONS ACCURACY

   NO CHANGE from the basic document.

12. PROCEDURES FOR CYCLIC TEST

   CHANGE clause 12.1.3 to read:

   12.1.3   In each complete cycle, apply test pressure to all parts of the pressure containing envelope being tested, in combination or sequentially, in a manner that simulates operating modes occurring in service. (See Section 14.)

13. PROCEDURES FOR STATIC TEST

    CHANGE clause 13.1.1 to read:

    13.1.1 Apply test pressure to all elements, individually or in combination, in a manner that simulates pressure containing envelope operating modes occurring in actual service. (See Section 14.)

14. ADDITIONAL PROVISIONS

    CHANGE the entire Section to read:

    14.1 Perform a separate cyclic and a separate static test with the valve to simulate each mode of operation (i.e., spool, plug, gate, poppet or slide in neutral, intermediate and extreme positions). In each test, pressurize simultaneously all ports that are rated for pressure (including exhaust ports unless the exhaust ports are not rated for pressure).

    14.2 Ensure that each chamber connected to a pressure rated port is pressurized during at least one test.

    EXAMPLES:

    1. A two-port shutoff valve requires two tests to simulate the two modes of valve operation.

       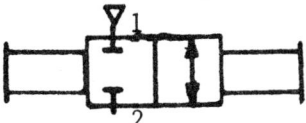

       Test 1: Valve in closed position with port 2 open to atmosphere. Pressurize port 1.

       Test 2: Valve in open position with port 1 connected to port 2. Pressurize ports 1 and 2 simultaneously.

    2. A three-port valve requires two tests to simulate the two possible modes of valve operation.

       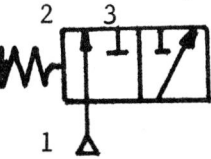

       Test 1: Valve positioned to connect ports 1 and 2. Pressurize ports 1 and 2. Open port 3 to atmosphere.

Test 2: Valve positioned to connect ports 1 and 3. Pressurize ports 1 and 3. Open port 2 to atmosphere.

3. A double pilot operated 4-way, 2 position, poppet valve in which all ports are rated for pressure requires four tests to simulate the four modes of operation.

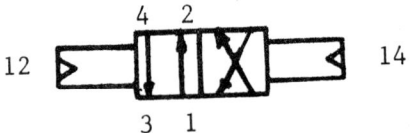

Test 1: Poppets in position corresponding to pressurization of pilot port 12. Pressurize ports 12, 1, 2, 3, and 4. Open port 14 to atmosphere. (This test would not be required if port 3 were not to be rated for pressure.)

Test 2: Poppets in position corresponding to pressurization of pilot port 12. Pressurize ports 12, 1 and 2. Open ports 3, 4 and 14 to atmosphere.

Test 3: Poppets in position corresponding to pressurization of pilot port 14. Pressurize ports 14, 1, 2, 3 and 4. Open port 12 to atmosphere. (This test would not be required if port 3 were not to be rated for pressure.)

Test 4: Poppets in position corresponding to pressurization of pilot port 14. Pressurize ports 14, 1 and 4. Open ports 2, 3 and 12 to atmosphere.

4. A pilot operated (internal pilot), closed center, 4-way, 3-position, 5-port spool valve in which all ports are rated for pressure requires five tests to simulate the five modes of operation.

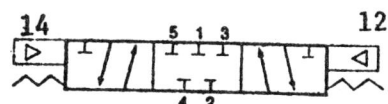

Test 1: Spool in center position. Pressurize port 1. Open ports 2, 3, 4, 5, 14 and 12 to atmosphere.

Test 2: Spool in position corresponding to pressurization of pilot port 12. Pressurize ports 12, 1, 2, 4 and 5. Open ports 3 and 14 to atmosphere. (This test would not be required if port 5 were not to be rated for pressure.)

Test 3: Spool in position corresponding to pressurization of pilot port 12. Pressurize ports 12, 1 and 2. Open ports 3, 4, 5 and 14 to atmosphere.

Test 4: Spool in position corresponding to pressurization of pilot port 14. Pressurize ports 14, 1, 2, 3 and 4. Open ports 5 and 12 to atmosphere. (This test would not be required if port 3 were not to be rated for pressure.)

Test 5: Spool in position corresponding to pressurization of pilot port 14. Pressurize ports 14, 1 and 4. open ports 2, 3, 5 and 12 to atmosphere.

14.3 Assure complete pressurization of the guide tube for at least one of the tests of a valve with a tube type solenoid in which the tube forms a portion of the pressure containing envelope.

14.4 Conduct separate tests for pilot circuits in pilot operated valves when a different rating is desired for pilot ports. These tests may be performed concurrently with the valve testing.

NFPA/T2.6.1 S2

14.5 Test a group of valves using a separate valve for each test which simulates a different mode of operation. The group of valves will constitute one test unit for determination of $K_V$.

EXAMPLE: A valve having three modes of operation will require tests of three individual valves, one for each operating mode.

NOTE: Simultaneous testing of a number of individual valves which are set up for the same or for different tests is permitted.

14.5.1 It is optionally permissible to use only one valve for the several tests applying the pressure pulses to each test in sequence or to each test separately. However, only the total number of cycles from one of the tests may be used for the determination of $K_N$ and $K_V$.

NOTE: It should be realized that a certain amount of conservatism will be introduced into the RFP verification by this procedure since some parts of the valve will experience perhaps as much as five times the number of cycles used for the test duration factor, $K_N$.

14.5.2 If it can be shown that two or more sections of the valve are nominally identical (within manufacturing tolerances) in dimensions and geometry, then it is necessary to test only one of the nominally identical sections provided that each of the nominally identical sections is tested during the pressure rating program.

NOTE: The body of the valve in Example 4 of clause 14.2 includes three basic sections: A center section with the inlet port; and two end sections each of which contains a pilot, a cylinder and an exhaust port. If the two end sections are nominally identical, only three of the five tests listed in Example 4 (tests 1, 2 and 3 or 1, 4 and 5) are required to constitute a test of the entire valve. However, all five tests must be performed within the group of samples subjected to test.

NFPA/T2.6.1 S2

14.6 Use conventional type caps and plugs to block ports and prevent leakage during testing.

14.7 Remove return or centering springs and use loose fitting slugs, spacers or other metal pieces to hold spools or poppets, etc., in the extreme positions required for the various tests. The contact between such parts and the pressure containing envelope must be identical to that of the parts being replaced.

14.8 Specify ratings of valve and subbase assemblies at an RSP and RFP no greater than the least verified value of the individual parts if tested separately. Obtain verification of these ratings as a result of static and cyclic tests with actual parts or test fixtures, assembling with all mounting bolts torqued to manufacturer's recommended values.

14.9 Retest a valve if changes have been made in the metallic elements of the fluid containing envelope which previously had been verified by test. It is required to rerun only those tests which are necessary to verify the changed elements to the original ratings.

14.10 Verification of an RFP may be made from the results of a cyclic endurance test by actually operating the valve through all operating modes provided that the basic provisions of this document have been met during the test.

14.11 If a diaphragm is used as a seal between portions of the valve body, the center of the diaphragm may be cut out, leaving that portion of the diaphragm that serves as a seal.

15. CRITERIA FOR RFP VERIFICATION

CHANGE clauses 15.2 up to and including 15.3 to read:

15.2 Consider any crack which is produced by metal fatigue due to pressure cycling and which results in leakage to constitute a failure. Alternate methods to determine failure by cracking such as fluorescent penetrant or magnetic particle techniques may be used if the material conditions both before and after testing are examined.

15.3 Consider excessive leakage at seals or sealing surfaces that results from deformation of the metallic portion of the pressure containing envelope to constitute a failure.

NOTE: It is permissible to retighten fasteners and replace seals after completion of a test to determine if any observed excessive leakage is corrected and, if so, to assume that such leakage was not due to deformation of the metallic portion of the pressure containing envelope.

16. CRITERIA FOR RSP VERIFICATION

CHANGE clauses 16.2 up to and including 16.3 to read:

16.2 Consider any crack which is produced by internal static pressure and which results in leakage to constitute a failure. Alternate methods to determine failure by cracking such as fluorescent penetrant or magnetic particle techniques may be used if the material conditions both before and after testing are examined.

16.3 Consider excessive leakage at seals or sealing surfaces which results from deformation of the metallic portion of the pressure containing envelope to constitute a failure.

NOTE: It is permissible to retighten fasteners and replace seals after completion of a test to determine if any observed excessive leakage is corrected and, if so, to assume that such leakage was not due to deformation of the metallic portion of the pressure containing envelope.

DELETE clause 16.4.

17. DATA PRESENTATION

ADD clause 17.1.3:

17.1.3 Identify clearly the specific portions of the valve which correspond to each value of RFP and RSP if different pressure ratings are to be applied to different portions of the valve. (See clauses 9.4, 14.1 and 14.4.)

18. SUMMARY OF DESIGNATED INFORMATION

NO CHANGE from the basic document.

19. JUSTIFICATION STATEMENT

CHANGE the entire section to read:

This recommended standard verification procedure is based upon the combined expert experiences of those who have participated in its preparation and review, and in the preparation and review of the basic document, NFPA/T2.6.1-1974, and its tutorial reference.

Additional justification data will be obtained and compiled as the procedures set forth in this document become more widely used. Recommendations for the further improvement of this recommended standard are always welcome. Forward them directly to NFPA at the address listed on the cover of this document.

20. TEST/PRODUCTION SIMILARITY

NO CHANGE from the basic document.

21. IDENTIFICATION STATEMENT

Use the following statement in test reports, catalogs and sales literature when electing to comply with this voluntary standard:

"Method of verifying rated fatigue and rated static pressures of pneumatic valves conforms to NFPA Recommended Standard NFPA/T2.6.1 S2 (T3.21.4)-1977, catagory_____*."

*NOTE: Catagory code numbers are as follows:

|  |  | ASSURANCE LEVEL | | |
| --- | --- | --- | --- | --- |
|  |  | 90% | 99% | 99.9% |
| CONFIDENCE LEVEL | 90% | 1/90 | 2/90 ** | 3/90 |
|  | 99% | 1/99 | 2/99 | 3/99 |

(For explanation of percentage values, see clause 12.2.2)

NFPA/T2.6.1 S2

EXAMPLE:

RFP (2/90) = ____ psi (bar). This means that the Pneumatic Valves Rated Fatigue Pressure is ____ psi (bar), based on a 99% assurance level and a 90% confidence level. (The use of code 2/90 in this example does not suggest that this level is mandatory).

NOTE: 2/90** rating is a minimum recommended level for RFP and RSP for pneumatic valves.

## 22. KEY WORDS

The following Key Words, useful in indexes and in information retrieval systems, are suggested for this recommended standard:

fluid power
pneumatic valve, fluid power
pressure, cyclic test
pressure, rated fatigue
pressure, rated static

pressure, static test
pressure rating, by similarity
pressure rating, by test
pressure rating, pneumatic valve

NFPA Recommended Standard
T2.6.1M S3 (T3.29.2M)-1976 (R1982)

**AN INDUSTRY STANDARD FOR FLUID POWER**

Pressure Switch Pressure Rating

Supplement No. 3 to

NFPA Recommended Standard for Verifying the

Fatigue and Static Pressure Ratings of the

Pressure Containing Envelope of a Metal Fluid Power Component

Approved as an NFPA Recommended Standard
25 February 1976

published by
**NATIONAL FLUID POWER ASSOCIATION, INC.**

3333 N. Mayfair Road  /  Milwaukee, WI 53222  /  414-778-3344  /  TLX 26898

## FOREWORD

(This Foreword is not part of Pressure Switch Pressure Rating Supplement No. 3 to NFPA Recommended Standard for Verifying the Fatigue and Static Pressure Ratings of the Pressure Containing Envelope of a Metal Fluid Power Component, NFPA/T2.6.1 S3 (T3.29.2)-1976.)

Early in 1974, the approval of NFPA/T2.6.1-1974 established a group of common requirements intended to provide an industry-wide philosophy and basic standard, providing a rationale for judging a component's ability as a pressure containing envelope. Although the specific applicability of NFPA/T2.6.1-1974 is limited, it immediately established a uniform base for subsequent, more specific proposed NFPA recommended standards for individual components.

The Pressure Switch Pressure Rating Project Group held its first meeting on 22 May 1973. The TSP was approved at the Technical Board meeting on 4 October 1973. On 20 November 1973 the Project Group met to discuss the development of the first draft, which was completed on 3 January 1974. Draft No. 1 was circulated among the Project Group members on 9 January 1974. Their comments were forwarded to Chairman Blickley, who prepared the Final Working Draft and submitted it to NFPA Headquarters on 6 September 1974. Headquarters' Technical Staff prepared the General Review Draft on 16 December 1974.

Comments resulting from General Review were resolved on 22 July 1975, and the document received approval to ballot from the Technical Board on 21 August 1975. The Ballot Draft was prepared on 28 August 1975. The one negative ballot received on this standard was resolved on 30 January 1976. One affirmative comment offering some suggested changes was also received; it was discussed at the Pressure Switch Section meeting on 6 November 1975, but none of the suggestions was incorporated in the document.

On 4 February 1976, the Technical Board unanimously voted to recommend that this document be approved as an NFPA Recommended Standard. The NFPA Board of Directors granted approval to NFPA/T2.6.1 S3 (T3.29.2)-1976 on 25 February 1976.

NFPA/T2.6.1 S3

## PROJECT GROUP MEMBERS WHO DEVELOPED THIS STANDARD

| | | |
|---|---|---|
| Blickley, George | Project Chairman and Section Chairman (1975 to present) | SOR, Inc. |
| Brown, James | Section Chairman (1972 to 1975) | DeLaval Turbine Inc. |
| Mueller, John | Technical Auditor | The Weatherhead Co. |
| Luecke, John R. | Director of National Technical Services | National Fluid Power Association |

Lyons, J. — Chemetron Corporation
Van Skike, O. — DeLaval Turbine Inc.

## REFERENCES

1. American National Standard Glossary of Terms for Fluid Power, ANSI/B93.2-1971, and Supplements thereto. (ISO/TC 131/SC 1 (USA-2) 3)

2. SI units and recommendations for the use of their multiples and of certain other units, ISO 1000-1973.

3. National Fluid Power Association Recommended Standard Method for Verifying the Fatigue and Static Pressure Ratings of the Pressure Containing Envelope of a Metal Fluid Power Component, NFPA/T2.6.1-1974.

## BACKGROUND REFERENCE

1. Proposed National Fluid Power Association Recommended Standard Glossary of Terms for Pressure Switches, NFPA/T3.29.1-19xx.

NFPA/T2.6.1 S3

## PRESSURE SWITCH PRESSURE RATING

SUPPLEMENT NO. 3 to

NFPA Recommended Standard for Verifying the

Fatigue and Static Pressure Ratings of the

Pressure Containing Envelope of a Metal Fluid Power Component

> HEADQUARTERS NOTE - The Project Group which developed this Supplement intended it for use only with the basic pressure rating document, NFPA/T2.6.1-1974. This Supplement provides a list of additions, deletions and changes to the basic document which are necessary for the establishment of pressure ratings for metal pressure switches. All clauses which have been modified are so indicated in this Supplement. Read all other clauses as they appear in the basic pressure rating document, NFPA/T2.6.1-1974.
>
> Since revisions of the basic pressure rating document will apply to this Supplement, the reader is cautioned to always use the most recent edition of the basic document. When reading this Supplement with the basic document, substitute the phrase "pressure switch(es)" whenever the term "component(s)" appears in the basic document.

### INTRODUCTION

CHANGE the first three paragraphs to read:

In fluid power systems, power is transmitted and controlled thru a fluid (liquid or gas) under pressure within an enclosed circuit.

A pressure switch is a device that measures positive or negative pressures (with respect to atmosphere) or a differential pressure with one or more pressure sensing elements which produce a motion and/or force which is utilized to operate a switching element. The switching element is generally an electrical switch or contact(s), but may also be a pneumatic or hydraulic element such as a valve.

The functions of a pressure switch in a fluid power system are many and varied, but they generally fall into one or more of the following three categories:

a) Alarm. In this case the pressure switch initiates a visual or audible alarm to inform an operator of a certain condition of pressure.

b) Control. In this case the pressure switch starts or stops a pump, or opens or closes a valve.

c) Safety. In this case the pressure switch initiates an action that protects the system from a failure mode of pressure.

A basic requirement of fluid power pressure switches is that they should be capable of completely containing the pressurized fluid.

The basic pressure rating document, NFPA/T2.6.1-1974 established a group of common requirements intended to provide an industry-wide philosophy and basic standard, providing a rationale for judging a component's ability as a pressure containing envelope. Although the specific applicability of NFPA/T2.6.1-1974 is limited, it immediately established a uniform base for subsequent, more specific NFPA Proposed Recommended Standards for individual fluid power components. This Recommended Standard implements NFPA/T2.6.1-1974 and specifically applies to pressure switches.

Ratings verified in accordance with this standard do not replace pressure ratings based on considerations such as performance, overrange, leakage and heat rejection. Instead, these new ratings are intended to supplement those that are provided by present practice and may be numerically different from prior ratings.

No attempt is made in this standard to deal with the electrical, pneumatic or hydraulic switching elements of pressure switches. The ratings for pneumatic and hydraulic switching elements (such as valves) are dealt with in other NFPA documents. The ratings for electrical switching elements (such as snap-action switches) are covered thoroughly by other industry standards.

The system designer should be aware that the cyclic tests in this standard do not include the switching element (either under load or with no load) and the life of the switching element may be quite different from the test cycles performed under this standard.

NFPA/T2.6.1 S3

1. SCOPE

   ADD the following clause:

   1.1.5   Made of the above listed materials (clause 1.1.4) but may contain seals or interfaces of elastomeric, organic or other non-metallic elements. Such elements may be part of the pressure containing envelope, but are not verified by these procedures.

   CHANGE clause 1.2 to read:

   1.2   These verification procedures do not include the functioning of a pressure switch in its ability to maintain a set point when either RFP or RSP is applied. It is expected that a future standard will cover the methods for determining Rated Overrange Pressure or that static pressure above which there can be expected a shift in, or malfunction of, the set point.

2. PURPOSE

   NO CHANGE from the basic document.

3. TERMS AND DEFINITIONS

   CHANGE clause 3.2 to read:

   3.2   Rated Static Pressure. A pressure that a component pressure containing envelope is represented to sustain for only one cycle without failure.

   CHANGE clause 3.9 to read:

   3.9   Pressure Containing Envelope Failure. A structural fracture, crack or any seal leakage caused by deformation or permanent deformation resulting from an RFP verification test; or a structural fracture, crack or any seal leakage caused by deformation or permanent deformation which interfered with the functioning of the pressure containing envelope of the pressure switch resulting from an RSP verification test.

CHANGE clause 3.10 to read:

3.10 Rated Overrange Pressure. The pressure to which a pressure switch can be subjected for extended time without change in operating characteristics, shift in set point, or damage to the device.

4. UNITS OF MEASUREMENT

NO CHANGE from the basic document.

5. LETTER SYMBOLS

NO CHANGE from the basic document.

6. OUTLINE OF PROCEDURES FOR RFP VERIFICATION BY TEST

NO CHANGE from the basic document.

7. RFP VERIFICATION BY SIMILARITY

CHANGE the entire Section to read:

Verification of the rated static pressure ratings of the pressure containing envelope of a pressure switch by similarity is allowed when the similar pressure switch differs from the tested pressure switch only in the size of the external port openings, provided those ports are of the same type and smaller than in the tested pressure switch.

8. OUTLINE OF PROCEDURES FOR RSP VERIFICATION BY TEST

NO CHANGE from the basic document.

NFPA/T2.6.1 S3

9. PREPARATIONS FOR TESTING

CHANGE clause 9.3 to read:

9.3   Bleed the entrapped air from the circuit and from the pressure containing envelope being tested.

CHANGE clause 9.8 to read:

9.8   Perform cyclic and static tests on separate pressure containing envelopes.

ADD the following clause:

9.12   If the sensing element is separable from the total assembly, and is not normally restrained in either force or motion by an external means, it may be removed from the total assembly for cyclic testing.

10. TEST EQUIPMENT

CHANGE clause 10.1.3 to read:

10.1.3   Mount the pressure measuring instrument directly into the pressure containing envelope, thru a pressurized port which is not being used to supply the test fluid.

CHANGE clause 10.1.4 to read:

10.1.4   When no such port (clause 10.1.3) is available, minimize restrictions between the instrument and the pressure containing envelope by mounting the pressure measuring instrument on the pressure supply line.

11. TEST CONDITIONS ACCURACY

NO CHANGE from the basic document.

12. PROCEDURES FOR CYCLIC TEST

    DELETE clause 12.1.5.

13. PROCEDURES FOR STATIC TEST

    CHANGE clause 13.1.2 to read:

    13.1.2   Apply test pressure at a rate no greater than one-tenth the selected RSP (per Section 8) per second. Maintain the selected RSP for one minute.

14. ADDITIONAL PROVISIONS

    CHANGE the entire Section to read:

    14.1   Restrain moving parts that are normally restrained at their point of maximum travel at RFP or zero pressure during cyclic testing.

    14.2   Remove springs, linkages or other mechanical parts for cyclic test purposes only if the removal is consistent with clause 14.1.

    14.3   Remove the electrical switching element for cyclic test purposes only if its removal is consistent with clause 14.1.

    14.4   All parts of the sensing element that move during normal functioning of the pressure switch must move in the same manner during cycling.

15. CRITERIA FOR RFP VERIFICATION

    ADD the following clause:

    15.4   Consider any leakage at threaded inlet ports due to distortion, a failure.

16. CRITERIA FOR RSP VERIFICATION

   CHANGE clause 16.4 to read:

   16.4   Consider any leakage at threaded inlet ports due to distortion, a failure.

   ADD the following clause:

   16.5   Consider any permanent deformation which interferes in any way with the proper functioning of the pressure containing envelope, but which is consistent with clause 1.2 of the Scope, a failure.

17. DATA PRESENTATION

   CHANGE clause 17.2.1 to read:

   17.2.1   All physical values obtained in Sections 6 and 8 as selected in clause 13.1.2.

   CHANGE clause 17.2.2 to read:

   17.2.2   Modifications to the pressure switch as permitted by clauses 9.11, 9.12, 12.1.1, 12.1.4, 14.1, 14.2 and 14.3.

   ADD the following clause:

   17.2.3   Accuracies and ranges of pressure gauges used to verify test pressures.

18. SUMMARY OF DESIGNATED INFORMATION

   NO CHANGE from the basic document.

19. JUSTIFICATION STATEMENT

    CHANGE the entire Section to read:

    This recommended standard verification procedure is based upon the combined expert experiences of those who have participated in its preparation and review, and in the preparation and review of the basic document, NFPA/T2.6.1-1974, and its tutorial reference.

    Additional justification data will be obtained and compiled as the procedures set forth in this document become more widely used. Recommendations for the further improvement of this recommended standard are always welcome. Forward them directly to NFPA at the following address:

    > National Fluid Power Association, Inc.
    > 3333 North Mayfair Road
    > Milwaukee, Wisconsin 53222

20. TEST/PRODUCTION SIMILARITY

    NO CHANGE from the basic document.

21. IDENTIFICATION STATEMENT

    CHANGE the quoted statement to read:

    > "Method of verifying the rated fatigue and rated static pressures of the pressure containing envelope of a pressure switch conforms to NFPA Recommended Standard, NFPA/T2.6.1 S3 (T3.29.2)-1976, category ____ *."

22. KEY WORDS

    The following Key Words, useful in indexes and in information retrieval systems, are suggested for this recommended standard:

    fluid power
    pressure, cyclic test
    pressure, rated fatigue
    pressure, rated static
    pressure, static test
    pressure rating, by similarity
    pressure rating, by test
    pressure rating, pressure switch
    pressure switch, fluid power

## NOTES

NFPA Recommended Standard
T2.6.1M S4 (T3.12.10M)-1976 (R1981)

**AN INDUSTRY STANDARD FOR FLUID POWER**

Air Line Filter, Regulator and/or Lubricator Pressure Rating

Supplement No. 4 to

NFPA Recommended Standard for Verifying the

Fatigue and Static Pressure Ratings of the

Pressure Containing Envelope of a Metal Fluid Power Component

Approved as an NFPA Recommended Standard
3 June 1976

published by
**NATIONAL FLUID POWER ASSOCIATION, INC.**

3333 N. Mayfair Road / Milwaukee, WI 53222 / 414-778-3344 / TLX 26898

# FOREWORD

(This Foreword is not part of Air Line Filter, Regulator and/or Lubricator Pressure Rating Supplement No. 4 to NFPA Recommended Standard for Verifying the Fatigue and Static Pressure Ratings of the Pressure Containing Envelope of a Metal Fluid Power Component, NFPA/T2.6.1 S4 (T3.12.10)-1976.)

Early in 1974, the approval of NFPA/T2.6.1-1974 established a group of common requirements intended to provide an industry-wide philosophy and basic standard, providing a rationale for judging a component's ability as a pressure containing envelope. Although the specific applicability of NFPA/T2.6.1-1974 is limited, it immediately established a uniform base for subsequent, more specific proposed NFPA recommended standards for individual components.

Draft No. 1 of this document was prepared by Chairman Colter (Watts) and was received at NFPA Headquarters on 14 January 1974. The FRL Section met on 6 February 1974 to commence merging all FRL pressure rating work into Project T3.12.10. Draft No. 1 of the document was reviewed at this meeting. Draft No. 2 was prepared by the NFPA Technical Staff on 5 April 1974, and was reviewed at the 20 May 1974 meeting of T3.12. Draft No. 3 was prepared on 9 October 1974; it was reviewed at the 30 October 1974 FRL Section meeting. A Final Working Draft was achieved at that meeting. The NFPA Technical Staff prepared the General Review Draft on 20 December 1974.

General Review comments were discussed and answered at the 12 March 1975 and 1 October 1975 T3.12 meetings. All comments were resolved by 6 October 1975. The Technical Board granted approval to ballot on 5 November 1975. The NFPA Technical Staff prepared the Ballot Draft on 5 December 1975. All negative ballots were resolved by 2 April 1976.

On 7 April 1976, the NFPA Technical Board voted to recommend approval of this document as an NFPA Recommended Standard. The NFPA Technical Staff prepared the edited Ballot Draft on 15 April 1976.

The NFPA Board of Directors approved this document as NFPA Recommended Standard T2.6.1 S4 (T3.12.10)-1976 on 3 June 1976.

NFPA/T2.6.1 S4

## PROJECT GROUP MEMBERS WHO DEVELOPED THIS STANDARD

| | | |
|---|---|---|
| Colter, John | Project Chairman and Section Chairman | Watts Regulator Co. |
| Bailey, Richard | Project Vice Chairman | C. A. Norgren Co. |
| Kay, Robert | Technical Auditor | Sperry Vickers |
| Luecke, John R. | Director of National Technical Services | National Fluid Power Association |

| | |
|---|---|
| Clark, A. | Aro Corporation |
| Demmler, F. | WABCO Fluid Power Division |
| Foltz, D. | Hankison Corporation |
| Hoffman, R. | Parker Hannifin Corporation |
| Kesselring, W. | Ross Operating Valve Company |
| Love, H. | Bellows International |
| Moore, R. | Monnier Brothers |
| Ogg, W. | Pall Corporation |
| Pousma, J. | C. A. Norgren Company |
| Ray, W. | Master Pneumatic-Detroit, Inc. |
| Russell, K. | Wilkerson Corporation |
| Sessoms, W. | Scovill Fluid Power Division |
| Thrasher, G. | Master Pneumatic-Detroit, Inc. |
| Ward, W. | Perfecting Service Division |
| Yoder, J. | Aro Corporation |

## REFERENCES

1. American National Standard Glossary of Terms for Fluid Power, ANSI/B93.2-1971, and Supplements thereto. (ISO/TC 131/SC 1 (USA-2) 3)

2. SI units and recommendations for the use of their multiples and of certain other units, ISO 1000-1973.

3. National Fluid Power Association Recommended Standard Method for Verifying the Fatigue and Static Pressure Ratings of the Pressure Containing Envelope of a Metal Fluid Power Component, NFPA/T2.6.1-1974.

NFPA/T2.6.1 S4

# AIR LINE FILTER, REGULATOR AND/OR LUBRICATOR PRESSURE RATING

## SUPPLEMENT NO. 4 to

NFPA Recommended Standard for Verifying the

Fatigue and Static Pressure Ratings of the

Pressure Containing Envelope of a Metal Fluid Power Component

---

HEADQUARTERS NOTE: The Project Group which developed this Supplement intended it for use only with the basic pressure rating document, NFPA/T2.6.1-1974. This Supplement provides a list of additions, deletions and changes to the basic document which are necessary for the establishment of pressure ratings for metal air line filters, regulators and/or lubricators. All clauses which have been modified are so indicated in this Supplement. Read all other clauses as they appear in the basic pressure rating document, NFPA/T2.6.1-1974.

Since revisions of the basic pressure rating document will apply to this Supplement, the reader is cautioned to always use the most recent edition of the basic document. When reading this Supplement with the basic document, substitute the phrase "air line filter(s), regulator(s) and/or lubricator(s)" whenever the term "component(s)" appears in the basic document.

---

## INTRODUCTION

CHANGE the first and second paragraphs to read:

In pneumatic fluid power systems, power is transmitted and controlled thru a gas under pressure within an enclosed circuit. A filter removes contaminants from the air line, a regulator reduces and controls the air pressure in the line, and a lubricator provides lubrication for downstream equipment.

A basic requirement of fluid power air line filters, regulators and/or lubricators is that they should be capable of adequately containing the pressurized gas.

NFPA/T2.6.1 S4

The basic pressure rating document, NFPA/T2.6.1-1974, established a group of common requirements intended to provide an industry-wide philosophy and basic standard, providing a rationale for judging a component's ability as a pressure containing envelope. Although the specific applicability of NFPA/T2.6.1-1974 is limited, it immediately established a uniform base for subsequent, more specific NFPA Recommended Standards for individual fluid power components. This Recommended Standard implements NFPA/T2.6.1-1974, and specifically applied to air line filters, regulators and/or lubricators.

1. SCOPE

   CHANGE clause 1.1 to read:

   1.1  To include methods for verifying ratings of the fatigue and static strengths of metal fluid power air line filters, regulators and/or lubricators or integral combinations thereof, pressure containing envelopes by test; or by analysis by geometric similarity (based on tests conducted in accordance with this recommended standard) that are:

   CHANGE clause 1.2 to read:

   1.2  This recommended standard does not establish a standard means for determining the magnitude of pressure transients in systems.

   NOTE - These magnitudes are usually determined by the system designer who then selects air line filters, regulators and/or lubricators with adequate strengths. However, it is generally recognized that there is very little authoritative data available to assist the system designer in making this determination. It is anticipated that the promulgation of this recommended standard will also stimulate the study of relationships between peak and operating pressures.

   CHANGE clause 1.3 to read:

   1.3  Fatigue and static pressure ratings do not apply to elastomeric or non-metallic diaphragms and/or pistons in regulators.

ADD the following clause:

1.4 This recommended standard specifically excludes pressure gauges used on metal fluid power air line filters, regulators and/or lubricators, or integral combinations thereof.

2. PURPOSE

CHANGE the entire section to read:

To provide standard methods for verifying the fatigue and static pressure ratings of a metal fluid power air line filter, regulator and/or lubricator pressure containing envelope.

3. TERMS AND DEFINITIONS

CHANGE clause 3.12 to read:

3.12 Elements. The integral pieces which make up a part or component (i.e., air line filter, regulator and/or lubricator), (e.g., body, bonnet, bowl, plug, etc.) of the pressure containing envelope.

4. UNITS OF MEASUREMENT

NO CHANGE from the basic document.

5. LETTER SYMBOLS

NO CHANGE from the basic document.

6. OUTLINE OF PROCEDURES FOR RFP VERIFICATION BY TEST

NO CHANGE from the basic document.

NFPA/T2.6.1 S4

7. RFP VERIFICATION BY SIMILARITY

CHANGE the entire section to read:

In the case of a series of air line filters, regulators and/or lubricators which differ only in the size of ports in the pressure containing envelope with all other dimensions, elements, materials and shapes being identical, RFP may be verified by testing the largest port sized component in the series.

8. OUTLINE OF PROCEDURES FOR RSP VERIFICATION BY TEST

NO CHANGE from the basic document.

9. PREPARATIONS FOR TESTING

CHANGE clause 9.3 to read:

9.3 When liquids are used, bleed the entrapped air from the circuit and from the pressure containing envelope being tested.

CHANGE clause 9.4 to read:

9.4 Apply different ratings to separate portions of the pressure containing envelope as required.

CHANGE clause 9.5 to read:

9.5 Allow ports that are not part of the pressure containing envelopes to remain at atmospheric pressure (i.e., regulator relief port).

10. TEST EQUIPMENT

CHANGE clause 10.1.3 to read:

10.1.3 Mount the pressure measuring instrument directly into the pressure containing envelope, thru a pressurized port which is not being used to supply the test fluid.

NFPA/T2.6.1 S4

CHANGE clause 10.1.4 to read:

10.1.4   When no such port (clause 10.1.3) is available, minimize restrictions between the instrument and the pressure containing envelope by mounting the pressure measuring instrument on the pressure supply line.

NOTE - It is recommended that it be verified that the pressure generated in the pressure containing envelope at the actual cycling rate be the intended value.

11. TEST CONDITIONS ACCURACY

NO CHANGE from the basic document.

12. PROCEDURES FOR CYCLIC TEST

NO CHANGE from the basic document.

13. PROCEDURES FOR STATIC TEST

NO CHANGE from the basic document.

14. ADDITIONAL PROVISIONS

CHANGE the entire section to read:

14.1   Integrate the pressure rating test procedures into performance tests, where possible.

14.2   Internal components may be removed or blocked open to speed up the test cycle, providing that maximum envelope stress is not affected.

15. CRITERIA FOR RFP VERIFICATION

CHANGE clause 15.2 to read:

15.2   Consider any crack produced by metal fatigue due to pressure cycling, as verified by magnetic particle or fluorescent penetrant or other appropriate techniques after testing, a failure.

NFPA/T2.6.1 S4

16. CRITERIA FOR RSP VERIFICATION

    CHANGE clause 16.2 to read:

    16.2 Consider any crack produced by internal static pressure, as verified by magnetic particle or fluorescent penetrant or other appropriate techniques after testing, a failure.

17. DATA PRESENTATION

    NO CHANGE from the basic document.

18. SUMMARY OF DESIGNATED INFORMATION

    NO CHANGE from the basic document.

19. JUSTIFICATION STATEMENT

    CHANGE the entire section to read:

    This recommended standard verification procedure is based upon the combined expert experiences of those who have participated in its preparation and review, and in the preparation and review of the basic document, NFPA/T2.6.1-1974, and its tutorial reference.

    Additional justification data will be obtained and compiled as the procedures set forth in this document become more widely used. Recommendations for the further improvement of this proposed recommended standard are always welcome. Forward them directly to NFPA at the following address:

    National Fluid Power Association, Inc.
    3333 North Mayfair Road
    Milwaukee, WI 53222

20. TEST/PRODUCTION SIMILARITY

    NO CHANGE from the basic document.

NFPA/T2.6.1 S4

21. IDENTIFICATION STATEMENT

CHANGE the quoted statement to read:

"Method of verifying rated fatigue and rated static pressures of an air line filter, regulator and/or lubricator conforms to NFPA Recommended Standard, NFPA/T2.6.1 S4 (T3.12.10)-1976, category \_\_\_\_\_ *."

22. KEY WORDS

The following Key Words, useful in indexes and in information retrieval systems, are suggested for this proposed recommended standard:

filter, air line

fluid power

lubricator, pneumatic fluid power

pressure, cyclic test

pressure, rated fatigue

pressure, rated static

pressure, static test

pressure rating, by similarity

pressure rating, by test

pressure rating, air line filter

pressure rating, FRL

pressure rating, lubricator

pressure rating, regulator

regulator, pneumatic fluid power

NFPA Recommended Standard
T2.6.1M S9 (T3.5.26M)-1977 (R1982)

## AN INDUSTRY STANDARD FOR FLUID POWER

Hydraulic Valve Pressure Rating

Supplement No. 9 to

NFPA Recommended Standard for Verifying the

Fatigue and Static Pressure Ratings of the

Pressure Containing Envelope of a Metal Fluid Power Component

Approved as an NFPA Recommended Standard
23 March 1977

published by

**NATIONAL FLUID POWER ASSOCIATION, INC.**

3333 N. Mayfair Road  /  Milwaukee, WI 53222  /  414-778-3344  /  TLX 26898

NFPA/T2.6.1 S9

## FOREWORD

This Foreword is not part of Hydraulic Valve Pressure Rating Supplement No. 9 to NFPA Recommended Standard for Verifying the Fatigue and Static Pressure Ratings of the Pressure Containing Envelope of a Metal Fluid Power Component, NFPA/T2.6.1 S9 (T3.5.26)-1977.

Early in 1974, the approval of NFPA/T2.6.1-1974 established a group of common requirements intended to provide an industry-wide philosophy and basic standard, providing a rationale for judging a component's ability as a pressure containing envelope. Although the specific applicability of NFPA/T2.6.1-1974 is limited, it immediately established a uniform base for subsequent, more specific proposed NFPA recommended standards for individual components.

The T3.5.26 Project Group met on 20 November 1973 to prepare the first draft of this document based on NFPA/T2.6.1-1974. On 4 January 1974, the Electrohydraulic Servovalve Section, T3.17, requested a joint pressure rating document with the Hydraulic Valve Section, T3.5. The first draft was completed on 8 April 1974, and was based on the results of the 20 November 1973 meeting, the worksheet provided by NFPA Headquarters, and the combination of T3.5 and T3.17 efforts into one document.

Project Group T3.5.26 met on 20 May 1974 to review the first draft and make revisions as required. On 21 May 1974, Section T3.5 was presented with the revisions to the first draft along with the Project Group's recommendation that the revised document be circulated for General Review. Section members concurred with this recommendation, and the Final Working Draft was forwarded to Headquarters. The General Review Draft was prepared on 31 May 1974.

As part of the 30 October 1974 meeting, Project Group and Section members discussed General Review comments and formulated revisions to the document. Early in 1975, Project Group Chairman Harold Jacoby (Webster Electric Co.) submitted replies to those who had offered General Review comments.

On 15 April 1975, the Project Group met to resolve the remaining General Review comments as well as the new comments received as a result of the replies submitted in January 1975. Letters indicating

continued . . .

NFPA/T2.6.1 S9

## FOREWORD (continued)

revisions to the document were submitted by Chairman Jacoby on 22 April 1975. Unresolved comments from General Review were presented to the Pressure Rating Coordinating Committee, T2.6, at the 12 June 1975 meeting. T2.6 comments were discussed at the 1 October 1975 Hydraulic Valve Section meeting, and suggestions were offered for revising the document.

During December 1975, Chairman Jacoby incorporated suggested changes in the document. On 6 January 1976, he forwarded the marked-up copy to NFPA Headquarters. Approval to ballot was granted by the Technical Board on 7 April 1976, and the NFPA Technical Staff prepared the Ballot Draft on 12 April 1976.

Negative comments received during ballot were answered by the Section on 28 September 1976. One comment, which requested that the Identification Statement of this document be modified to allow a manufacturer to cite either Rated Fatigue Pressure or Rated Static Pressure, or both, was referred to the NFPA Pressure Rating Coordinating Committee; this committee will consider including the requested change in a future revision of the basic NFPA Pressure Rating Standard. All negative votes were withdrawn by 28 January 1977.

On 2 February 1977, the NFPA Technical Board voted to recommend to the Board of Directors that this document be approved as an NFPA Recommended Standard. The NFPA Board of Directors granted final approval to NFPA/T2.6.1 S9 (T3.5.26)-1977 on 23 March 1977.

NFPA/T2.6.1 S9

## PROJECT GROUP MEMBERS WHO DEVELOPED THIS STANDARD

| | | |
|---|---|---|
| Jacoby, Harold | Project Chairman | Webster Electric Co., Inc. |
| Prevallet, David | Project Secretary and Section Vice Chairman | Gresen Manufacturing Co. |
| Olen, Robert | Section Chairman (1976 - present) | DeLaval Turbine Inc. |
| Krehbiel, Rob | Section Chairman (1975) | Cessna Fluid Power Division |
| Mills, Nathan | Section Chairman (1971 - 1975) | Sperry Vickers* |
| Pippenger, John | Technical Auditor | Hydraulic Products, Inc. |
| Luecke, John R. | Director of National Technical Services | National Fluid Power Association |

| | |
|---|---|
| Becker, L. | Hydreco |
| Bowman, R. | Continental Hydraulics Div. |
| Gagnon, J. | HUSCO Division |
| George, P. | Parker Hannifin Corp. |
| Hedge, J. | Abex Corp. |
| Jackson, A. | Cessna Fluid Power Div. |
| Keir, J. | Sperry Vickers |
| McNeil, R. | Moog, Inc. |
| Obergefell, R. | Fluid Controls, Inc. |
| Pauken, D. | Double A Products Co. |
| Tenkku, W. | Fluid Controls, Inc. |
| Thomas, V. | Tomco, Inc. |
| Thorson, C. | Eaton Corp. |
| Turko, J. | Commercial Shearing, Inc. |

*Retired

## REFERENCES

1. American National Standard Glossary of Terms for Fluid Power, ANSI/B93.2-1971, and Supplements thereto. (ISO Draft Proposal 5598)

2. SI units and recommendations for the use of their multiples and of certain other units, ISO 1000-1973.

3. NFPA Recommended Standard for Verifying the Fatigue and Static Pressure Ratings of the Pressure Containing Envelope of a Metal Fluid Power Component, NFPA/T2.6.1-1974.

# HYDRAULIC VALVE PRESSURE RATING

SUPPLEMENT NO. 9 to

NFPA Recommended Standard for Verifying the

Fatigue and Static Pressure Ratings of the

Pressure Containing Envelope of a Metal Fluid Power Component

---

HEADQUARTERS NOTE: The Project Group which developed this Supplement intended it for use only with the basic pressure rating document, NFPA/T2.6.1-1974. This Supplement provides a list of additions, deletions and changes to the basic document which are necessary for the establishment of pressure ratings for metal hydraulic valves. All clauses which have been modified are so indicated in this Supplement. Read all other clauses as they appear in the basic pressure rating document, NFPA/T2.6.1-1974.

Since revisions of the basic pressure rating document will apply to this Supplement, the reader is cautioned always to use the most recent edition of the basic document. When reading this Supplement with the basic document, substitute the phrase "hydraulic valve(s)" whenever the term "component(s)" appears in the basic document.

---

## INTRODUCTION

CHANGE the first three paragraphs to read:

In hydraulic fluid power systems, power is transmitted and controlled thru a liquid under pressure within an enclosed circuit. A hydraulic valve is a device which controls fluid flow direction, pressure, or flow rate. A basic requirement of hydraulic fluid power valves is that they should be capable of adequately containing the pressurized liquid.

The basic pressure rating document, NFPA/T2.6.1-1974, established a group of common requirements intended to provide an industry-wide philosophy and basic standard, providing a rationale for judging a component's ability as a pressure containing envelope. Although the specific applicability of NFPA/T2.6.1-1974 is limited, it immediately established a uniform base for subsequent, more specific proposed NFPA recommended standards for individual fluid power components.

NFPA/T2.6.1 S9

This recommended standard implements NFPA/T2.6.1-1974, and specifically applies to metal hydraulic fluid power valves.

Ratings verified in accordance with this standard do not replace pressure ratings based on considerations such as performance and internal leakage. Instead, these new ratings are intended to supplement those that are provided by present practice and may be numerically different from prior ratings.

1. SCOPE

   CHANGE the entire Section to read:

   1.1 To include methods for verifying ratings of the fatigue and static strengths of metal hydraulic fluid power valve pressure containing envelopes by test; or by geometric similarity analysis (based on tests conducted in accordance with this recommended standard) that are:

   1.1.1 Subjected to pressure induced stresses.

   1.1.2 Used within the $-20^\circ$ to $200^\circ F$ ($-29^\circ$ to $93^\circ C$) fluid temperature range.

   1.1.3 Not subject to loss of strength due to corrosion or other chemical action.

   1.1.4 Made of aluminum, magnesium, steel, iron, copper-based alloys and stainless steels. (Other metals will be included in subsequent revisions as soon as suitable data are available.)

   1.2 Hydraulic valves may contain one or more pressure zones. Each pressure zone will carry static and fatigue pressure ratings based on a maximum differential pressure to an adjacent lower pressure zone, or atmosphere.

   1.3 Attachments can be rated as an integral part or as an independent component.

NFPA/T2.6.1 S9

1.4  This recommended standard does not establish a standard means for determining the magnitude of pressure transients in systems.

NOTE: These magnitudes are usually determined by the system designer who then selects hydraulic valves with adequate strengths. However, it is generally recognized that there is very little authoritative data available to assist the system designer in making this determination. It is anticipated that the promulgation of this recommended standard will also stimulate the study of relationships between peak and operating pressures.

2. PURPOSE

CHANGE the entire Section to read:

To provide standard methods for verifying the fatigue and static pressure ratings of a metal hydraulic fluid power valve pressurized zones with regard to sustaining cyclic and steady pressure loads that can be used as part of specific hydraulic valve standards.

3. TERMS AND DEFINITIONS

CHANGE the entire Section to read:

(For definition of other fluid power terms used, see Reference No. 1.)

3.1  Pressure Zone. The pressure containing envelope or a separate portion or portions having by design or function, separate fatigue and static strength values (e.g., pressure, cylinder and tank passages of mobile directional control valves).

3.2  Rated Fatigue Pressure. A pressure that a component (i.e., hydraulic valve) pressure containing envelope or zone is represented to sustain 10 million times without failure.

3.3  Rated Static Pressure. A pressure that a component (i.e., hydraulic valve) pressure containing envelope or zone is represented to sustain for only one cycle without failure.

3.4  Cyclic Test Pressure. A pressure range applied in cyclic tests that are performed to verify a Rated Fatigue Pressure.

3.5  Static Test Pressure. A pressure applied in static tests performed to verify a Rated Static Pressure.

3.6  Test Duration Factor. A multiplier applied to the Rated Fatigue Pressure for calculating the Cyclic Test Pressure which will permit shortening the qualification test from 10 million to one million cycles.

3.7  Variability Factor. A multiplier applied to the Rated Fatigue Pressure for calculating the Cyclic Test Pressure to account for the variability in fatigue strength of metals. It is also applied to the Rated Static Pressure for calculating the Static Test Pressure.

3.8  Assurance Level. The minimum percentage of pressure containing envelopes or pressure zones of a verified design that will sustain 10 million applications of its Rated Fatigue Pressure.

3.9  Confidence Level. The degree to which a manufacturer can be assured that the desired assurance level is attained.

3.10  Pressure Containing Envelope or Pressure Zone Failure. A structural fracture, crack or excessive seal leakage caused by deformation resulting from an RFP verification test; or a structural fracture, crack or leakage caused by deformation or permanent deformation resulting from an RSP verification test.

NOTE: It is the responsibility of the hydraulic valve manufacturer to define "excessive seal leakage".

3.11  Attachment. Any device appending to the hydraulic valve in communication with a pressure zone.

NFPA/T2.6.1 S9

The following terms, common in fluid power terminology, are defined in general usage dictionaries, so therefore do not appear in Reference No. 1. Because these terms have more than one accepted meaning, we have set forth below the meaning applied to their use in this document.

3.12 Pressure Containing Envelope. That part of the component (i.e., hydraulic valve) which contains the pressurized fluid.

3.13 Elements. The integral pieces which make up a part or component (i.e., hydraulic valve), (e.g., body, spool, etc.).

The following terms, not included in this document but found in related reference material, are defined with the meaning that the Committee applied when using the reference.

3.14 Peak Pressure. The maximum pressure actually experienced by a component (i.e., hydraulic fluid power valve) in an application. It is caused by any number of events, individually or in combination, such as (but not limited to) impact, inertia, shock, heat, etc.

3.15 Nominal Pressure. A pressure value assigned to a component (i.e., hydraulic valve) or a system for the purpose of convenient designation. This may be an average of normal values without peaks, or a design figure for stress calculations.

4. UNITS OF MEASUREMENT

NO CHANGE from the basic document.

5. LETTER SYMBOLS

NO CHANGE from the basic document.

6. OUTLINE OF PROCEDURES FOR RFP VERIFICATION BY TEST

CHANGE clauses 6.5 and 6.6 to read:

6.5 Test the specified pressure containing envelopes or pressure zones in the number of hydraulic valves upon which the factor $K_V$ was selected in clause 6.2.

6.6 Provide pressure pulses per clause 12.3.

CHANGE Section 7 to read:

7. OUTLINE OF PROCEDURES FOR RSP VERIFICATION BY TEST

7.1 Select a $K_V$ factor per clause 12.2.2.

7.2 Calculate the STP per clause 13.2.

7.3 Use test equipment per clause 10.2.

7.4 Test the specific pressure containing envelopes or pressure zones in the number of hydraulic valves upon which the factor $K_V$ was selected in clause 7.1.

7.5 Conduct static tests per Section 13.

7.6 Verify the RSP per Section 16.

7.7 Record data per Section 17.

CHANGE Section 8 to read:

8. RFP AND RSP VERIFICATION BY SIMILARITY

8.1 Rated Fatigue Pressure and Rated Static Pressure by geometric similarity are permitted for hydraulic valves which utilize the same geometric element as the hydraulic valve tested and differ only in the size of the external port openings, provided those ports are of the same type and smaller than in the hydraulic valve tested as the basis for the similarity analysis.

8.2 The "same geometric element" for hydraulic valves means that the pressure containing envelope or pressure zone must have the same dimensional configuration and be of the same material.

8.3 Verification by similarity does not apply when the similar geometric element is modified in a manner which could reduce the pressure capability from the hydraulic valve tested as the basis for the similarity analysis.

8.4 Assurance and Confidence Levels of verification made by similarity must be the same as the values established for the hydraulic valve tested as the basis for the similarity analysis.

## 9. PREPARATIONS FOR TESTING

CHANGE the entire Section to read:

9.1 Use liquid to pressurize pressure containing envelopes or pressure zones being tested.

9.2 CAUTION: Take special precautions, while conducting any of the pressure tests, in the design of test rigs, and in training the operators to maximize operator safety.

9.3 Bleed the entrapped air from the hydraulic circuit and from the pressure containing envelope or pressure zone being tested.

9.4 Apply different ratings to separate pressure zones as required by operating conditions.

9.5 Allow all drains and low pressure ports that are not part of the pressure containing envelopes or zones to drain freely and keep them at atmospheric pressure.

9.6 Test pressure containing envelopes or pressure zones whose geometries, material properties and manufacturing methods are known and representative of normal production.

9.7 Do not perform manufacturer's proof tests prior to cyclic testing, unless such tests are made on all production hydraulic valves.

9.8 Perform cyclic and static tests on separate pressure containing envelopes or pressure zones when a sufficient number of test pieces is available.

9.9 Perform the cyclic tests first, when both RFP and RSP verification tests must be made on the same pressure containing envelope or pressure zone.

9.10 Complete machining and processing of test hydraulic valve(s) to the degree necessary for duplicating operating stress distributions and final strength in the pressure containing envelope or pressure zone.

9.11 It is permissible to make modifications to the test piece to facilitate cyclic or static tests providing that such modifications do not increase the pressure capabilities of the pressure containing envelope or pressure zone being tested.

## 10. TEST EQUIPMENT

CHANGE clauses 10.1 up to and including 10.1.4 to read:

10.1 Cyclic Test

10.1.1 Use a circuit on the test stand which will produce repeatable pressure pulses (per clause 12.3) which cycle between approximately zero and a specified maximum pressure.

NOTE: The specific design of the test stand and circuit is optional.

10.1.2 Use any suitable non-corrosive test fluid.

10.1.3 Mount the pressure measuring instrument directly into the pressure containing envelope or pressure zone being tested thru a pressurized port which is not being used to supply the test fluid.

10.1.4   When no such port (clause 10.1.3) is available, minimize restrictions between the instrument and the pressure containing envelope or pressure zone by mounting the pressure measuring instrument on the pressure supply line.

NOTE: It is recommended that it be verified that the pressure generated in the pressure containing envelope or pressure zone at the actual cycling rate be the intended value.

## 11. TEST CONDITIONS ACCURACY

NO CHANGE from the basic document.

## 12. PROCEDURES FOR CYCLIC TEST

CHANGE clauses 12.1.1 up to and including 12.1.3 to read:

12.1.1   Place metal shot or loosely fitting metal pieces in the pressure containing envelope or pressure zone being tested to reduce the volume of the pressurized fluid when higher cycling rates are desired.

12.1.2   Subject all like pressure containing envelopes or pressure zones to the same pressure loadings when testing multiple samples.

12.1.3   In each complete test cycle for hydraulic valves with adjacent pressure zones (having a common boundary), alternately apply the maximum differential test pressure to each zone.

CHANGE clause 12.1.5 to read:

12.1.5   Verify with a strain measuring device that the ratio of induced stresses to pressure under static load is attained at the test cycling rate where pressures must penetrate between close fitting parts, very large hydraulic valves are tested, or where hysteresis in joints can significantly affect stresses.

ADD the following clause:

> 12.1.6 If threaded fluid connections used for test purposes, such as fittings, loosen during cyclic tests, they may be tightened without constituting a failure.

CHANGE clause 12.2.1.2 to read:

> 12.2.1.2 Use factors specified for non-ferrous metals in testing all pressure containing envelopes or pressure zones with non-ferrous elements.

CHANGE clause 12.2.2.2 to read:

> 12.2.2.2 Use the Variability Factor for the materials contained in the pressure containing envelope or pressure zone being tested which have the highest value.

CHANGE clauses 12.3.1 up to and including 12.3.2 to read:

> 12.3.1 Provide a pressure pulse that decreases to below 0.05 CTP in each cycle.

> 12.3.2 Provide a maximum pressure pulse that exceeds the minimum pressure pulse (per clause 12.3.1) by an amount equal to the CTP.

13. PROCEDURES FOR STATIC TEST

CHANGE clause 13.1.1 to read:

> 13.1.1 Apply test pressure to the pressure containing envelope or pressure zones, individually or in combination, to induce the most critical loading.

14. ADDITIONAL PROVISIONS

DELETE clauses 14.1 up to and including 14.10.

NFPA/T2.6.1 S9

15. CRITERIA FOR RFP VERIFICATION

CHANGE clauses 15.2 up to and including 15.3 to read:

15.2 Consider any crack produced by metal fatigue due to pressure cycling, that results in leakage or as verified by magnetic particle or fluorescent penetrant or other appropriate technique after testing, a failure.

15.3 Consider excessive leakage at seals or sealing surfaces, that results from deformation of the metallic portion of the pressure containing envelope or pressure zone, including fastening devices, a failure.

16. CRITERIA FOR RSP VERIFICATION

CHANGE clauses 16.2 up to and including 16.4 to read:

16.2 Consider any crack produced by internal static pressure, that results in leakage or as verified by magnetic particle or fluorescent penetrant or other appropriate technique after testing, a failure.

16.3 Consider leakage at seals or sealing surfaces, that results from deformation of the metallic portion of the pressure containing envelope or pressure zone, including fastening devices, a failure.

16.4 Consider any permanent deformation, which interferes in any way with the proper functioning of the pressure containing envelope or pressure zone, a failure.

17. DATA PRESENTATION

ADD the following clause:

17.1.3 Each pressure zone of a hydraulic valve must have the above ratings specified when tested as separate pressure zones.

ADD the following clause:

17.2.3 All data and information necessary to verify the basis used for rating by geometric similarity.

NFPA/T2.6.1 S9

18. SUMMARY OF DESIGNATED INFORMATION

   NO CHANGE from the basic document.

19. JUSTIFICATION STATEMENT

   CHANGE the entire Section to read:

   This recommended standard verification procedure is based upon the combined expert experiences of those who have participated in its preparation and review, and in the preparation and review of the basic document, NFPA/T2.6.1-1974, and its tutorial reference.

   Additional justification data will be obtained and compiled as the procedures set forth in this document become more widely used. Recommendations for the further improvement of this recommended standard are always welcome. Forward them to NFPA at the following address:

   > National Fluid Power Association, Inc.
   > 3333 North Mayfair Road
   > Milwaukee, Wisconsin 53222

20. TEST/PRODUCTION SIMILARITY

   NO CHANGE from the basic document.

21. IDENTIFICATION STATEMENT

   CHANGE the quoted statement to read:

   > "Method of verifying rated fatigue and rated static pressures of hydraulic valves conforms to NFPA Recommended Standard, NFPA/T2.6.1 S9 (T3.5.26)-1977, category _____ *."

- 136 -

NFPA/T2.6.1 S9

ADD the following Section:

22. KEY WORDS

The following Key Words, useful in indexes and in information retrieval systems, are suggested for this recommended standard:

fluid power
pressure, cyclic test
pressure, rated fatigue
pressure, rated static
pressure, static test
pressure rating, by similarity
pressure rating, by test
pressure rating, hydraulic valve
similarity, geometric
testing, hydraulic valve
testing, pressure
valve, hydraulic fluid power
zone, pressure

# NOTES

ANSI/B93.7M-1986

## AN INDUSTRY STANDARD FOR FLUID POWER

American National Standard

Hydraulic fluid power -

Valves -

Mounting interfaces

Approved as an ANSI Standard
4 June 1986

**Descriptors:** fluid power; hydraulic fluid power; directional control valves, compensated flow control valve, check valve, pilot operated check valves, pressure control valves, sequence valves, throttle valves, unloading valves, pressure relief valves; valve interchangeability; dimensional surfaces, dimensional criteria; mounting interfaces.

published by
# NATIONAL FLUID POWER ASSOCIATION, INC.
3333 N. Mayfair Road / Milwaukee, WI 53222 / 414-778-3344 / TLX 26898

## FOREWORD

This Foreword is not part of American National Standard Hydraulic fluid power - Valves - Mounting interfaces, ANSI/B93.7M-1986 (revision of ANSI/B93.7-1968 and ANSI/B93.7A - 1981).

On 20 April 1977 the NFPA Valve Section acknowledged the need to revise NFPA/T3.5.1M (ANSI/B93.7M-1968 [R1979]). The T3.5 Section recommended all work (NFPA proposal, two ANSI Standards, and an ISO proposal) be consolidated into one convenient reference to promote international interchangeability of hydraulic valves. Headquarters assigned the undertaking the project number T3.5.1M R1.

A Title, Scope and Purpose was developed and forwarded to the Technical Board for approval. The TSP was granted approval 11 May 1977.

A preliminary draft document was developed and distributed among Project Group members.

On 9 May 1979, the T3.5 Valve Section recommended the reaffirmation of ANSI/B93.7M with the knowledge that a revision was underway.

The Project Group also met on 9 May 1979. Project Group Chairman Coleman restated that their primary purpose was to consolidate standards, ANSI/B93.40M, NFPA/T3.5.1M S1 R1, and also consider ISO 4401.

When the Project Group met on 16 August 1979, each member agreed to check a particular portion of the preliminary draft for errors (i.e., valve and pilot operated check valve interfaces, flow control interfaces, directional control interfaces, and pressure control interfaces).

On 24 October 1979 the Project Group members were asked to continue their work on checking the accuracy of specific interface patterns. There was considerable discussion concerning subplate interfaces.

The work completed was reviewed at the Project Group meeting 17 April 1980. It was recommended that the evolving document be forwarded to Headquarters for distribution among Project Group members.

On 14 August 1980 the Project Group recommended that because existing and emerging NFPA and ANSI standards include codes based on dimensional units (D01, D02, etc.) and ISO standards being similar in form but; differently based and playing a significant role in the US marketplace, the NFPA and ANSI series in NFPA/T3.5.1M R1 be changed to coincide with the ISO series. This recommendation was circulated for comment to NFPA Valve Section (T3.5), Terms Coordinating Committee (T2.1) and Metrication Coordinating Committee (T2.10) by the Project Group on 29 August 1980.

The Project Group reviewed and noted responses at their meeting of 9 October 1980. It was agreed to use the codification numbers conforming to ISO 5783 when the General Review Draft was submitted for comments.

It was reported to the T3.5 Valve Section on 24 March 1981 that necessary information to complete the Draft Document would be forwarded to Headquarters.

The report filed with Valve Section T3.5 on 19 October 1981 indicated a revised draft of NFPA/T3.5.1 R1 was completed and would be forwarded to Headquarters for preparation as a General Review Draft.

Headquarters Staff prepared the document for General Review on 8 April 1982.

The General Review closed 10 May 1982. Numerous comments were received. Comments were discussed at length, reviewed and resolved at the 28 September 1982 Project Group meeting.

Project Group Chairman Coleman forwarded the document to the Technical Board where it received approval to Ballot on 2 February 1983.

Headquarters Technical Staff prepared T3.5.1M R1 for ballot on 25 March 1983. Balloting closed 25 April 1983 with one negative comment. The negative was withdrawn upon the condition that a genuine effort to remove all ambiguities of valve port markings be addressed thru the NFPA/T3.5.2M R1 document.

On 28 March 1984 it was reported to the T3.5 Valve Section that the ballot was complete and recommended the document be forwarded to the NFPA Technical Board for approval.

The Technical Board unanimously granted final approval on 17 May 1984 and recommended to the Board of Directors that the document be approved as an NFPA Recommended Standard and that after approval the document be submitted to ANSC B93 for promulgation as an ANSI standard.

The NFPA Board of Directors supported the Technical Board recommendations and granted final approval on 6 June 1984.

Project Group members who developed this document:

**Larry Coleman**
Project Group Chairman
Continental Hydraulics

**Richard Woodring**
Project Secretary
Parker Hannifin Corp.

**David Prevallet**
Section Chairman
Dana Corp./Gresen

**Harold Jacoby**
Section Vice Chairman
Dana Corp./Racine Hydraulics Div.

**Wayland Tenkku** *
Section Secretary
Fluid Controls Inc.

**Pete Wolf**
Technical Auditor
Eaton Corporation

**Allen E. Tucker** *
Director of Technical Services
National Fluid Power Association

**Jeff Cooper**
Double A Products

**Tom Clark** **
Dana Corp./Racine Hydraulics Div.

**Tom Frankenfield**
Rexroth Corporation

**Frank Hedge**
Abex/Denison Division

**Robert Olen**
Dana Corp./Gresen

**Jack Walrad**
Vickers Inc.

---

\* Company affiliation has changed.
\*\* Retired

mkm

On 27 November 1985, ANSI/B93.7M was submitted to ANSI Committee B93 for Ballot. There were no negative comments.

ANSI/B93.7M-1986 (revision of ANSI/B93.7-1968 and ANSI/B93.7A-1981), was approved by ANSI's Board of Standards Review on 4 June 1986.

The membership roster of Standards Committee B93 at the time of ballot:

**Jack McPherson**
Chairman

**Daniel B. Shore**
Vice Chairman

**Allen E. Tucker**
Co-Secretary

**Herb Kaufman**
Co-Secretary

**Compressed Air & Gas Institute**
David E. Bonn
John Addington (alternate)

**Construction Industry Manufacturers Association**
Glenn Stewart

**Fluid Controls Institute**
Jude Pauli
E.C. Rutter (alternate)

**Fluid Power Distributors Association**
Thomas Neff

**Fluid Power Society**
Nick Beaver
J. Otto Byers
Robert L. Firth
Paul Gies
Harry R. Holsen
Dale H. Killen
Verne L. Middleton
David Prevallet
James C. White

**Fluid Sealing Association**
Alex Pilecki

**Material Handling Institute**
Jack C. McPherson
Willard Chichester (alternate)

**National Fluid Power Association**
Richard N. Bailey
Walter Forster
Robert Kay
Z.J. Lansky
Paul Schacht

**National Machine Tool Builders' Association**
John B. Deam

**Society of Automotive Engineers**
William A. Hertel
John T. Parrett
Daniel B. Shore
W.L. Snyder
Robert W. White

**US Department of Defense**
S. Nguyen

**Company Members**
Logan Mathis
Lloyd L. Schmaltz
Don McGeachy
John Welker (alternate)
Logan Mathis

**Individual Members**
Dr. E.C. Fitch
Carroll Grigsby
John J. Pippenger
A.O. Roberts
Jack Walrad
Tom Wanke
Frank Yeaple

# Hydraulic fluid power - Valves - Mounting interfaces

## 0 INTRODUCTION

In hydraulic fluid power systems, power is transmitted and controlled thru a liquid, under pressure, within an enclosed circuit. Typical components found in such systems are hydraulic valves. These devices control flow direction, pressure, or flow rate of liquids in the enclosed circuit.

## 1 SCOPE AND FIELD OF APPLICATION

**1.1** This Standard includes mounting surfaces for the following:

- directional control valves;
- compensated flow control valves;
- check valves;
- pilot operated check valves;
- pressure control valves;
- sequence valves;
- throttle valves;
- unloading valves;
- pressure relief valves;

**1.2** This Standard includes the following dimensional criteria:

- minimum surface dimensions;
- sizes and locations of tapped holes for mounting bolts;
- sizes and locations of ports;
- sizes and locations of dowel or rest pins where required.

**1.3** This Standard provides the following general criteria:

- surface finish and flatness;
- indication of tolerances where pertinent;
- indication of appropriate radii.

**1.4** This Standard only applies to the dimensional criteria of products manufactured in conformance with this standard. It does not apply to their functional characteristics.

## 2 REFERENCES

ANSI/B93.2 and ANSI/B93.2A, *American National Standard Glossary of Terms for Fluid Power.*

ANSI/B93.9, *American National Standard Symbols for Marking Electrical Leads and Ports on Fluid Power Valves.*

ANSI/B93.40, *American National Standard Series of Mounting Interfaces for 4567 Maximum psi (315 bar) Four Port Hydraulic Fluid Power Directional Valves.*

ANSI/Y14.5M, *Dimensioning and Tolerancing.*

ANSI/Y14.36, *Surface Texture Symbols.*

NFPA/T2.1.1 R1, *National Fluid Power Association Recommended Standard Fluid power systems and products - Glossary.*

NFPA/T2.10.1M, *National Fluid Power Association Metric units for fluid power applications.*

## 3 TERMS AND DEFINITIONS

For definitions of terms used, see NFPA/T2.1.1 R1, ANSI/B93.2 and ANSI/B93.2A.

## 4 UNITS OF MEASUREMENT

**4.1** Inches and decimal parts of an inch are used.

**4.2** Approximate conversions to millimetres are given and appear above their inch counterparts.

## 5 SYMBOLS

For the purpose of this standard the following letters apply:

**5.1** A, B, P, T, L, V, X and Y identify ports.

**5.2** $F_1$, $F_2$, $F_3$, $F_4$, $F_5$, and $F_6$, identify thread holes for fixing bolts.

**5.3** $G_1$ and $G_2$ identify location pin holes.

## 6 SUBPLATE IDENTIFICATION CODING

Drawing identification coding corresponds to the following prefixes:

D - Directional control valves;

F, 2F, 3F - Compensated flow control valves;

C - Check valves;

FB - CAM actuated two-stage compensated flow control valves;

POC - Pilot operated valves;

P - Pressure control valves (other than relief valves), sequence valves, throttle valves and check valves;

R - Pressure relief valves.

## 7 MOUNTING SURFACE SELECTION

Mounting surface for subplate hydraulic fluid power valves may be selected from the following:

**7.1** Directional control valves:

D03 - see figure 1, table 1

D05 - see figure 2, table 2

D06 - see figure 3, table 3

D07 - see figure 4, table 4

D08 - see figure 5, table 5

D10 - see figure 6, table 6

**7.2** Compensated flow control valves:

F03 - see figure 7, table 7

2F06 - see figure 8, table 8

3F06 - see figure 9, table 9

2F07 - see figure 10, table 10

3F07 - see figure 11, table 11

2FB07 - see figure 12, table 12

2F08 - see figure 13, table 13

2F09 - see figure 14, table 14

**7.3** Check valves:

C06 - see figure 15, table 15

C08 - see figure 16, table 16

C09 - see figure 17, table 17

**7.4** Pilot operated check valves:

POC06 - see figure 18, table 18

POC08 - see figure 19, table 19

POC09 - see figure 20, table 20

**7.5** Pressure control valves (other than relief valves), sequence valves, throttle valves and check valves.

P03 - see figure 21, table 21

P06 - see figure 22, table 22

P08 - see figure 23, table 23

P10 - see figure 24, table 24

**7.6** Pressure relief valves

R03 - see figure 25, table 25

R06 - see figure 26, table 26

R08 - see figure 27, table 27

R10 - see figure 28, table 28

RP06 - see figure 29, table 29

RP08 - see figure 30, table 30

RP10 - see figure 31, table 31

## 8 TOLERANCES

The following tolerances apply except when minimum or maximum appear in the tabulation for dimensions.

**8.1** Surface roughness (Ra):

$$\frac{0.8 \text{ micrometre}}{32 \text{ microinch}}$$ (N6 in accordance with ISO 1302)

See ANSI Y14.36

**8.2** Surface flatness:

0.01 mm over a distance of 100 mm (.0004 inches over a distance of 4 inches)

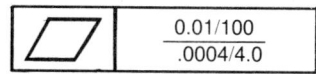

See ANSI/Y14.5M-1982

**8.3** Locating pin hole:

tolerance for diameters

$$\frac{\pm\ 0.2}{\pm\ .008}$$

**8.4** Observe the following tolerances along the X and Y axes with respect to the origin:

- for pin holes:

$$\frac{\pm\ 0.1}{\pm\ .004}$$

- for bolt holes:

$$\frac{\pm\ 0.1}{\pm\ .004}$$

- for port holes:

$$\frac{\pm\ 0.2}{\pm\ .008}$$

**NOTE:** See figures and tables for dimensions.

## 9  IDENTIFICATION STATEMENT

Use the following statement in catalogs and sales literature when electing to comply with this voluntary standard:

"Mounting surface dimensions conform to ANSI/B93.7M, *Hydraulic fluid power - Valves - Mounting surfaces.*"

ANSI/B93.7M

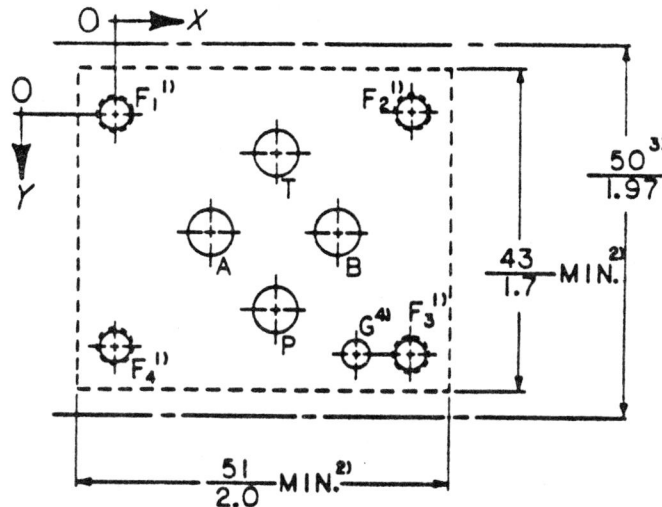

Figure 1 - DO3 mounting surface for four-port hydraulic directional control valves with 6.3/.25 maximum port diameter.

Table 1 - DO3 mounting surface for four-port hydraulic directional control valves with 6.3/.25 maximum port diameter.[5]

|   | P | A | T | B | G | $F_1$ | $F_2$ | $F_3$ | $F_4$ |
|---|---|---|---|---|---|---|---|---|---|
| X | 21.5/.85 | 12.7/.5 | 21.5/.85 | 30.2/1.19 | 33/1.3 | 0/0 | 40.5/1.594 | 40.5/1.594 | 0/0 |
| Y | 25.9/1.02 | 15.5/.61 | 5.1/.2 | 15.5/.61 | 31.75/1.25 | 0/0 | -0.75/-.03 | 31.75/1.25 | 31/1.22 |
| $\phi$ | 6.3/.25 Max. | 6.3/.25 Max. | 6.3/.25 Max. | 6.3/.25 Max. | 4/.16 | M5/.19 | M5/.19 | M5/.19 | M5/.19 |

1. The minimum thread depth is 1.5 bolt diameter. The recommended engagement of fixing bolt thread for ferrous mounting is 1.25 bolt diameter.

2. The dimensions specifying the area within the dashed lines are the minimum dimensions for the mounting surface. The corners of the rectangle may be radiused to a maximum radius equal to the thread diameter of the fixing bolts.

   Along each axis the fixing holes are at equal distance to the mounting surface edges.

3. This dimension gives the minimum spacing distance between the valve and adjacent obstructions, for example, another valve or a wall. This dimension is, therefore, the minimum distance from centerline to centerline of two similar mounting surfaces placed on a manifold block. The fixing holes are at equal distances to this dimension.

   The valve manufacturer's attention is drawn to the fact that no part of the width of the complete valve assembly is to exceed this dimension.

4. Blind hole in the mounting surface to accommodate the locating pin on the valve. The minimum depth is 4/.16.

5. The maximum limit of the working pressure for subplates and manifold blocks with this mounting surface will be supplied by the manufacturer.

# ANSI/B93.7M

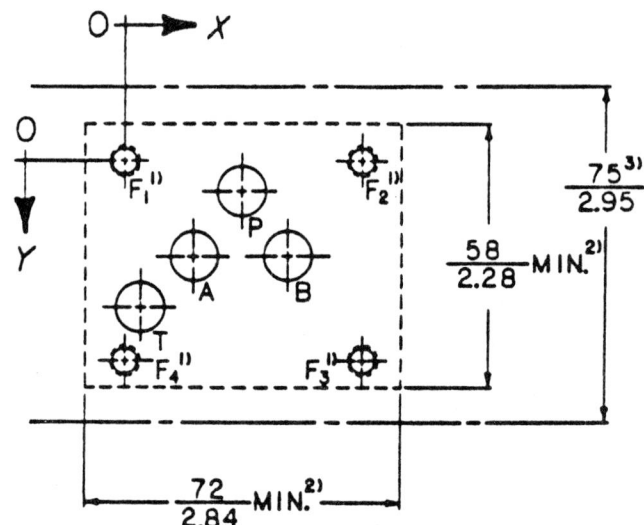

Figure 2 - **D05 mounting surface for four-port hydraulic directional control valves with 11.2/.44 maximum port diameter.**

Table 2 - **D05 mounting surface for four-port hydraulic directional control valves with 11.2/.44 maximum port diameter.**[4]

|   | P | A | T | B | $F_1$ | $F_2$ | $F_3$ | $F_4$ |
|---|---|---|---|---|---|---|---|---|
| X | 27 / 1.06 | 16.7 / .66 | 3.2 / .13 | 37.3 / 1.47 | 0 / 0 | 54 / 2.125 | 54 / 2.125 | 0 / 0 |
| Y | 6.3 / .25 | 21.4 / 0.84 | 32.5 / 1.28 | 21.4 / .84 | 0 / 0 | 0 / 0 | 46 / 1.812 | 46 / 1.812 |
| $\phi$ | 11.2 / .44 Max. | 11.2 / .44 Max. | 11.2 / .44 Max. | 11.2 / .44 Max. | M6 / .25 | M6 / .25 | M6 / .25 | M6 / .25 |

1. The minimum thread depth is 1.5 bolt diameter. The recommended engagement of fixing bolt thread for ferrous mounting is 1.25 bolt diameter.

2. The dimensions specifying the area within the dashed lines are the minimum dimensions for the mounting surface. The corners of the rectangle may be radiused to a maximum radius equal to the thread diameter of the fixing bolts.

   Along each axis the fixing holes are at equal distance to the mounting surface edges.

3. This dimension gives the minimum spacing distance between the valve and adjacent obstructions, for example, another valve or a wall. This dimension is, therefore, the minimum distance from centerline to centerline of two similar mounting surfaces placed on a manifold block. The fixing holes are at equal distances to this dimension.

   The valve manufacturer's attention is drawn to the fact that no part of the width of the complete valve assembly is to exceed this dimension.

4. The maximum limit of the working pressure for subplates and manifold blocks with this mounting surface will be supplied by the manufacturer.

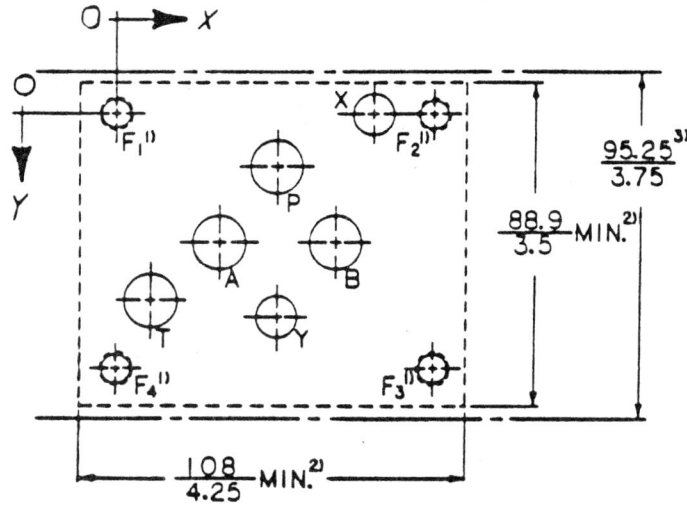

Figure 3 - **D06 mounting surface for four-port hydraulic directional control valves with 14.7/.58 maximum port diameter**

Table 3 - **D06 mounting surface for four-port hydraulic directional control valves with 14.7/.58 maximum port diameter.**[4]

|   | P | A | T | B | X | Y | $F_1$ | $F_2$ | $F_3$ | $F_4$ |
|---|---|---|---|---|---|---|---|---|---|---|
| X | 44.45 / 1.75 | 28.45 / 1.12 | 9.65 / .38 | 60.45 / 2.38 | 71.4 / 2.81 | 44.45 / 1.75 | 0 / 0 | 88.9 / 3.5 | 88.9 / 3.5 | 0 / 0 |
| Y | 14.25 / .56 | 35.05 / 1.38 | 50.8 / 2.0 | 35.05 / 1.38 | 0 / 0 | 55.37 / 2.18 | 0 / 0 | 0 / 0 | 69.85 / 2.75 | 69.85 / 2.75 |
| $\phi$ | 14.7 / .58 Max. | 14.7 / .58 Max. | 14.7 / .58 Max. | 14.7 / .58 Max. | 11.2 / .44 Max. | 11.2 / .44 Max. | MIO / .375 | MIO / .375 | MIO / .375 | MIO / .375 |

1. The minimum thread depth is 1.5 bolt diameter. The recommended engagement of fixing bolt thread for ferrous mounting is 1.25 bolt diameter.

2. The dimensions specifying the area within the dashed lines are the minimum dimensions for the mounting surface. The corners of the rectangle may be radiused to a maximum radius equal to the thread diameter of the fixing bolts.

   Along each axis the fixing holes are at equal distance to the mounting surface edges.

3. This dimension gives the minimum spacing distance between the valve and adjacent obstructions, for example, another valve or a wall. This dimension is, therefore, the minimum distance from centerline to centerline of two similar mounting surfaces placed on a manifold block. The fixing holes are at equal distances to this dimension.

   The valve manufacturer's attention is drawn to the fact that no part of the width of the complete valve assembly is to exceed this dimension.

4. The maximum limit of the working pressure for subplates and manifold blocks with this mounting surface will be supplied by the manufacturer.

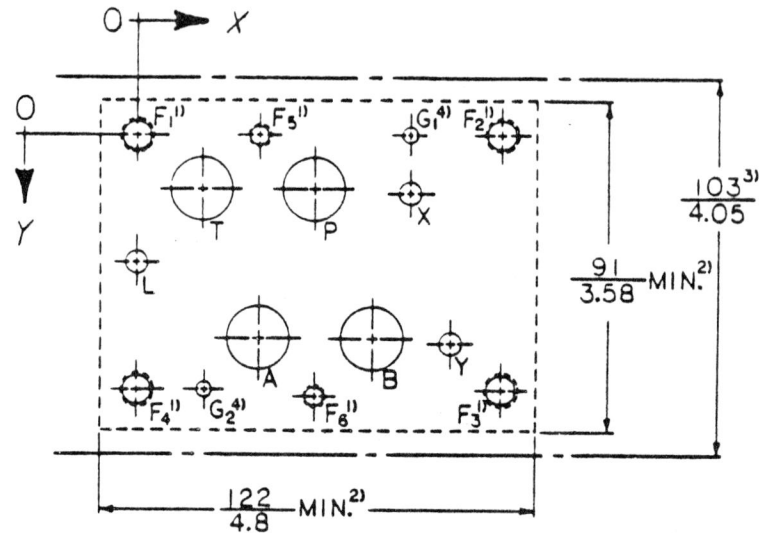

Figure 4 - **D07 mounting surface for four-port hydraulic directional control valves with 17.5/.69 maximum port diameter.**

Table 4 - **D07 mounting surfce for four-port hydraulic directional control valves with 17.5/.69 maximum port diameter.**[5]

| | P | A | T | B | L | X | Y | $G_1$ | $G_2$ | $F_1$ | $F_2$ | $F_3$ | $F_4$ | $F_5$ | $F_6$ |
|---|---|---|---|---|---|---|---|---|---|---|---|---|---|---|---|
| X | $\frac{50}{1.97}$ | $\frac{34.1}{1.34}$ | $\frac{18.3}{.72}$ | $\frac{65.9}{2.6}$ | $\frac{0}{0}$ | $\frac{76.6}{3.016}$ | $\frac{88.1}{3.47}$ | $\frac{76.6}{3.016}$ | $\frac{18.3}{.72}$ | $\frac{0}{0}$ | $\frac{101.6}{4.0}$ | $\frac{101.6}{4.0}$ | $\frac{0}{0}$ | $\frac{34.1}{1.342}$ | $\frac{50}{1.97}$ |
| Y | $\frac{14.3}{.56}$ | $\frac{55.6}{2.19}$ | $\frac{14.3}{.56}$ | $\frac{55.6}{2.19}$ | $\frac{34.9}{1.37}$ | $\frac{15.9}{.62}$ | $\frac{57.2}{2.25}$ | $\frac{0}{0}$ | $\frac{69.9}{2.75}$ | $\frac{0}{0}$ | $\frac{0}{0}$ | $\frac{69.9}{2.75}$ | $\frac{69.9}{2.75}$ | $\frac{-1.6}{-.063}$ | $\frac{71.5}{2.815}$ |
| $\phi$ | $\frac{17.5}{.69}$ Max. | $\frac{17.5}{.69}$ Max. | $\frac{17.5}{.69}$ Max. | $\frac{17.5}{.69}$ Max. | $\frac{6.3}{.25}$ Max. | $\frac{6.3}{.25}$ Max. | $\frac{6.3}{.25}$ Max. | $\frac{4}{.16}$ | $\frac{4}{.16}$ | $\frac{M10}{.375}$ | $\frac{M10}{.375}$ | $\frac{M10}{.375}$ | $\frac{M10}{.375}$ | $\frac{M6}{.25}$ | $\frac{M6}{.25}$ |

1. The minimum thread depth is 1.5 bolt diameter. The recommended engagement of fixing bolt thread for ferrous mounting is 1.25 bolt diameter.

2. The dimensions specifying the area within the dashed lines are the minimum dimensions for the mounting surface. The corners of the rectangle may be radiused to a maximum radius equal to the thread diameter of the fixing bolts.

   Along each axis the fixing holes are at equal distance to the mounting surface edges.

3. This dimension gives the minimum spacing distance between the valve and adjacent obstructions, for example, another valve or a wall. This dimension is, therefore, the minimum distance from centerline to centerline of two similar mounting surfaces placed on a manifold block. The fixing holes are at equal distances to this dimension.

   The valve manufacturer's attention is drawn to the fact that no part of the width of the complete valve assembly is to exceed this dimension.

4. Blind hole in the mounting surface to accommodate the locating pin on the valve. The minimum depth is 8/.31.

5. The maximum limit of the working pressure for subplates and manifold blocks with this mounting surface will be supplied by the manufacturer.

ANSI/B93.7M

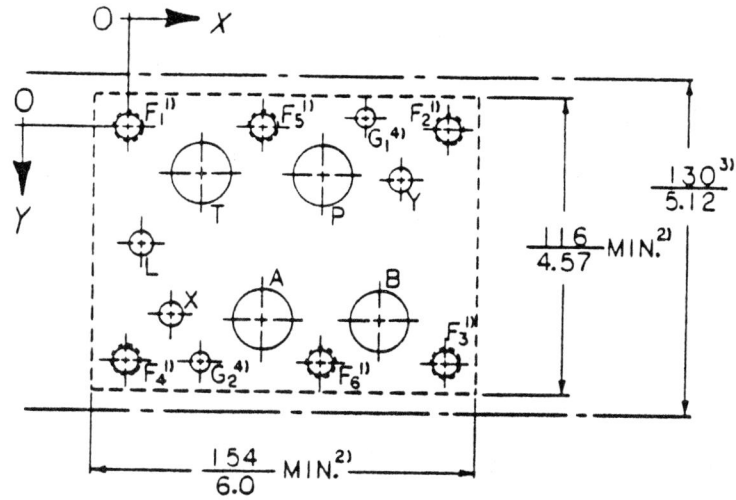

Figure 5 - D08 mounting surface for four-port hydraulic directional control valves with 23.4/.92 maximum port diameter.

Table 5 - D08 mounting surface for four-port hydraulic directional control valves with 23.4/.92 maximum port diameter.[5]

| | P | A | T | B | L | X | Y | $G_1$ | $G_2$ | $F_1$ | $F_2$ | $F_3$ | $F_4$ | $F_5$ | $F_6$ |
|---|---|---|---|---|---|---|---|---|---|---|---|---|---|---|---|
| X | 77 / 3.03 | 53.2 / 2.09 | 29.4 / 1.16 | 100.8 / 3.97 | 5.6 / .22 | 17.5 / .69 | 112.7 / 4.44 | 94.5 / 3.719 | 29.4 / 1.156 | 0 / 0 | 130.2 / 5.125 | 130.2 / 5.125 | 0 / 0 | 53.2 / 2.094 | 77 / 3.031 |
| Y | 17.5 / .69 | 74.6 / 2.93 | 17.5 / .69 | 74.6 / 2.93 | 46 / 1.81 | 73 / 2.88 | 19 / .75 | -4.8 / -.187 | 92.1 / 3.625 | 0 / 0 | 0 / 0 | 92.1 / 3.625 | 92.1 / 3.625 | 0 / 0 | 92.1 / 3.625 |
| φ | 23.4 / .92 Max. | 23.4 / .92 Max. | 23.4 D / .92 Max. | 23.4 / .92 Max. | 11.2 / .44 Max. | 11.2 / .44 Max. | 11.2 / .44 Max. | 7.5 / .28 | 7.5 / .28 | M12 / .5 | M12 / .5 | M12 / .5 | M12 / .5 | M12 / .5 | M12 / .5 |

1. The minimum thread depth is 1.5 bolt diameter. The recommended engagement of fixing bolt thread for ferrous mounting is 1.25 bolt diameter.

2. The dimensions specifying the area within the dashed lines are the minimum dimensions for the mounting surface. The corners of the rectangle may be radiused to a maximum radius equal to the thread diameter of the fixing bolts.

   Along each axis the fixing holes are at equal distance to the mounting surface edges.

3. This dimension gives the minimum spacing distance between the valve and adjacent obstructions, for example, another valve or a wall. This dimension is, therefore, the minimum distance from centerline to centerline of two similar mounting surfaces placed on a manifold block. The fixing holes are at equal distances to this dimension.

   The valve manufacturer's attention is drawn to the fact that no part of the width of the complete valve assembly is to exceed this dimension.

4. Blind hole in the mounting surface to accommodate the locating pin on the valve. The minimum depth is 8/.31.

5. The maximum limit of the working pressure for subplates and manifold blocks with this mounting surface will be supplied by the manufacturer.

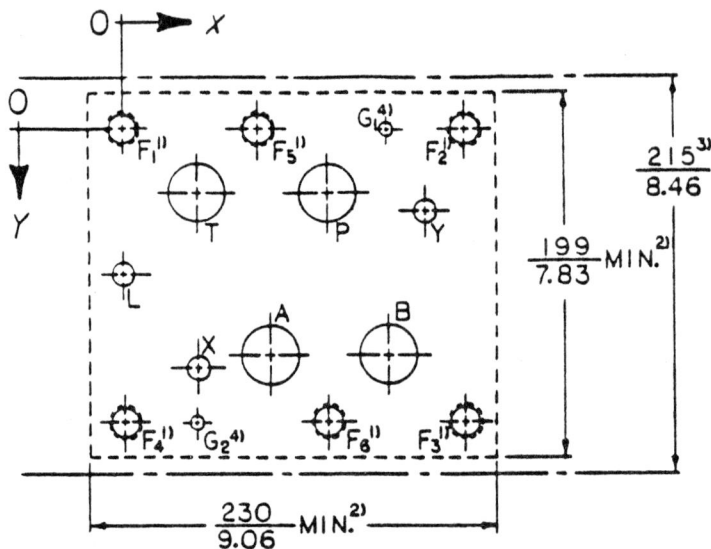

Figure 6 - D10 mounting surface for four-port hydraulic directional control valves with 32/1.25 maximum port diameter.

Table 6 - D10 mounting surface for four-port hydraulic directional control valves with 32/1.25 maximum port diameter.[5),]

|   | P | A | T | B | L | X | Y | $G_1$ | $G_2$ | $F_1$ | $F_2$ | $F_3$ | $F_4$ | $F_5$ | $F_6$ |
|---|---|---|---|---|---|---|---|---|---|---|---|---|---|---|---|
| X | 114.3 / 4.5 | 82.5 / 3.25 | 41.3 / 1.63 | 147.6 / 5.81 | 0 / 0 | 41.3 / 1.63 | 168.3 / 6.63 | 147.6 / 5.812 | 41.3 / 1.625 | 0 / 0 | 190.5 / 7.5 | 190.5 / 7.5 | 0 / 0 | 76.2 / 3.0 | 114.3 / 4.5 |
| Y | 35 / 1.38 | 123.8 / 4.87 | 35 / 1.38 | 123.8 / 4.87 | 79.4 / 3.13 | 130.2 / 5.13 | 44.5 / 1.75 | 0 / 0 | 158.8 / 6.25 | 0 / 0 | 0 / 0 | 158.8 / 6.25 | 158.8 / 6.25 | 0 / 0 | 158.8 / 6.25 |
| $\phi$ | 32 / 1.25 Max. | 32 / 1.25 Max. | 32 / 1.25 Max. | 32 / 1.25 Max. | 11.2 / .44 Max. | 11.2 / .44 Max. | 11.2 / .44 Max. | 7.5 / .28 | 7.5 / .28 | M20 / .75 | M20 / .75 | M20 / .75 | M20 / .75 | M20 / .75 | M20 / .75 |

1. The minimum thread depth is 1.5 bolt diameter. The recommended engagement of fixing bolt thread for ferrous mounting is 1.25 bolt diameter.

2. The dimensions specifying the area within the dashed lines are the minimum dimensions for the mounting surface. The corners of the rectangle may be radiused to a maximum radius equal to the thread diameter of the fixing bolts.

   Along each axis the fixing holes are at equal distance to the mounting surface edges.

3. This dimension gives the minimum spacing distance between the valve and adjacent obstructions, for example, another valve or a wall. This dimension is, therefore, the minimum distance from centerline to centerline of two similar mounting surfaces placed on a manifold block. The fixing holes are at equal distances to this dimension.

   The valve manufacturer's attention is drawn to the fact that no part of the width of the complete valve assembly is to exceed this dimension.

4. Blind hole in the mounting surface to accommodate the locating pin on the valve. The minimum depth is 8/.31.

5. The maximum limit of the working pressure for subplates and manifold blocks with this mounting surface will be supplied by the manufacturer.

ANSI/B93.7M

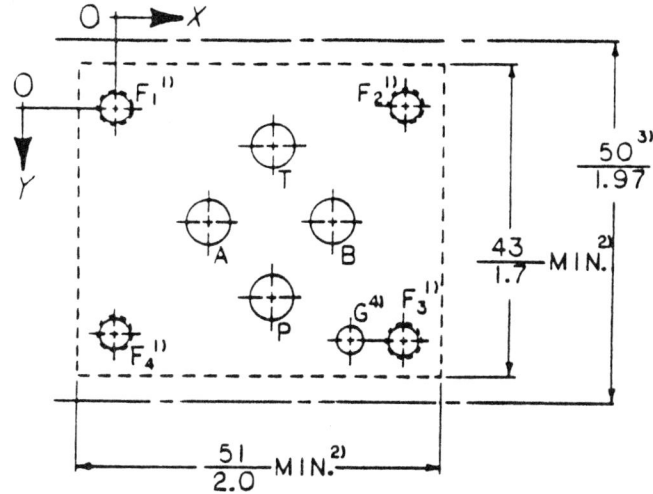

Figure 7 - **F03 mounting surface for compensated flow control valves with main ports of 6.3/.25 maximum port diameter.**

Table 7 - **F03 mounting surface for compensated flow control valves with main ports of 6.3/.25 maximum port diameter.**[5]

|   | P | A | T | B | G | $F_1$ | $F_2$ | $F_3$ | $F_4$ |
|---|---|---|---|---|---|---|---|---|---|
| X | 21.5 / .85 | 12.7 / .5 | 21.5 / .85 | 30.2 / 1.19 | 33 / 1.3 | 0 / 0 | 40.5 / 1.594 | 40.5 / 1.594 | 0 / 0 |
| Y | 25.9 / 1.02 | 15.5 / .61 | 5.1 / .2 | 15.5 / .61 | 31.75 / 1.25 | 0 / 0 | -0.75 / -.03 | 31.75 / 1.25 | 31 / 1.22 |
| $\phi$ | 6.3 / .25 Max. | 6.3 / .25 Max. | 6.3 / .25 Max. | 6.3 / .25 Max. | 4 / .16 | M5 / .19 | M5 / .19 | M5 / .19 | M5 / .19 |

1. The minimum thread depth is 1.5 bolt diameter. The recommended engagement of fixing bolt thread for ferrous mounting is 1.25 bolt diameter.

2. The dimensions specifying the area within the dashed lines are the minimum dimensions for the mounting surface. The corners of the rectangle may be radiused to a maximum radius equal to the thread diameter of the fixing bolts.

   Along each axis the fixing holes are at equal distance to the mounting surface edges.

3. This dimension gives the minimum spacing distance between the valve and adjacent obstructions, for example, another valve or a wall. This dimension is, therefore, the minimum distance from centerline to centerline of two similar mounting surfaces placed on a manifold block. The fixing holes are at equal distances to this dimension.

   The valve manufacturer's attention is drawn to the fact that no part of the width of the complete valve assembly is to exceed this dimension.

4. Blind hole in the mounting surface to accommodate the locating pin on the valve. The minimum depth is 4/.16.

5. The maximum limit of the working pressure for subplates and manifold blocks with this mounting surface will be supplied by the manufacturer.

# ANSI/B93.7M

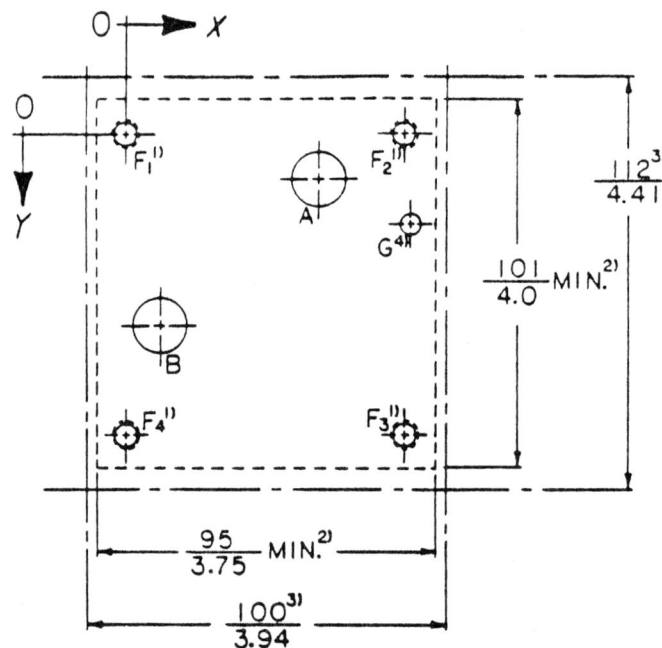

Figure 8 - **2F06 mounting surface for compensated flow control valves with two main ports of 14.7/.58 maximim port diameter.**

Table 8 - **2F06 mounting surface for compensated flow control valves with two main ports of 14.7/.58 maximum port diameter.**[5]

|   | A | B | G | $F_1$ | $F_2$ | $F_3$ | $F_4$ |
|---|---|---|---|---|---|---|---|
| X | $\dfrac{54}{2.125}$ | $\dfrac{9.5}{.375}$ | $\dfrac{79.4}{3.125}$ | $\dfrac{0}{0}$ | $\dfrac{76.2}{3.0}$ | $\dfrac{76.2}{3.0}$ | $\dfrac{0}{0}$ |
| Y | $\dfrac{11.1}{.437}$ | $\dfrac{52.4}{2.062}$ | $\dfrac{23.8}{.937}$ | $\dfrac{0}{0}$ | $\dfrac{0}{0}$ | $\dfrac{82.6}{3.25}$ | $\dfrac{82.6}{3.25}$ |
| $\phi$ | $\dfrac{14.7}{.58}$ Max. | $\dfrac{14.7}{.58}$ Max. | $\dfrac{7.1}{.28}$ | $\dfrac{M8}{.312}$ | $\dfrac{M8}{.312}$ | $\dfrac{M8}{.312}$ | $\dfrac{M8}{.312}$ |

1. The minimum thread depth is 1.5 bolt diameter. The recommended engagement of fixing bolt thread for ferrous mounting is 1.25 bolt diameter.

2. The dimensions specifying the area within the dashed lines are the minimum dimensions for the mounting surface. The corners of the rectangle may be radiused to a maximum radius equal to the thread diameter of the fixing bolts.

   Along each axis the fixing holes are at equal distance to the mounting surface edges.

3. This dimension gives the minimum spacing distance between the valve and adjacent obstructions, for example, another valve or a wall. This dimension is, therefore, the minimum distance from centerline to centerline of two similar mounting surfaces placed on a manifold block. The fixing holes are at equal distances to this dimension.

   The valve manufacturer's attention is drawn to the fact that no part of the width of the complete valve assembly is to exceed this dimension.

4. Blind hole in the mounting surface to accommodate the locating pin on the valve. The minimum depth is 8/.31.

5. The maximum limit of the working pressure for subplates and manifold blocks with this mounting surface will be supplied by the manufacturer.

ANSI/B93.7M

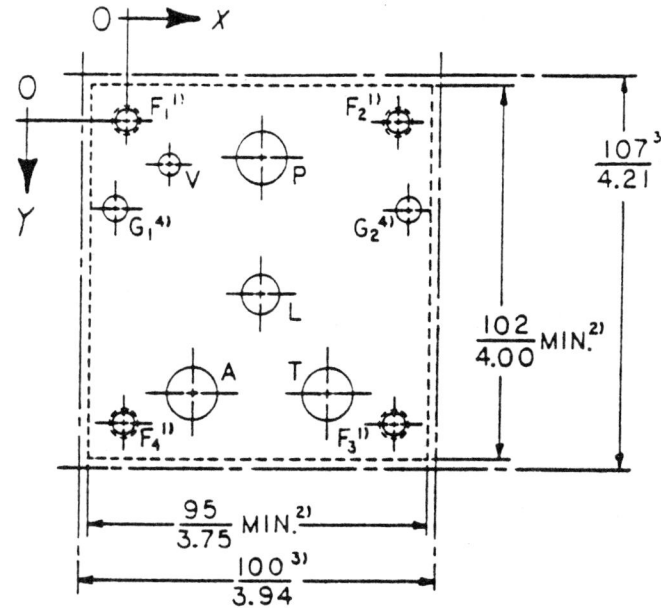

Figure 9 - 3F06 mounting surface for compensated flow control valves with three main ports of 14.7/.58 maximum port diameter.

Table 9 - 3F06 mounting surface for compensated flow control valves with three main ports of 14.7/.58 maximum port diameter.[5]

|   | P | A | T | L | V | $G_1$ | $G_2$ | $F_1$ | $F_2$ | $F_3$ | $F_4$ |
|---|---|---|---|---|---|---|---|---|---|---|---|
| X | 38<br>1.5 | 19<br>.75 | 57<br>2.25 | 38<br>1.5 | 11.9<br>.47 | -3.2<br>-.125 | 79.4<br>3.125 | 0<br>0 | 76.2<br>3.0 | 76.2<br>3.0 | 0<br>0 |
| Y | 9.5<br>.375 | 73.8<br>2.906 | 73.8<br>2.906 | 46.8<br>1.843 | 11.9<br>.47 | 23.8<br>.937 | 23.8<br>.937 | 0<br>0 | 0<br>0 | 82.6<br>3.25 | 82.6<br>3.25 |
| $\phi$ | 14.7<br>.58<br>Max. | 14.7<br>.58<br>Max. | 14.7<br>.58<br>Max. | 11.1<br>.44<br>Max. | 6.3<br>.25<br>Max. | 7.1<br>.28 | 7.1<br>.28 | M8<br>.312 | M8<br>.312 | M8<br>.312 | M8<br>.312 |

1. The minimum thread depth is 1.5 bolt diameter. The recommended engagement of fixing bolt thread for ferrous mounting is 1.25 bolt diameter.

2. The dimensions specifying the area within the dashed lines are the minimum dimensions for the mounting surface. The corners of the rectangle may be radiused to a maximum radius equal to the thread diameter of the fixing bolts.

   Along each axis the fixing holes are at equal distance to the mounting surface edges.

3. This dimension gives the minimum spacing distance between the valve and adjacent obstructions, for example, another valve or a wall. This dimension is, therefore, the minimum distance from centerline to centerline of two similar mounting surfaces placed on a manifold block. The fixing holes are at equal distances to this dimension.

   The valve manufacturer's attention is drawn to the fact that no part of the width of the complete valve assembly is to exceed this dimension.

4. Blind hole in the mounting surface to accommodate the locating pin on the valve. The minimum depth is 8/.31.

5. The maximum limit of the working pressure for subplates and manifold blocks with this mounting surface will be supplied by the manufacturer.

ANSI/B93.7M

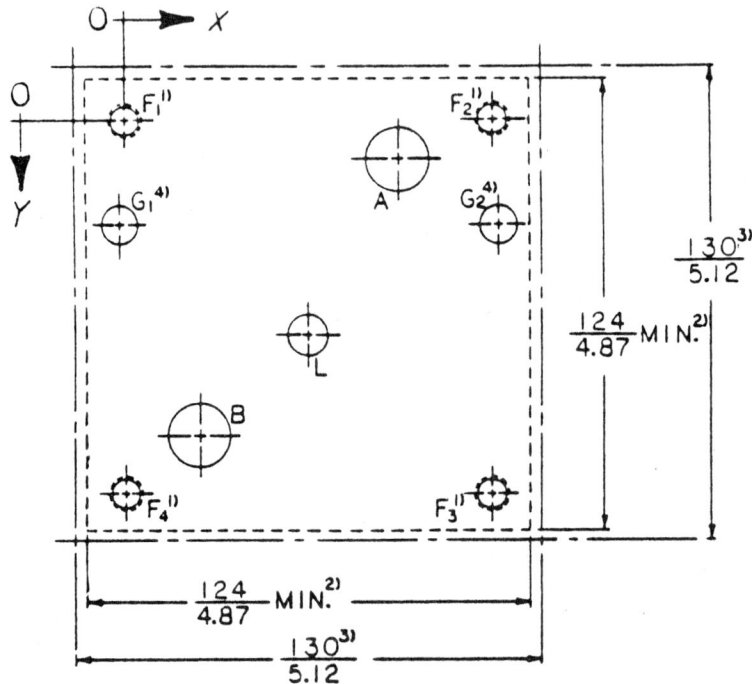

Figure 10 - 2F07 mounting surface for compensated flow control valves with two main ports of 17.5/.69 maximum port diameter.

Table 10 - 2F07 mounting surface for compensated flow control valves with two main ports of 17.5/.69 maximum port diameter.[5]

|   | A | B | L | $G_1$ | $G_2$ | $F_1$ | $F_2$ | $F_3$ | $F_4$ |
|---|---|---|---|---|---|---|---|---|---|
| X | 75 / 2.953 | 20.6 / .812 | 50.8 / 2.0 | - .8 / - .031 | 2.4 / 4.031 | 0 / 0 | 101.6 / 4.0 | 101.6 / 4.0 | 0 / 0 |
| Y | 11.1 / .437 | 86.5 / 3.406 | 58.7 / 2.312 | 28.6 / 1.125 | 28.6 / 1.125 | 0 / 0 | 0 / 0 | 101.6 / 4.0 | 101.6 / 4.0 |
| $\phi$ | 17.5 / .69 Max. | 17.5 / .69 Max. | 11.1 / .44 Max. | 10.4 / .41 | 10.4 / .41 | M10 / .375 | M10 / .375 | M10 / .375 | M10 / .375 |

1. The minimum thread depth is 1.5 bolt diameter. The recommended engagement of fixing bolt thread for ferrous mounting is 1.25 bolt diameter.

2. The dimensions specifying the area within the dashed lines are the minimum dimensions for the mounting surface. The corners of the rectangle may be radiused to a maximum radius equal to the thread diameter of the fixing bolts.

   Along each axis the fixing holes are at equal distance to the mounting surface edges.

3. This dimension gives the minimum spacing distance between the valve and adjacent obstructions, for example, another valve or a wall. This dimension is, therefore, the minimum distance from centerline to centerline of two similar mounting surfaces placed on a manifold block. The fixing holes are at equal distances to this dimension.

   The valve manufacturer's attention is drawn to the fact that no part of the width of the complete valve assembly is to exceed this dimension.

4. Blind hole in the mounting surface to accommodate the locating pin on the valve. The minimum depth is 8/.31.

5. The maximum limit of the working pressure for subplates and manifold blocks with this mounting surface will be supplied by the manufacturer.

ANSI/B93.7M

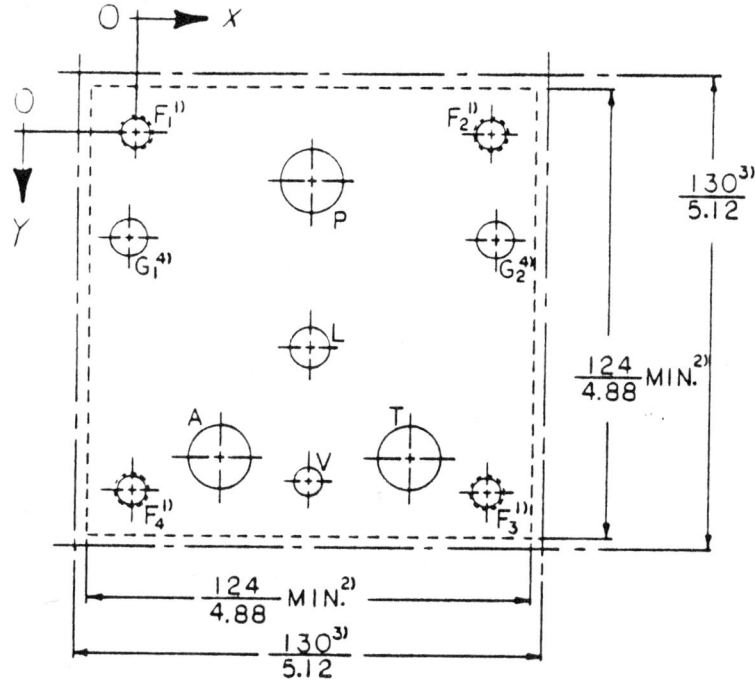

Figure 11 - 3F07 mounting surface for compensated flow control valves with three main ports of 17.5/.69 maximum port diameter.

Table 11 - 3F07 mounting surface for compensated flow control valves with three main ports of 17.5/.69 maximum port diameter.[5]

|   | P | A | T | L | V | $G_1$ | $G_2$ | $F_1$ | $F_2$ | $F_3$ | $F_4$ |
|---|---|---|---|---|---|---|---|---|---|---|---|
| X | 50.8 / 2.0 | 23.8 / .937 | 77.8 / 3.062 | 50.8 / 2.0 | 50.8 / 2.0 | -.8 / -.031 | 102.4 / 4.031 | 0 / 0 | 101.6 / 4.0 | 101.6 / 4.0 | 0 / 0 |
| Y | 12.7 / .5 | 88.9 / 3.5 | 88.9 / 3.5 | 58.7 / 2.312 | 95.3 / 3.75 | 28.6 / 1.125 | 28.6 / 1.125 | 0 / 0 | 0 / 0 | 101.6 / 4.0 | 101.6 / 4.0 |
| φ | 17.5 / .69 Max. | 17.5 / .69 Max. | 17.5 / .69 Max. | 11.1 / .44 Max. | 7.9 / .31 Max. | 10.4 / .41 | 10.4 / .41 | M10 / .375 | M10 / .375 | M10 / .375 | M10 / .375 |

1. The minimum thread depth is 1.5 bolt diameter. The recommended engagement of fixing bolt thread for ferrous mounting is 1.25 bolt diameter.

2. The dimensions specifying the area within the dashed lines are the minimum dimensions for the mounting surface. The corners of the rectangle may be radiused to a maximum radius equal to the thread diameter of the fixing bolts.

   Along each axis the fixing holes are at equal distance to the mounting surface edges.

3. This dimension gives the minimum spacing distance between the valve and adjacent obstructions, for example, another valve or a wall. This dimension is, therefore, the minimum distance from centerline to centerline of two similar mounting surfaces placed on a manifold block. The fixing holes are at equal distances to this dimension.

   The valve manufacturer's attention is drawn to the fact that no part of the width of the complete valve assembly is to exceed this dimension.

4. Blind hole in the mounting surface to accommodate the locating pin on the valve. The minimum depth is 8/.31.

5. The maximum limit of the working pressure for subplates and manifold blocks with this mounting surface will be supplied by the manufacturer.

# ANSI/B93.7M

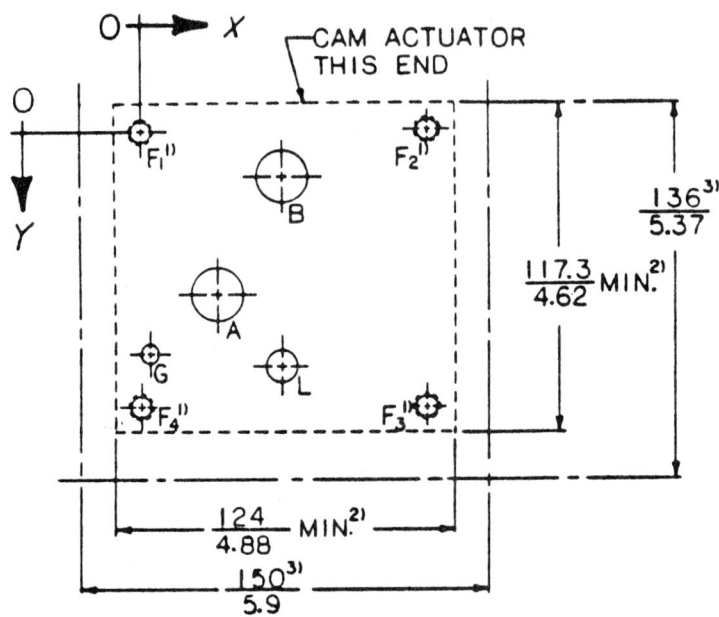

Figure 12 - **2FB07 mounting surface for compensated two-stage cam actuated flow control valves with two main ports of 19.1/.75 maximum port diameter.**

Table 12 - **2FB07 mounting surface for compensated two-stage cam actuated flow control valves with two main ports of 19.1/.75 maximum port diameter.**[5]

|   | A | B | L | G | $F_1$ | $F_2$ | $F_3$ | $F_4$ |
|---|---|---|---|---|---|---|---|---|
| X | 28.6 / 1.125 | 48.4 / 1.907 | 48.4 / 1.907 | 4.1 / .16 | 0 / 0 | 104.8 / 4.125 | 104.8 / 4.125 | 0 / 0 |
| Y | 58.7 / 2.312 | 16.7 / .656 | 85.7 / 3.375 | 79.4 / 3.125 | 0 / 0 | 0 / 0 | 98.4 / 3.875 | 98.4 / 3.875 |
| $\phi$ | 19.1 / .75 Max. | 19.1 / .75 Max. | 11.1 / .44 Max. | 7.1 / .28 | M10 / .375 | M10 / .375 | M10 / .375 | M10 / .375 |

1. The minimum thread depth is 1.5 bolt diameter. The recommended engagement of fixing bolt thread for ferrous mounting is 1.25 bolt diameter.

2. The dimensions specifying the area within the dashed lines are the minimum dimensions for the mounting surface. The corners of the rectangle may be radiused to a maximum radius equal to the thread diameter of the fixing bolts.

   Along each axis the fixing holes are at equal distance to the mounting surface edges.

3. This dimension gives the minimum spacing distance between the valve and adjacent obstructions, for example, another valve or a wall. This dimension is, therefore, the minimum distance from centerline to centerline of two similar mounting surfaces placed on a manifold block. The fixing holes are at equal distances to this dimension.

   The valve manufacturer's attention is drawn to the fact that no part of the width of the complete valve assembly is to exceed this dimension.

4. Blind hole in the mounting surface to accommodate the locating pin on the valve. The minimum depth is 8/.31.

5. The maximum limit of the working pressure for subplates and manifold blocks with this mounting surface will be supplied by the manufacturer.

ANSI/B93.7M

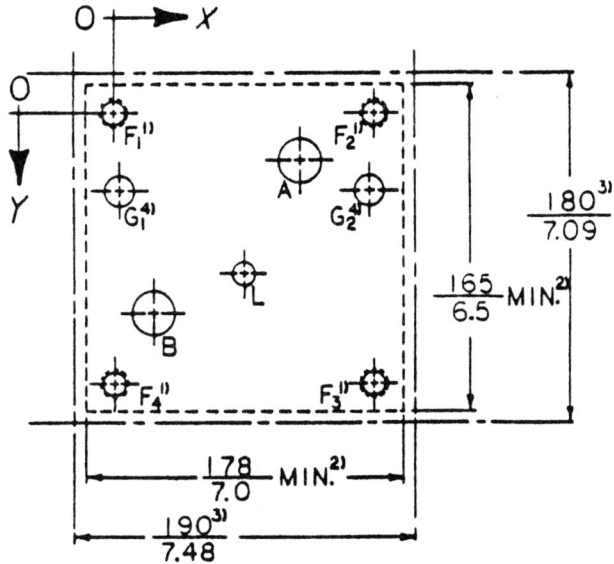

Figure 13 - **2F08 mounting surface for compensated flow control valves with two main ports of 23.4/.92 maximum port diameter.**

Table 13 - **2F08 mounting surface for compensated flow control valves with two main ports of 23.4/.92 maximum port diameter.**[5]

|   | A | B | L | $G_1$ | $G_2$ | $F_1$ | $F_2$ | $F_3$ | $F_4$ |
|---|---|---|---|---|---|---|---|---|---|
| X | 104.8/4.125 | 22.2/.875 | 73/2.875 | 1.6/.062 | 144.5/5.687 | 0/0 | 146/5.75 | 146/5.75 | 0/0 |
| Y | 12.7/.5 | 104.8/4.125 | 85.7/3.375 | 41.3/1.625 | 41.3/1.625 | 0/0 | 0/0 | 133.4/5.25 | 133.4/5.25 |
| $\phi$ | 23.4/.92 Max. | 23.4/.92 Max. | 11.1/.44 Max. | 16.5/.65 | 16.5/.65 | M16/.625 | M16/.625 | M16/.625 | M16/.625 |

1. The minimum thread depth is 1.5 bolt diameter. The recommended engagement of fixing bolt thread for ferrous mounting is 1.25 bolt diameter.

2. The dimensions specifying the area within the dashed lines are the minimum dimensions for the mounting surface. The corners of the rectangle may be radiused to a maximum radius equal to the thread diameter of the fixing bolts.

   Along each axis the fixing holes are at equal distance to the mounting surface edges.

3. This dimension gives the minimum spacing distance between the valve and adjacent obstructions, for example, another valve or a wall. This dimension is, therefore, the minimum distance from centerline to centerline of two similar mounting surfaces placed on a manifold block. The fixing holes are at equal distances to this dimension.

   The valve manufacturer's attention is drawn to the fact that no part of the width of the complete valve assembly is to exceed this dimension.

4. Blind hole in the mounting surface to **accommodate** the locating pin on the valve. The minimum depth is 8/.31.

5. The maximum limit of the working pressure for subplates and manifold blocks with this mounting surface will be supplied by the manufacturer.

# ANSI/B93.7M

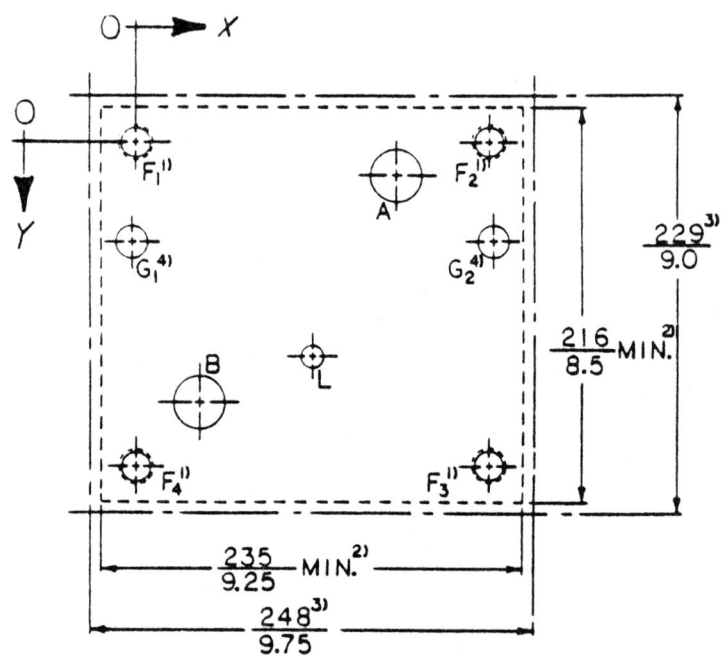

Figure 14 - **2F09 mounting surface for compensated flow control valves with two main ports of 28.5/1.12 maximum port diameter.**

Table 14 - **2F09 mounting surface for compensated flow control valves with two main ports of 28.5/1.12 maximum port diameter.**[5]

|   | A | B | L | $F_1$ | $F_2$ | $F_3$ | $F_4$ | $G_1$ | $G_2$ |
|---|---|---|---|---|---|---|---|---|---|
| X | 144.4 / 5.687 | 34.9 / 1.375 | 98.4 / 3.875 | 0 / 0 | 196.9 / 7.75 | 196.9 / 7.75 | 0 / 0 | -1.6 / -.062 | 198.4 / 7.812 |
| Y | 17.4 / .687 | 144.4 / 5.687 | 119 / 4.687 | 0 / 0 | 0 / 0 | 177.8 / 7.0 | 177.8 / 7.0 | 55.5 / 2.187 | 55.5 / 2.187 |
| $\phi$ | 28.5 / 1.12 Max. | 28.5 / 1.12 Max. | 11.2 / .44 Max. | M20 / .75 | M20 / .75 | M20 / .75 | M20 / .75 | 19.8 / .78 | 19.8 / .78 |

1. The minimum thread depth is 1.5 bolt diameter. The recommended engagement of fixing bolt thread for ferrous mounting is 1.25 bolt diameter.

2. The dimensions specifying the area within the dashed lines are the minimum dimensions for the mounting surface. The corners of the rectangle may be radiused to a maximum radius equal to the thread diameter of the fixing bolts.

   Along each axis the fixing holes are at equal distance to the mounting surface edges.

3. This dimension gives the minimum spacing distance between the valve and adjacent obstructions, for example, another valve or a wall. This dimension is, therefore, the minimum distance from centerline to centerline of two similar mounting surfaces placed on a manifold block. The fixing holes are at equal distances to this dimension.

   The valve manufacturer's attention is drawn to the fact that no part of the width of the complete valve assembly is to exceed this dimension.

4. Blind hole in the mounting surface to accommodate the locating pin on the valve. The minimum depth is 8/.31.

5. The maximum limit of the working pressure for subplates and manifold blocks with this mounting surface will be supplied by the manufacturer.

**Recommended uses of the ports for various types of flow control valves.**

**Examples of symbols for compensated flow control valves with two main ports of 6.3/.25 maximum diameter, refer to figure 7.**

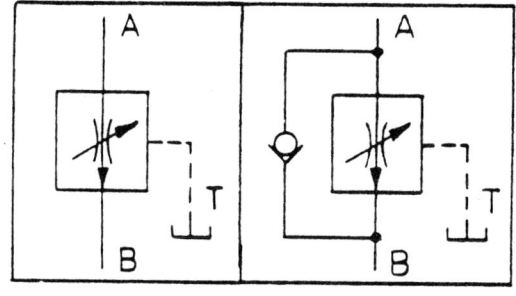

External drain

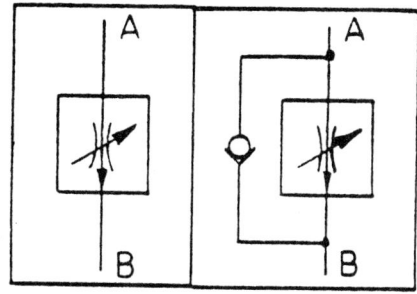

Internal drain

**Examples of symbols for compensated flow control valves with two main ports of 14.7/.58 maximum diameter, refer to figure 8.**

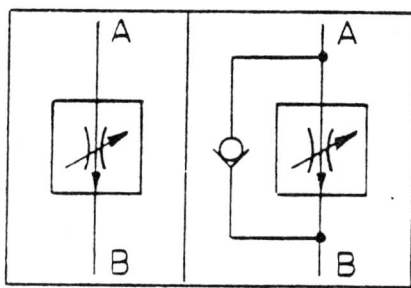

Internal drain

**Examples of symbols for compensated flow control valves with two main ports of 17.5/.69 and 23.4/.92 maximum diameter, refer to figures 10 and 13.**

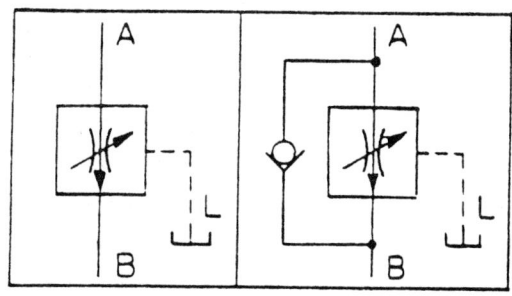

External drain

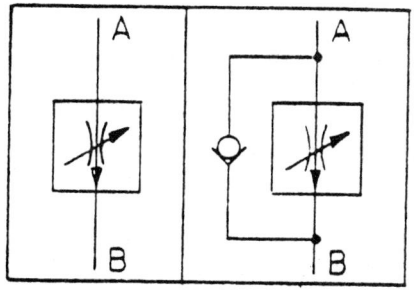

Internal drain

# ANSI/B93.7M

## Recommended uses of the ports for various types of flow control valves.

**Examples of symbols for compensated flow control valves with three main ports of 14.7/.58 and 17.5/.69 maximum diameter, refer to figures 9 and 11.**

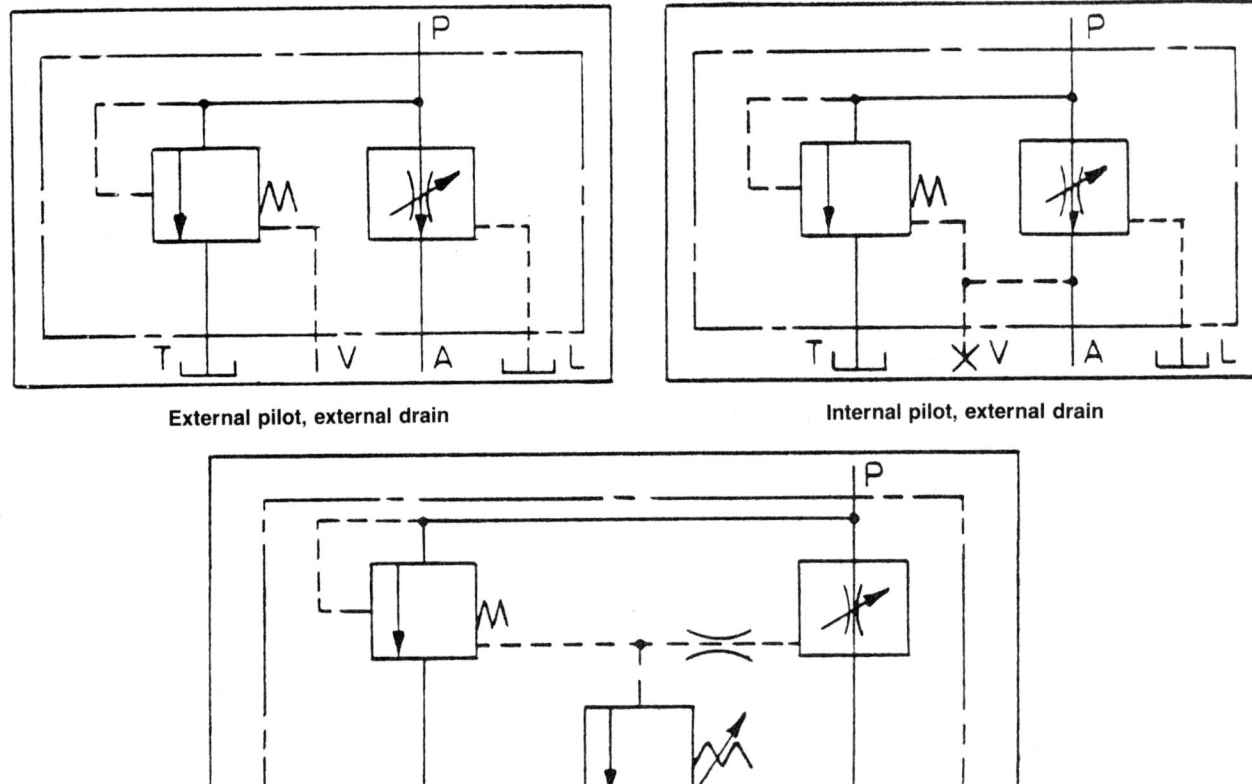

External pilot, external drain

Internal pilot, external drain

Internal pilot, internal drain.

**Examples of symbols for compensated two-stage cam actuated flow control valves with two main ports of 19.1/.75 maximum diameter, refer to figure 12.**

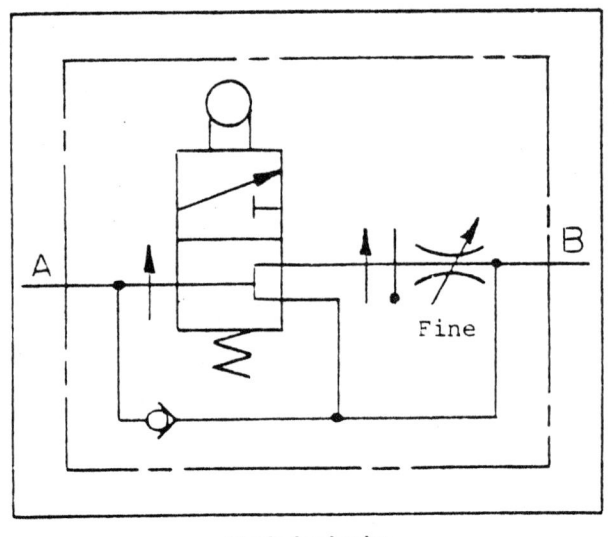

Single feed valve

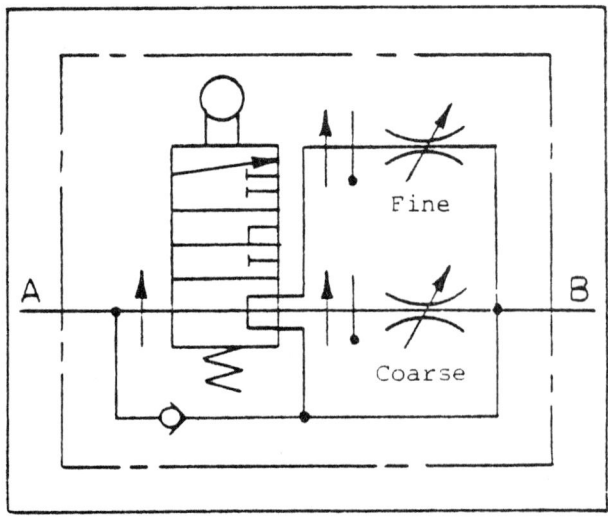

Dual feed valve

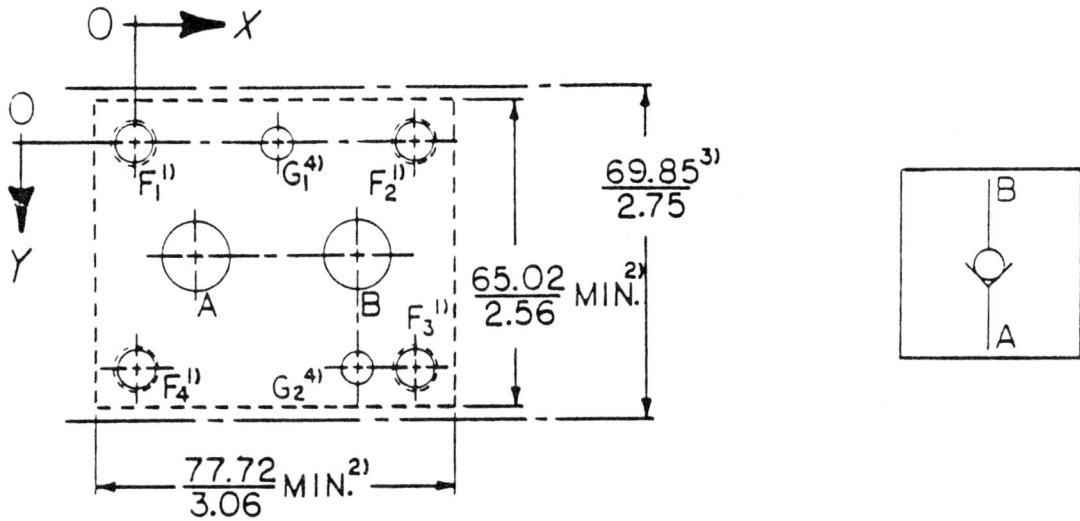

Figure 15 - **C06 mounting surface for check valves with main ports of 14.7/.58 maximum port diameter.**

Table 15 - **C06 mounting surface for check valves with main ports of 14.7/.58 maximum port diameter.**[5]

|   | A | B | $G_1$ | $G_2$ | $F_1$ | $F_2$ | $F_3$ | $F_4$ |
|---|---|---|---|---|---|---|---|---|
| X | 12.7/.5 | 47.6/1.875 | 30.1/1.187 | 47.6/1.875 | 0/0 | 60.3/2.375 | 60.3/2.375 | 0/0 |
| Y | 23.8/.938 | 23.8/.938 | 0/0 | 47.6/1.875 | 0/0 | 0/0 | 47.6/1.875 | 47.6/1.875 |
| $\phi$ | 14.7/.58 Max. | 14.7/.58 Max. | 7.1/.28 | 7.1/.28 | M10/.375 | M10/.375 | M10/.375 | M10/.375 |

1. The minimum thread depth is 1.5 bolt diameter. The recommended engagement of fixing bolt thread for ferrous mounting is 1.25 bolt diameter.

2. The dimensions specifying the area within the dashed lines are the minimum dimensions for the mounting surface. The corners of the rectangle may be radiused to a maximum radius equal to the thread diameter of the fixing bolts.

   Along each axis the fixing holes are at equal distance to the mounting surface edges.

3. This dimension gives the minimum spacing distance between the valve and adjacent obstructions, for example, another valve or a wall. This dimension is, therefore, the minimum distance from centerline to centerline of two similar mounting surfaces placed on a manifold block. The fixing holes are at equal distances to this dimension.

   The valve manufacturer's attention is drawn to the fact that no part of the width of the complete valve assembly is to exceed this dimension.

4. Blind hole in the mounting surface to **accommodate** the locating pin on the valve. The minimum depth is 8/.31.

5. The maximum limit of the working pressure for subplates and manifold blocks with this mounting surface will be supplied by the manufacturer.

# ANSI/B93.7M

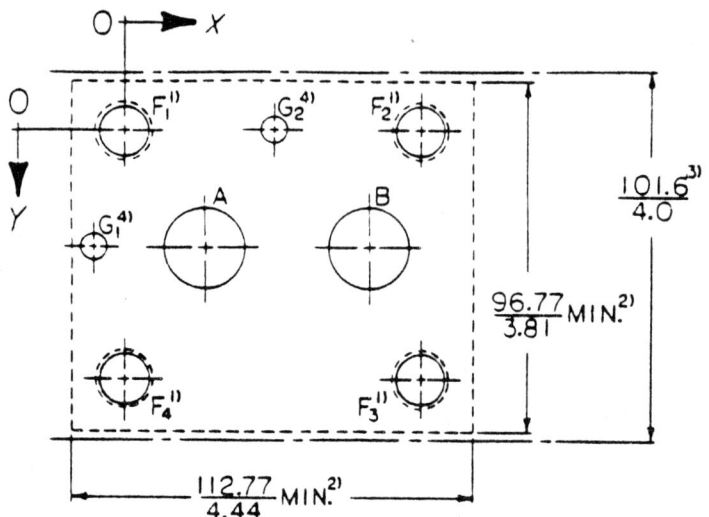
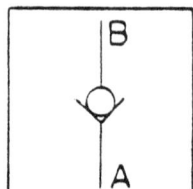

Figure 16 - **C08 mounting surface for check valves with main ports of 23.4/.92 maximum port diameter.**

Table 16 - **C08 mounting surface for check valves with main ports of 23.4/.92 maximum port diameter.**[5]

|   | A | B | $G_1$ | $G_2$ | $F_1$ | $F_2$ | $F_3$ | $F_4$ |
|---|---|---|---|---|---|---|---|---|
| X | 22.2 / .875 | 68.2 / 2.687 | -8.7 / -.344 | 40.5 / 1.594 | 0 / 0 | 81 / 3.187 | 81 / 3.187 | 0 / 0 |
| Y | 32.5 / 1.281 | 32.5 / 1.281 | 32.5 / 1.281 | 0 / 0 | 0 / 0 | 0 / 0 | 65.1 / 2.562 | 65.1 / 2.562 |
| $\phi$ | 23.4 / .92 Max. | 23.4 / .92 Max. | 7.1 / .28 | 7.1 / .28 | M16 / .625 | M16 / .625 | M16 / .625 | M16 / .625 |

1. The minimum thread depth is 1.5 bolt diameter. The recommended engagement of fixing bolt thread for ferrous mounting is 1.25 bolt diameter.

2. The dimensions specifying the area within the dashed lines are the minimum dimensions for the mounting surface. The corners of the rectangle may be radiused to a maximum radius equal to the thread diameter of the fixing bolts.

   Along each axis the fixing holes are at equal distance to the mounting surface edges.

3. This dimension gives the minimum spacing distance between the valve and adjacent obstructions, for example, another valve or a wall. This dimension is, therefore, the minimum distance from centerline to centerline of two similar mounting surfaces placed on a manifold block. The fixing holes are at equal distances to this dimension.

   The valve manufacturer's attention is drawn to the fact that no part of the width of the complete valve assembly is to exceed this dimension.

4. Blind hole in the mounting surface to accommodate the locating pin on the valve. The minimum depth is 8/.31.

5. The maximum limit of the working pressure for subplates and manifold blocks with this mounting surface will be supplied by the manufacturer.

ANSI/B93.7M

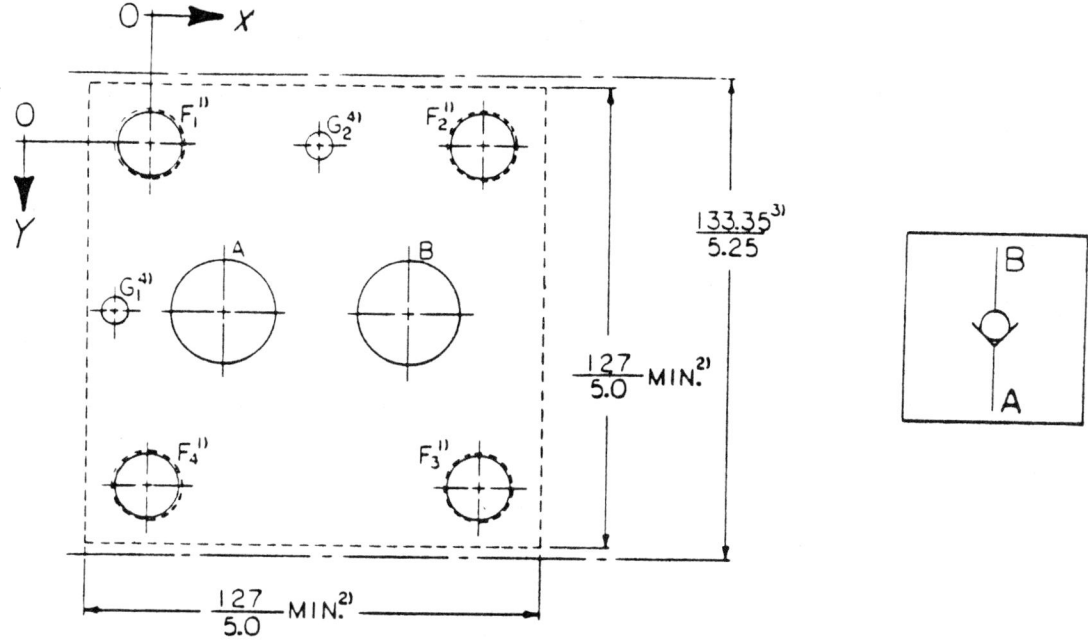

Figure 17 - **C09 mounting surface for check valves with main ports of 28.7/1.13 maximum port diameter.**

Table 17 - **C09 mounting surface for check valve with main ports of 28.7/1.13 maximum port diameter.**[5]

|   | A | B | $G_1$ | $G_2$ | $F_1$ | $F_2$ | $F_3$ | $F_4$ |
|---|---|---|---|---|---|---|---|---|
| X | 20.6 / .812 | 71.4 / 2.812 | -9.5 / -.375 | 46 / 1.812 | 0 / 0 | 92.1 / 3.625 | 92.1 / 3.625 | 0 / 0 |
| Y | 46 / 1.812 | 46 / 1.812 | 46 / 1.812 | 0 / 0 | 0 / 0 | 0 / 0 | 92.1 / 3.625 | 92.1 / 3.625 |
| $\phi$ | 28.7 / 1.13 Max. | 28.7 / 1.13 Max. | 7.1 / .28 | 7.1 / .28 | M20 / .75 | M20 / .75 | M20 / .75 | M20 / .75 |

1. The minimum thread depth is 1.5 bolt diameter. The recommended engagement of fixing bolt thread for ferrous mounting is 1.25 bolt diameter.

2. The dimensions specifying the area within the dashed lines are the minimum dimensions for the mounting surface. The corners of the rectangle may be radiused to a maximum radius equal to the thread diameter of the fixing bolts.

   Along each axis the fixing holes are at equal distance to the mounting surface edges.

3. This dimension gives the minimum spacing distance between the valve and adjacent obstructions, for example, another valve or a wall. This dimension is, therefore, the minimum distance from centerline to centerline of two similar mounting surfaces placed on a manifold block. The fixing holes are at equal distances to this dimension.

   The valve manufacturer's attention is drawn to the fact that no part of the width of the complete valve assembly is to exceed this dimension.

4. Blind hole in the mounting surface to **accommodate** the locating pin on the valve. The minimum depth is 8/.31.

5. The maximum limit of the working pressure for subplates and manifold blocks with this mounting surface will be supplied by the manufacturer.

# ANSI/B93.7M

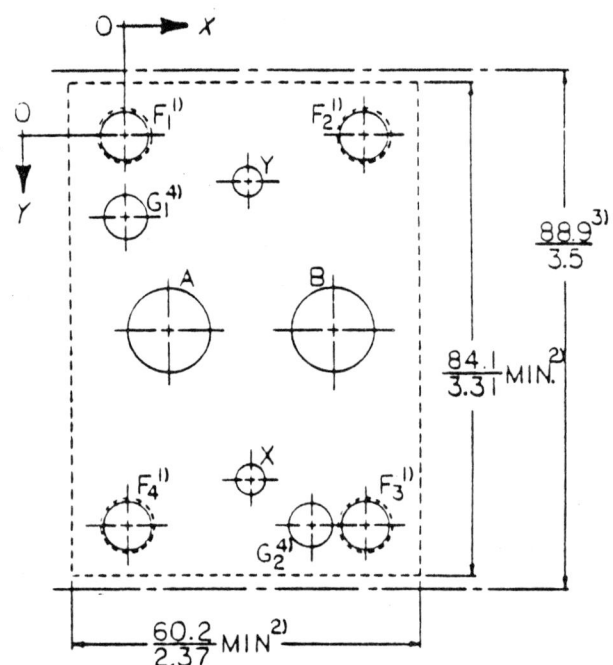

Figure 18 - POC06 mounting surface for pilot operated check valves with main ports of 14.7/.58 maximum port diameter.

Table 18 - POC06 mounting surface for pilot operated check valves with main ports of 14.7/.58 maximum port diameter.[5]

|   | A | B | X | Y | $G_1$ | $G_2$ | $F_1$ | $F_2$ | $F_3$ | $F_4$ |
|---|---|---|---|---|---|---|---|---|---|---|
| X | 7.1 / .281 | 35.7 / 1.406 | 21.4 / .843 | 21.4 / .843 | 0 / 0 | 31.8 / 1.25 | 0 / 0 | 42.9 / 1.688 | 42.9 / 1.688 | 0 / 0 |
| Y | 33.3 / 1.313 | 33.3 / 1.313 | 58.7 / 2.313 | 7.9 / .312 | 14.3 / .563 | 66.7 / 2.625 | 0 / 0 | 0 / 0 | 66.7 / 2.625 | 66.7 / 2.625 |
| $\phi$ | 14.7 / .58 Max. | 14.7 / .58 Max. | 4.8 / .19 Max. | 4.8 / .19 Max. | 7.1 / .28 | 7.1 / .28 | M10 / .375 | M10 / .375 | M10 / .375 | M10 / .375 |

1. The minimum thread depth is 1.5 bolt diameter. The recommended engagement of fixing bolt thread for ferrous mounting is 1.25 bolt diameter.

2. The dimensions specifying the area within the dashed lines are the minimum dimensions for the mounting surface. The corners of the rectangle may be radiused to a maximum radius equal to the thread diameter of the fixing bolts.

   Along each axis the fixing holes are at equal distance to the mounting surface edges.

3. This dimension gives the minimum spacing distance between the valve and adjacent obstructions, for example, another valve or a wall. This dimension is, therefore, the minimum distance from centerline to centerline of two similar mounting surfaces placed on a manifold block. The fixing holes are at equal distances to this dimension.

   The valve manufacturer's attention is drawn to the fact that no part of the width of the complete valve assembly is to exceed this dimension.

4. Blind hole in the mounting surface to accommodate the locating pin on the valve. The minimum depth is 8/.31.

5. The maximum limit of the working pressure for subplates and manifold blocks with this mounting surface will be supplied by the manufacturer.

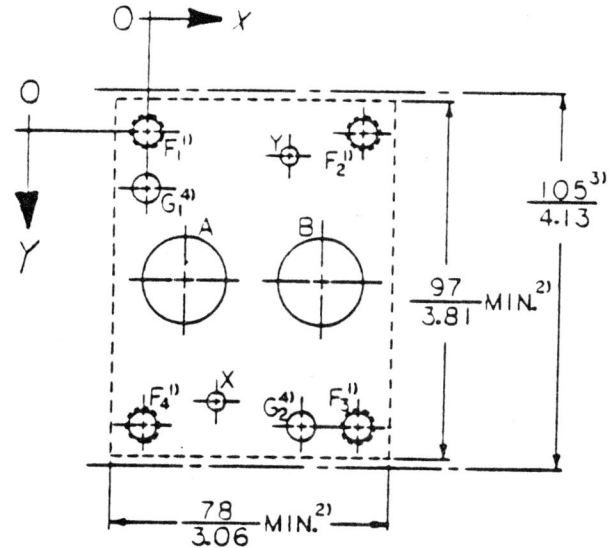
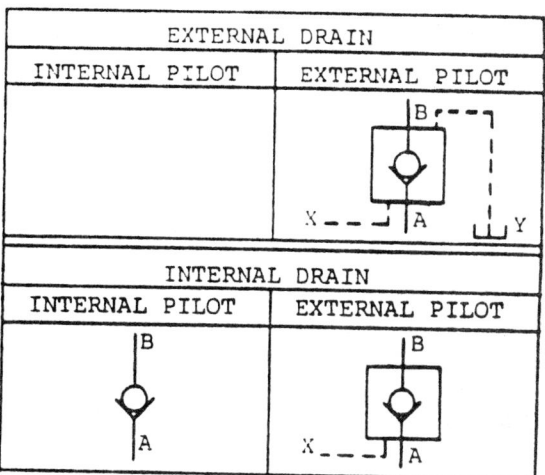

Figure 19 - **POC08 mounting surface for pilot operated check valves with main ports of 23.4/.92 maximum port diameter.**

Table 19 - **POC08 mounting surface for pilot operated check valves with main ports of 23.4/.92 maximum port diameter.[5]**

|   | A | B | X | Y | $G_1$ | $G_2$ | $F_1$ | $F_2$ | $F_3$ | $F_4$ |
|---|---|---|---|---|---|---|---|---|---|---|
| X | 11.1 / .438 | 49.2 / 1.938 | 20.6 / .813 | 39.7 / 1.563 | 0 / 0 | 44.5 / 1.75 | 0 / 0 | 60.3 / 2.375 | 60.3 / 2.375 | 0 / 0 |
| Y | 39.7 / 1.563 | 39.7 / 1.563 | 73 / 2.875 | 6.4 / .252 | 15.9 / .625 | 79.4 / 3.125 | 0 / 0 | 0 / 0 | 79.4 / 3.125 | 79.4 / 3.125 |
| φ | 23.4 / .92 Max. | 23.4 / .92 Max. | 4.8 / .19 Max. | 4.8 / .19 Max. | 7.1 / .28 | 7.1 / .28 | M10 / .375 | M10 / .375 | M10 / .375 | M10 / .375 |

1. The minimum thread depth is 1.5 bolt diameter. The recommended engagement of fixing bolt thread for ferrous mounting is 1.25 bolt diameter.

2. The dimensions specifying the area within the dashed lines are the minimum dimensions for the mounting surface. The corners of the rectangle may be radiused to a maximum radius equal to the thread diameter of the fixing bolts.

   Along each axis the fixing holes are at equal distance to the mounting surface edges.

3. This dimension gives the minimum spacing distance between the valve and adjacent obstructions, for example, another valve or a wall. This dimension is, therefore, the minimum distance from centerline to centerline of two similar mounting surfaces placed on a manifold block. The fixing holes are at equal distances to this dimension.

   The valve manufacturer's attention is drawn to the fact that no part of the width of the complete valve assembly is to exceed this dimension.

4. Blind hole in the mounting surface to accommodate the locating pin on the valve. The minimum depth is 8/.31.

5. The maximum limit of the working pressure for subplates and manifold blocks with this mounting surface will be supplied by the manufacturer.

# ANSI/B93.7M

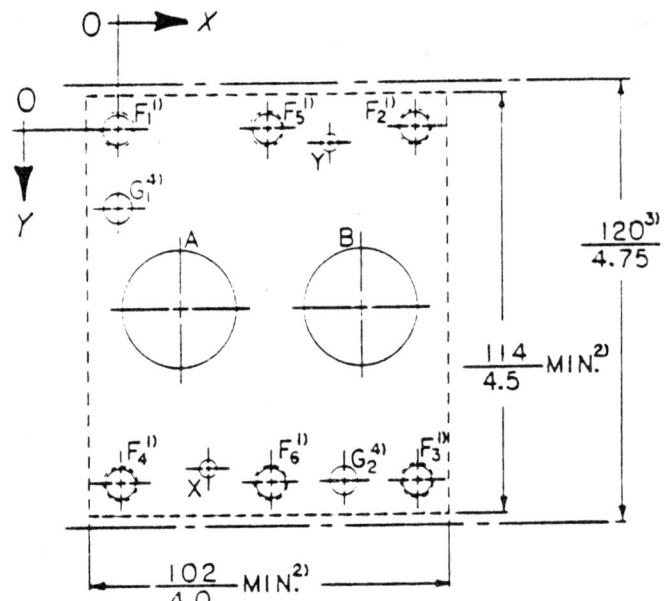

Figure 20 - **POC09 mounting surface for pilot operated check valves with main ports of 28.7/1.13 maximum port diameter.**

Table 20 - **POC09 mounting surface for pilot operated check valves with main ports of 28.7/1.13 maximum port diameter.**[5]

|   | A | B | X | Y | $G_1$ | $G_2$ | $F_1$ | $F_2$ | $F_3$ | $F_4$ | $F_5$ | $F_6$ |
|---|---|---|---|---|---|---|---|---|---|---|---|---|
| X | 16.7 / .656 | 67.5 / 2.656 | 24.6 / .969 | 59.5 / 2.344 | 0 / 0 | 62.7 / 2.469 | 0 / 0 | 84.1 / 3.312 | 84.1 / 3.312 | 0 / 0 | 42.1 / 1.656 | 42.1 / 1.656 |
| Y | 48.4 / 1.906 | 48.4 / 1.906 | 92.8 / 3.656 | 4 / .156 | 21.4 / .844 | 96.8 / 3.812 | 0 / 0 | 0 / 0 | 96.8 / 3.812 | 96.8 / 3.812 | 0 / 0 | 96.8 / 3.812 |
| $\phi$ | 28.7 / 1.13 Max. | 28.7 / 1.13 Max. | 4.8 / .19 Max. | 4.8 / .19 Max. | 7.1 / .28 | 7.1 / .28 | M10 / .375 | M10 / .375 | M10 / .375 | M10 / .375 | M10 / .375 | M10 / .375 |

1. The minimum thread depth is 1.5 bolt diameter. The recommended engagement of fixing bolt thread for ferrous mounting is 1.25 bolt diameter.

2. The dimensions specifying the area within the dashed lines are the minimum dimensions for the mounting surface. The corners of the rectangle may be radiused to a maximum radius equal to the thread diameter of the fixing bolts.

   Along each axis the fixing holes are at equal distance to the mounting surface edges.

3. This dimension gives the minimum spacing distance between the valve and adjacent obstructions, for example, another valve or a wall. This dimension is, therefore, the minimum distance from centerline to centerline of two similar mounting surfaces placed on a manifold block. The fixing holes are at equal distances to this dimension.

   The valve manufacturer's attention is drawn to the fact that no part of the width of the complete valve assembly is to exceed this dimension.

4. Blind hole in the mounting surface to accommodate the locating pin on the valve. The minimum depth is 8/.31.

5. The maximum limit of the working pressure for subplates and manifold blocks with this mounting surface will be supplied by the manufacturer.

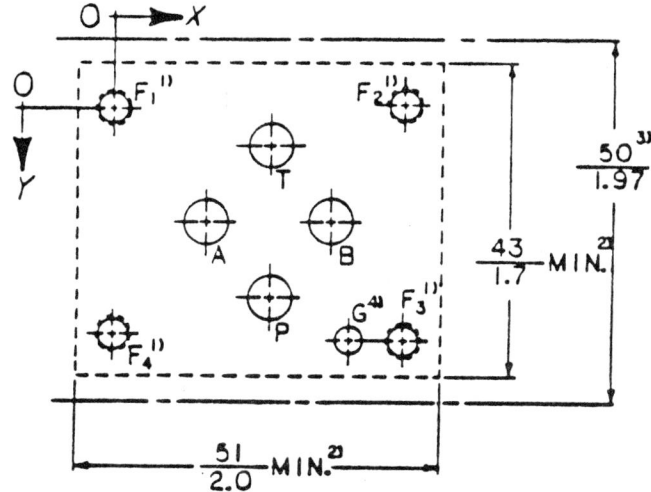

Figure 21 - P03 mounting surface for pressure control valves (excluding pressure relief valves), sequence valves, unloading valves, throttle valves and check valves with main ports of 6.3/.25 maximum port diameter.

Table 21 - P03 mounting surface for pressure control valves (excluding pressure relief valves), sequence valves, unloading valves, throttle valves and check valves with main ports of 6.3/.25 maximum port diameter.[5]

|   | P | A | T | B | G | $F_1$ | $F_2$ | $F_3$ | $F_4$ |
|---|---|---|---|---|---|---|---|---|---|
| X | 21.5 / .85 | 12.7 / .5 | 21.5 / .85 | 30.2 / 1.19 | 33 / 1.3 | 0 / 0 | 40.5 / 1.594 | 40.5 / 1.594 | 0 / 0 |
| Y | 25.9 / 1.02 | 15.5 / .61 | 5.1 / .2 | 15.5 / .61 | 31.75 / 1.25 | 0 / 0 | -0.75 / -.03 | 31.75 / 1.25 | 31 / 1.22 |
| $\phi$ | 6.3 / .25 Max. | 6.3 / .25 Max. | 6.3 / .25 Max. | 6.3 / .25 Max. | 4 / .16 | M5 / .19 | M5 / .19 | M5 / .19 | M5 / .19 |

1. The minimum thread depth is 1.5 bolt diameter. The recommended engagement of fixing bolt thread for ferrous mounting is 1.25 bolt diameter.

2. The dimensions specifying the area within the dashed lines are the minimum dimensions for the mounting surface. The corners of the rectangle may be radiused to a maximum radius equal to the thread diameter of the fixing bolts.

   Along each axis the fixing holes are at equal distance to the mounting surface edges.

3. This dimension gives the minimum spacing distance between the valve and adjacent obstructions, for example, another valve or a wall. This dimension is, therefore, the minimum distance from centerline to centerline of two similar mounting surfaces placed on a manifold block. The fixing holes are at equal distances to this dimension.

   The valve manufacturer's attention is drawn to the fact that no part of the width of the complete valve assembly is to exceed this dimension.

4. Blind hole in the mounting surface to **accommodate** the locating pin on the valve. The minimum depth is 4/.16.

5. The maximum limit of the working pressure for subplates and manifold blocks with this mounting surface will be supplied by the manufacturer.

# ANSI/B93.7M

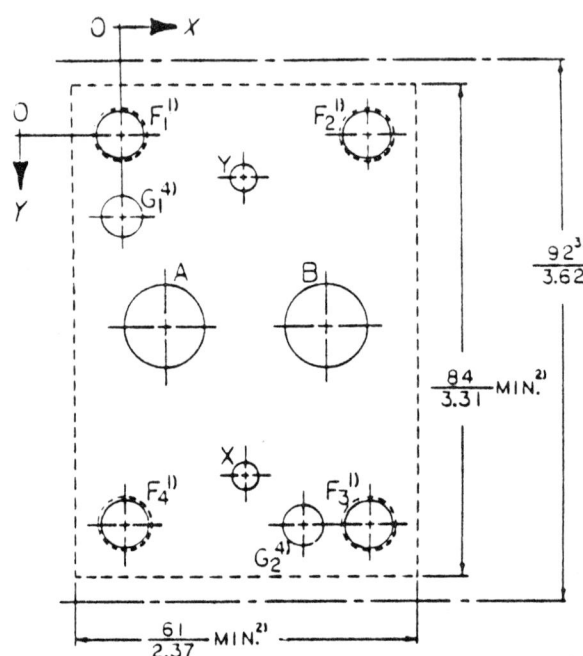

Figure 22 - P06 mounting surface for pressure control valves (excluding pressure relief valves), sequence valves, unloading valves, throttle valves and check valves with main ports of 14.7/.58 maximum port diameter.

Table 22 - P06 mounting surface for pressure control valves (excluding pressure relief valves), sequence valves, unloading valves, throttle valves and check valves with main ports of 14.7/.58 maximum port diameter.[5]

|   | A | B | X | Y | $G_1$ | $G_2$ | $F_1$ | $F_2$ | $F_3$ | $F_4$ |
|---|---|---|---|---|---|---|---|---|---|---|
| X | 7.1 / .281 | 35.7 / 1.406 | 21.4 / .844 | 21.4 / .844 | 0 / 0 | 31.8 / 1.25 | 0 / 0 | 42.9 / 1.687 | 42.9 / 1.687 | 0 / 0 |
| Y | 33.3 / 1.312 | 33.3 / 1.312 | 58.7 / 2.312 | 7.9 / .312 | 14.3 / .562 | 66.7 / 2.625 | 0 / 0 | 0 / 0 | 66.7 / 2.625 | 66.7 / 2.625 |
| $\phi$ | 14.7 / .58 Max. | 14.7 / .58 Max. | 4.8 / .19 Max. | 4.8 / .19 Max. | 7.1 / .28 | 7.1 / .28 | M10 / .375 | M10 / .375 | M10 / .375 | M10 / .375 |

1. The minimum thread depth is 1.5 bolt diameter. The recommended engagement of fixing bolt thread for ferrous mounting is 1.25 bolt diameter.

2. The dimensions specifying the area within the dashed lines are the minimum dimensions for the mounting surface. The corners of the rectangle may be radiused to a maximum radius equal to the thread diameter of the fixing bolts.

   Along each axis the fixing holes are at equal distance to the mounting surface edges.

3. This dimension gives the minimum spacing distance between the valve and adjacent obstructions, for example, another valve or a wall. This dimension is, therefore, the minimum distance from centerline to centerline of two similar mounting surfaces placed on a manifold block. The fixing holes are at equal distances to this dimension.

   The valve manufacturer's attention is drawn to the fact that no part of the width of the complete valve assembly is to exceed this dimension.

4. Blind hole in the mounting surface to accommodate the locating pin on the valve. The minimum depth is 8/.31.

5. The maximum limit of the working pressure for subplates and manifold blocks with this mounting surface will be supplied by the manufacturer.

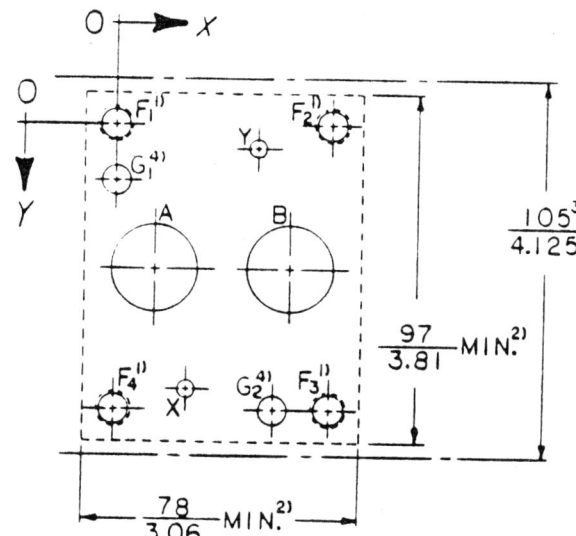

**NOTE:** Valves are known to exist in the general market having port usage different from the recommendation shown on pages 172 and 173 for this mounting surface.

Figure 23 - P08 mounting surface for pressure control valves (excluding pressure relief valves), sequence valves, unloading valves, throttle valves and check valves with main ports of 23.4/.92 maximum port diameter.

Table 23 - P08 mounting surface for pressure control valves (excluding pressure relief valves), sequence valves, unloading valves, throttle valves and check valves with main ports of 23.4/.92 maximum port diameter.[5]

|   | A | B | X | Y | $G_1$ | $G_2$ | $F_1$ | $F_2$ | $F_3$ | $F_4$ |
|---|---|---|---|---|---|---|---|---|---|---|
| X | 11.1 / .438 | 49.2 / 1.938 | 20.6 / .813 | 39.7 / 1.563 | 0 / 0 | 44.5 / 1.75 | 0 / 0 | 60.3 / 2.375 | 60.3 / 2.375 | 0 / 0 |
| Y | 39.7 / 1.563 | 39.7 / 1.563 | 73 / 2.875 | 6.4 / .25 | 15.9 / .625 | 79.4 / 3.125 | 0 / 0 | 0 / 0 | 79.4 / 3.125 | 79.4 / 3.125 |
| $\phi$ | 23.4 / .92 Max. | 23.4 / .92 Max. | 4.8 / .19 Max. | 4.8 / .19 Max. | 7.1 / .28 | 7.1 / .28 | M10 / .375 | M10 / .375 | M10 / .375 | M10 / .375 |

1. The minimum thread depth is 1.5 bolt diameter. The recommended engagement of fixing bolt thread for ferrous mounting is 1.25 bolt diameter.

2. The dimensions specifying the area within the dashed lines are the minimum dimensions for the mounting surface. The corners of the rectangle may be radiused to a maximum radius equal to the thread diameter of the fixing bolts.

   Along each axis the fixing holes are at equal distance to the mounting surface edges.

3. This dimension gives the minimum spacing distance between the valve and adjacent obstructions, for example, another valve or a wall. This dimension is, therefore, the minimum distance from centerline to centerline of two similar mounting surfaces placed on a manifold block. The fixing holes are at equal distances to this dimension.

   The valve manufacturer's attention is drawn to the fact that no part of the width of the complete valve assembly is to exceed this dimension.

4. Blind hole in the mounting surface to accommodate the locating pin on the valve. The minimum depth is 8/.31.

5. The maximum limit of the working pressure for subplates and manifold blocks with this mounting surface will be supplied by the manufacturer.

# ANSI/B93.7M

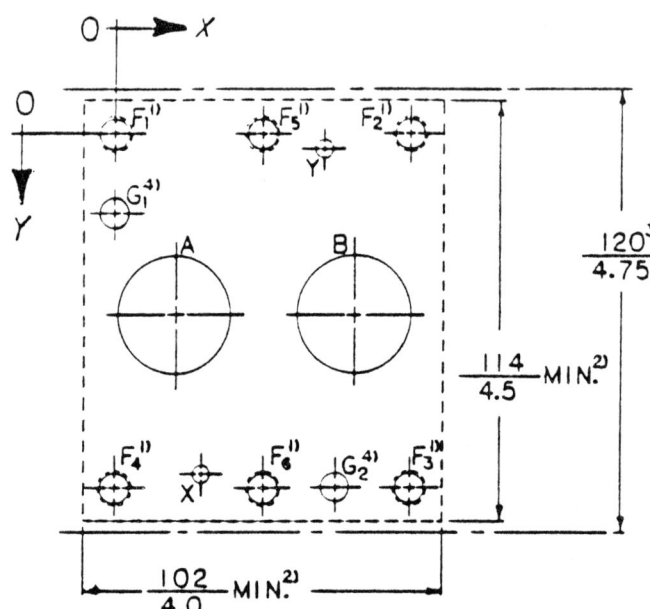

NOTE: Valves are known to exist in the general having port usage different from the recommen shown on pages 172 and 173 for this mounting s

Figure 24 - P10 mounting surface for pressure control valves (excluding pressure relief valves), sequence valves, unloading valves, throttle valves, and check valves with main ports of 32/1.25 maximum port diameter.

Table 24 - P10 mounting surface for pressure control valves (excluding pressure relief valves), sequence valves, unloading valves, throttle valves, and check valves with main ports of 32/1.25 maximum port diameter.[5]

|   | A | B | X | Y | $G_1$ | $G_2$ | $F_1$ | $F_2$ | $F_3$ | $F_4$ | $F_5$ | $F_6$ |
|---|---|---|---|---|---|---|---|---|---|---|---|---|
| X | 16.7/.656 | 67.5/2.656 | 24.6/.969 | 59.6/2.345 | 0/0 | 62.7/2.469 | 0/0 | 84.1/3.312 | 84.1/3.312 | 0/0 | 42.1/1.656 | 42.1/1.656 |
| Y | 48.4/1.906 | 48.4/1.906 | 92.9/3.656 | 4/.156 | 21.4/.844 | 96.8/3.812 | 0/0 | 0/0 | 96.8/3.812 | 96.8/3.812 | 0/0 | 96.8/3.812 |
| $\phi$ | 32/1.25 Max. | 32/1.25 Max. | 4.8/.19 Max. | 4.8/.19 Max. | 7.1/.28 | 7.1/.28 | M10/.375 | M10/.375 | M10/.375 | M10/.375 | M10/.375 | M10/.375 |

1. The minimum thread depth is 1.5 bolt diameter. The recommended engagement of fixing bolt thread for ferrous mounting is 1.25 bolt diameter.

2. The dimensions specifying the area within the dashed lines are the minimum dimensions for the mounting surface. The corners of the rectangle may be radiused to a maximum radius equal to the thread diameter of the fixing bolts.

   Along each axis the fixing holes are at equal distance to the mounting surface edges.

3. This dimension gives the minimum spacing distance between the valve and adjacent obstructions, for example, another valve or a wall. This dimension is, therefore, the minimum distance from centerline to centerline of two similar mounting surfaces placed on a manifold block. The fixing holes are at equal distances to this dimension.

   The valve manufacturer's attention is drawn to the fact that no part of the width of the complete valve assembly is to exceed this dimension.

4. Blind hole in the mounting surface to accommodate the locating pin on the valve. The minimum depth is 8/.

5. The maximum limit of the working pressure for subplates and manifold blocks with this mounting surface will be supplied by the manufacturer.

ANSI/B93.7M

Recommended uses of the ports for various two-port hydraulic valves.

| Description | External drain | | Internal drain | |
|---|---|---|---|---|
| | Internal pilot | External pilot | Internal pilot | External pilot |
| Pressure control valve | symbol with ports A, P, T | symbol with ports A, B, P, T | | |
| Pressure control valve with bypass check valve | symbol with bypass check and ports A, P, T | symbol with bypass check and ports A, B, P, T | | |
| Sequence valve | symbol with ports A, P, T | symbol with ports A, B, P, T | | |
| Sequence valve with bypass check valve | symbol with bypass check and ports A, P, T | symbol with bypass check and ports A, B, P, T | | |
| Unloading valve | | symbol with ports A, B, P, T | | symbol with ports A, B, P |
| Unloading valve with bypass check valve | | symbol with bypass check and ports A, B, P, T | | symbol with bypass check and ports A, B, P |
| Throttle valve | A —⋈— B, T | | A —⋈— B | |
| Throttle valve with bypass check valve | A —⋈— B with bypass check, T | | A —⋈— B with bypass check | |
| Check valve | | | | B —◯— A |

Examples of symbols for direct operated two-port valves with main ports of 6.3/.25 maximum diameter, refer to figure 21.

## ANSI/B93.7M
Recommended uses of the ports for various two-port hydraulic valves.

| Description | External drain | | Internal drain | |
|---|---|---|---|---|
| | Internal pilot | External pilot | Internal pilot | External pilot |
| Pressure control valves | 1.) [symbol with ports B, A, X] | | | |
| Pressure control valve with bypass check valve | 1.) [symbol with bypass check, ports B, A, X] | | | |
| Sequence valve | [symbol with ports B, A, Y] | [symbol with ports X, B, A, Y] | | |
| Sequence valve with bypass check valve | [symbol with bypass check, ports B, A, Y] | [symbol with bypass check, ports X, B, A, Y] | | |
| Unloading valve | | [symbol with ports X, B, A, Y] | | [symbol with ports X, B, A] |
| Unloading valve with bypass check valve | | [symbol with bypass check, ports X, B, A, Y] | | [symbol with bypass check, ports X, B, A] |
| Throttle valve | A —⋈— B, Y | | A —⋈— B | |
| Throttle valve with bypass check valve | A —[check/throttle]— B, Y | | A —[check/throttle]— B | |
| Check valve | | | B —◯— A | |

1. Valves are known to exist in the general market having port usage different from the recommendation shown on pages 31 and 32 for the mounting surface.

Examples of symbols for direct operated two-port valves with main ports of 14.7/.58, 23.4/.92, 32/1.25 maximum diameter, refer to figures 22, and 24.

ANSI/B93.7M

Recommended uses of the ports for various two-port hydraulic valves.

| Description | External drain | | Internal drain | |
|---|---|---|---|---|
| | Internal pilot | External pilot | Internal pilot | External pilot |
| Pressure control valves | [symbol with B, A, Y 1), X, note 2)] | | | |
| Pressure control valve with bypass check valve | [symbol with B, A, Y 1), X, note 2)] | | | |
| Sequence valve | [symbol with B, A, X 1), Y] | [symbol with B, X, A, Y] | | |
| Sequence valve with bypass check valve | [symbol with B, A, X 1), Y] | [symbol with B, X, A, Y] | | |
| Unloading valve | | [symbol with B, X, A, Y] | | [symbol with B, X, A] |
| Unloading valve with bypass check valve | | [symbol with B, X, A, Y] | | [symbol with B, X, A] |

. Port for supplementary remote control; may be blocked if not needed.

. Valves are known to exist in the general market having port usage different from the recommendation shown on pages 31 and 32 for this mounting surface.

xamples of symbols for pilot operated two-port valves with main ports of 14.7/.58, 23.4/.92, 32/1.25 maximum diameter, refer to figures 22, 23 and 4.

# ANSI/B93.7M

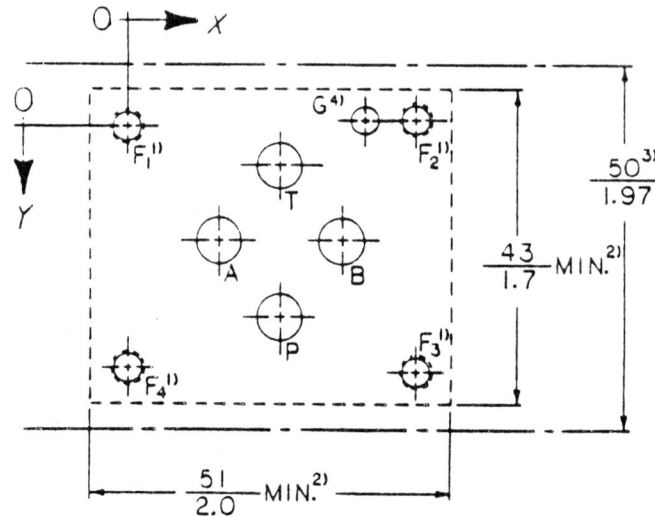

Figure 25 - **R03 mounting surface for pressure relief valves with main ports of 6.3/.25 maximum port diameter.**

Table 25 - **R03 mounting surface for pressure relief valves with main ports of 6.3/.25 maximum port diameter.**[5]

|   | P | A | T | B | G | $F_1$ | $F_2$ | $F_3$ | $F_4$ |
|---|---|---|---|---|---|---|---|---|---|
| X | 21.5 / .85 | 12.7 / .5 | 21.5 / .85 | 30.2 / 1.19 | 33 / 1.3 | 0 / 0 | 40.5 / 1.594 | 40.5 / 1.594 | 0 / 0 |
| Y | 25.9 / 1.02 | 15.5 / .61 | 5.1 / .2 | 15.5 / .61 | -0.75 / -.03 | 0 / 0 | -0.75 / -.03 | 31.75 / 1.25 | 31 / 1.22 |
| φ | 6.3 / .25 Max. | 6.3 / .25 Max. | 6.3 / .25 Max. | 6.3 / .25 Max. | 4 / .16 | M5 / .19 | M5 / .19 | M5 / .19 | M5 / .19 |

1. The minimum thread depth is 1.5 bolt diameter. The recommended engagement of fixing bolt thread for ferrous mounting is 1.25 bolt diameter.

2. The dimensions specifying the area within the dashed lines are the minimum dimensions for the mounting surface. The corners of the rectangle may be radiused to a maximum radius equal to the thread diameter of the fixing bolts.

   Along each axis the fixing holes are at equal distance to the mounting surface edges.

3. This dimension gives the minimum spacing distance between the valve and adjacent obstructions, for example, another valve or a wall. This dimension is, therefore, the minimum distance from centerline to centerline of two similar mounting surfaces placed on a manifold block. The fixing holes are at equal distances to this dimension.

   The valve manufacturer's attention is drawn to the fact that no part of the width of the complete valve assembly is to exceed this dimension.

4. Blind hole in the mounting surface to **accommodate** the locating pin on the valve. The minimum depth is 4/.16.

5. The maximum limit of the working pressure for subplates and manifold blocks with this mounting surface will be supplied by the manufacturer.

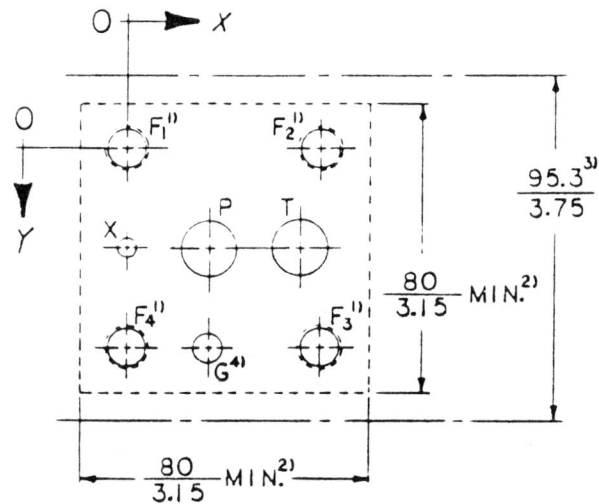

Figure 26 - **R06 mounting surface for pilot operated pressure relief valves with main ports of 14.7/.58 maximum port diameter.**

Table 26 - **R06 mounting surface for pilot operated pressure relief valves with main ports of 14.7/.58 maximum port diameter.**[5]

|   | P | T | X | G | $F_1$ | $F_2$ | $F_3$ | $F_4$ |
|---|---|---|---|---|---|---|---|---|
| X | 22.2 / .875 | 47.5 / 1.875 | 0 / 0 | 22.1 / .875 | 0 / 0 | 53.8 / 2.125 | 53.8 / 2.125 | 0 / 0 |
| Y | 26.9 / 1.063 | 26.9 / 1.063 | 26.9 / 1.063 | 53.8 / 2.125 | 0 / 0 | 0 / 0 | 53.8 / 2.125 | 53.8 / 2.125 |
| $\phi$ | 14.7 / .58 Max. | 14.7 / .58 Max. | 4.8 / .19 Max. | 7.1 / .28 | M12 / .5 | M12 / .5 | M12 / .5 | M12 / .5 |

1. The minimum thread depth is 1.5 bolt diameter. The recommended engagement of fixing bolt thread for ferrous mounting is 1.25 bolt diameter.

2. The dimensions specifying the area within the dashed lines are the minimum dimensions for the mounting surface. The corners of the rectangle may be radiused to a maximum radius equal to the thread diameter of the fixing bolts.

   Along each axis the fixing holes are at equal distance to the mounting surface edges.

3. This dimension gives the minimum spacing distance between the valve and adjacent obstructions, for example, another valve or a wall. This dimension is, therefore, the minimum distance from centerline to centerline of two similar mounting surfaces placed on a manifold block. The fixing holes are at equal distances to this dimension.

   The valve manufacturer's attention is drawn to the fact that no part of the width of the complete valve assembly is to exceed this dimension.

4. Blind hole in the mounting surface to **accommodate** the locating pin on the valve. The minimum depth is 8/.31.

5. The maximum limit of the working pressure for subplates and manifold blocks with this mounting surface will be supplied by the manufacturer.

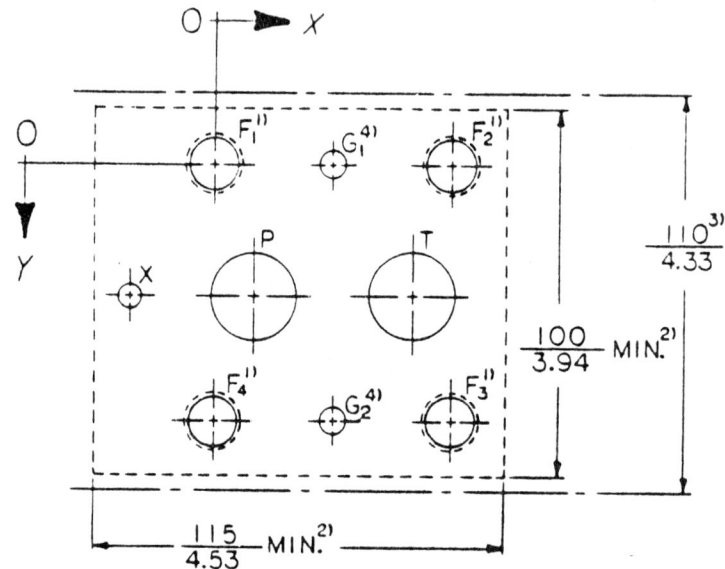

Figure 27 - **R08 mounting surface for pilot operated pressure relief valves with main ports of 23.4/.92 maximum port diameter.**

Table 27 - **R08 mounting surface for pilot operated pressure relief valves with main ports of 23.4/.92 maximum port diameter.**[5]

|   | P | T | X | G$_1$ | G$_2$ | F$_1$ | F$_2$ | F$_3$ | F$_4$ |
|---|---|---|---|---|---|---|---|---|---|
| X | 11.1 / .438 | 55.6 / 2.188 | -23.8 / -.938 | 33.4 / 1.313 | 33.4 / 1.313 | 0 / 0 | 66.7 / 2.625 | 66.7 / 2.625 | 0 / 0 |
| Y | 35 / 1.375 | 35 / 1.375 | 35 / 1.375 | 0 / 0 | 70 / 2.75 | 0 / 0 | 0 / 0 | 70 / 2.75 | 70 / 2.75 |
| $\phi$ | 23.4 / .92 Max. | 23.4 / .92 Max. | 6.3 / .25 Max. | 7.1 / .28 | 7.1 / .28 | M16 / .625 | M16 / .625 | M16 / .625 | M16 / .625 |

1. The minimum thread depth is 1.5 bolt diameter. The recommended engagement of fixing bolt thread for ferrous mounting is 1.25 bolt diameter.

2. The dimensions specifying the area within the dashed lines are the minimum dimensions for the mounting surface. The corners of the rectangle may be radiused to a maximum radius equal to the thread diameter of the fixing bolts.

   Along each axis the fixing holes are at equal distance to the mounting surface edges.

3. This dimension gives the minimum spacing distance between the valve and adjacent obstructions, for example, another valve or a wall. This dimension is, therefore, the minimum distance from centerline to centerline of two similar mounting surfaces placed on a manifold block. The fixing holes are at equal distances to this dimension.

   The valve manufacturer's attention is drawn to the fact that no part of the width of the complete valve assembly is to exceed this dimension.

4. Blind hole in the mounting surface to **accommodate** the locating pin on the valve. The minimum depth is 8/.31.

5. The maximum limit of the working pressure for subplates and manifold blocks with this mounting surface will be supplied by the manufacturer.

ANSI/B93.7M

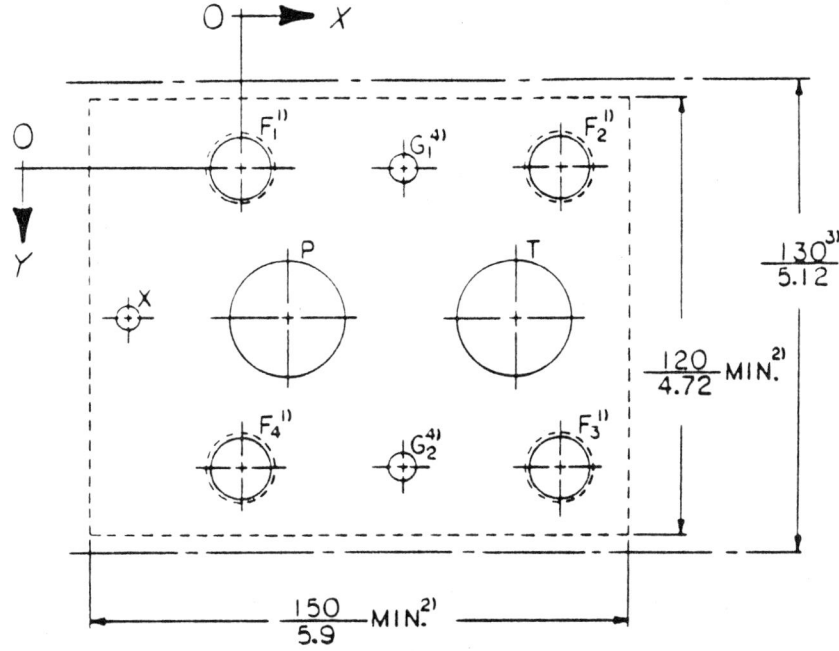

Figure 28 - **R10 mounting surface for pilot operated pressure relief valves with main ports of 32/1.26 maximum port diameter.**

Table 28 - **R10 mounting surface for pilot operated pressure relief valves with main ports of 32/1.26 maximum port diameter.**[5]

|   | P | T | X | $G_1$ | $G_2$ | $F_1$ | $F_2$ | $F_3$ | $F_4$ |
|---|---|---|---|---|---|---|---|---|---|
| X | 12.7 / .5 | 76.2 / 3.0 | -31.8 / -1.25 | 44.5 / 1.75 | 44.5 / 1.75 | 0 / 0 | 88.9 / 3.5 | 88.9 / 3.5 | 0 / 0 |
| Y | 41.3 / 1.625 | 41.3 / 1.625 | 41.3 / 1.625 | 0 / 0 | 82.6 / 3.25 | 0 / 0 | 0 / 0 | 82.6 / 3.25 | 82.6 / 3.25 |
| $\phi$ | 32 / 1.26 Max. | 32 / 1.26 Max. | 6.3 / .25 Max. | 7.1 / .28 | 7.1 / .28 | M18 / .75 | M18 / .75 | M18 / .75 | M18 / .75 |

1. The minimum thread depth is 1.5 bolt diameter. The recommended engagement of fixing bolt thread for ferrous mounting is 1.25 bolt diameter.

2. The dimensions specifying the area within the dashed lines are the minimum dimensions for the mounting surface. The corners of the rectangle may be radiused to a maximum radius equal to the thread diameter of the fixing bolts.

   Along each axis the fixing holes are at equal distance to the mounting surface edges.

3. This dimension gives the minimum spacing distance between the valve and adjacent obstructions, for example, another valve or a wall. This dimension is, therefore, the minimum distance from centerline to centerline of two similar mounting surfaces placed on a manifold block. The fixing holes are at equal distances to this dimension.

   The valve manufacturer's attention is drawn to the fact that no part of the width of the complete valve assembly is to exceed this dimension.

4. Blind hole in the mounting surface to **accommodate** the locating pin on the valve. The minimum depth is 8/.31.

5. The maximum limit of the working pressure for subplates and manifold blocks with this mounting surface will be supplied by the manufacturer.

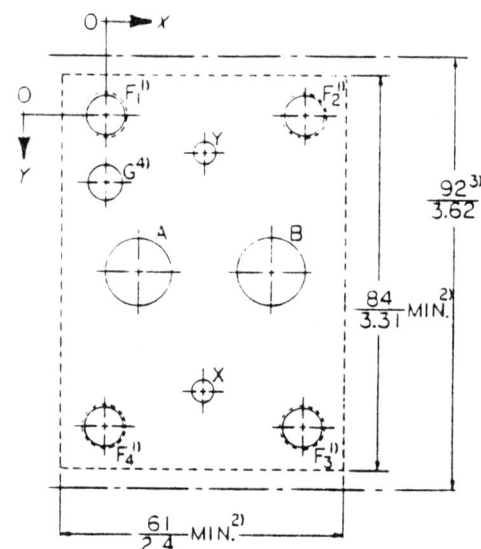

Figure 29 - **RP06 mounting surface for pilot operated pressure relief valves with main ports of 14.7/.58 maximum port diameter.**

Table 29 - **RP06 mounting surface for pilot operated pressure relief valves with main ports of 14.7/.58 maximum port diameter.**[5]

|   | A | B | X | Y | G | $F_1$ | $F_2$ | $F_3$ | $F_4$ |
|---|---|---|---|---|---|---|---|---|---|
| X | 7.1 / .28 | 35.7 / 1.406 | 21.4 / .843 | 21.4 / .843 | 0 / 0 | 0 / 0 | 42.9 / 1.688 | 42.9 / 1.688 | 0 / 0 |
| Y | 33.3 / 1.312 | 33.3 / 1.312 | 58.7 / 2.312 | 7.9 / .312 | 14.3 / .563 | 0 / 0 | 0 / 0 | 66.7 / 2.625 | 66.7 / 2.625 |
| $\phi$ | 14.7 / .578 Max. | 14.7 / .578 Max. | 4.8 / .19 Max. | 4.8 / .19 Max. | 7.1 / .28 | M10 / .375 | M10 / .375 | M10 / .375 | M10 / .375 |

1. The minimum thread depth is 1.5 bolt diameter. The recommended engagement of fixing bolt thread for ferrous mounting is 1.25 bolt diameter.

2. The dimensions specifying the area within the dashed lines are the minimum dimensions for the mounting surface. The corners of the rectangle may be radiused to a maximum radius equal to the thread diameter of the fixing bolts.

   Along each axis the fixing holes are at equal distance to the mounting surface edges.

3. This dimension gives the minimum spacing distance between the valve and adjacent obstructions, for example, another valve or a wall. This dimension is, therefore, the minimum distance from centerline to centerline of two similar mounting surfaces placed on a manifold block. The fixing holes are at equal distances to this dimension.

   The valve manufacturer's attention is drawn to the fact that no part of the width of the complete valve assembly is to exceed this dimension.

4. Blind hole in the mounting surface to accommodate the locating pin on the valve. The minimum depth is 8/.31.

5. The maximum limit of the working pressure for subplates and manifold blocks with this mounting surface will be supplied by the manufacturer.

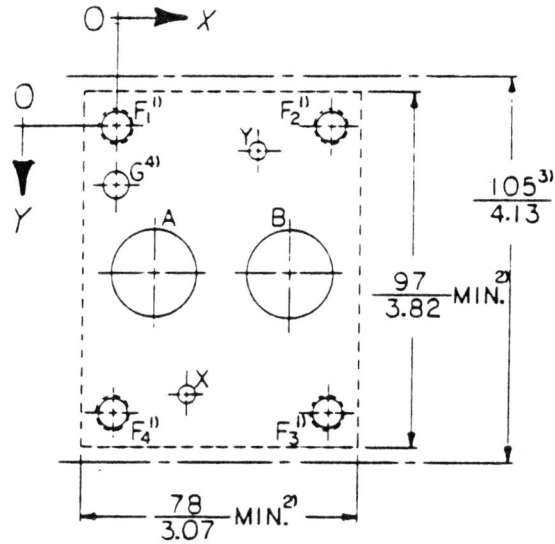

Figure 30 - **RP08 mounting surface for pilot operated pressure relief valves with main ports of 23.4/.92 maximum port diameter.**

Table 30 - **RP08 mounting surface for pilot operated pressure relief valves with main ports of 23.4/.92 maximum port diameter.**[5]

|   | A | B | X | Y | G | $F_1$ | $F_2$ | $F_3$ | $F_4$ |
|---|---|---|---|---|---|---|---|---|---|
| X | 11.1 / .438 | 49.2 / 1.938 | 20.6 / .812 | 39.7 / 1.563 | 0 / 0 | 0 / 0 | 60.3 / 2.375 | 60.3 / 2.375 | 0 / 0 |
| Y | 39.7 / 1.563 | 39.7 / 1.563 | 73 / 2.875 | 6.4 / .25 | 15.9 / .625 | 0 / 0 | 0 / 0 | 79.4 / 3.125 | 79.4 / 3.125 |
| $\phi$ | 23.4 / .92 Max. | 23.4 / .92 Max. | 4.8 / .19 Max. | 4.8 / .19 Max. | 7.1 / .28 | M10 / .375 | M10 / .375 | M10 / .375 | M10 / .375 |

1. The minimum thread depth is 1.5 bolt diameter. The recommended engagement of fixing bolt thread for ferrous mounting is 1.25 bolt diameter.

2. The dimensions specifying the area within the dashed lines are the minimum dimensions for the mounting surface. The corners of the rectangle may be radiused to a maximum radius equal to the thread diameter of the fixing bolts.

   Along each axis the fixing holes are at equal distance to the mounting surface edges.

3. This dimension gives the minimum spacing distance between the valve and adjacent obstructions, for example, another valve or a wall. This dimension is, therefore, the minimum distance from centerline to centerline of two similar mounting surfaces placed on a manifold block. The fixing holes are at equal distances to this dimension.

   The valve manufacturer's attention is drawn to the fact that no part of the width of the complete valve assembly is to exceed this dimension.

4. Blind hole in the mounting surface to accommodate the locating pin on the valve. The minimum depth is 8/.31.

5. The maximum limit of the working pressure for subplates and manifold blocks with this mounting surface will be supplied by the manufacturer.

# ANSI/B93.7M

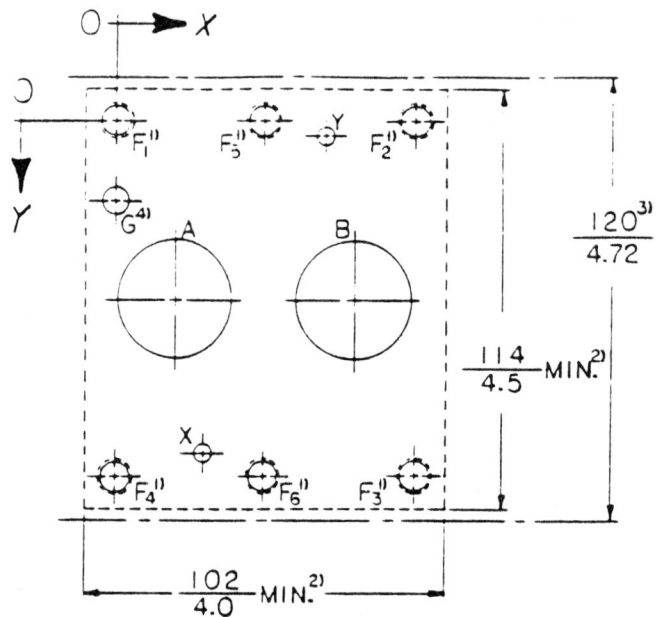

Figure 31 - **RP10** mounting surface for pilot operated pressure relief valves with main ports of 32/1.26 maximum port diameter.

Figure 31 - **RP10** mounting surface for pilot operated pressure relief valves with main ports of 32/1.26 maximum port diameter.[5]

|   | A | B | X | Y | G | $F_1$ | $F_2$ | $F_3$ | $F_4$ | $F_5$ | $F_6$ |
|---|---|---|---|---|---|---|---|---|---|---|---|
| X | 16.7 / .656 | 67.5 / 2.656 | 24.6 / .968 | 59.6 / 2.344 | 0 / 0 | 0 / 0 | 84.1 / 3.312 | 84.1 / 3.312 | 0 / 0 | 42.1 / 1.656 | 42.1 / 1.656 |
| Y | 48.4 / 1.906 | 48.4 / 1.906 | 92.9 / 3.656 | 4 / .156 | 21.4 / .844 | 0 / 0 | 0 / 0 | 96.8 / 3.812 | 96.8 / 3.812 | 0 / 0 | 96.8 / 3.812 |
| $\phi$ | 32 / 1.26 Max. | 32 / 1.26 Max. | 4.8 / .19 Max. | 4.8 / .19 Max. | 7.1 / .28 | M10 / .375 | M10 / .375 | M10 / .375 | M10 / .375 | M10 / .375 | M10 / .375 |

1. The minimum thread depth is 1.5 bolt diameter. The recommended engagement of fixing bolt thread for ferrous mounting is 1.25 bolt diameter.

2. The dimensions specifying the area within the dashed lines are the minimum dimensions for the mounting surface. The corners of the rectangle may be radiused to a maximum radius equal to the thread diameter of the fixing bolts.

   Along each axis the fixing holes are at equal distance to the mounting surface edges.

3. This dimension gives the minimum spacing distance between the valve and adjacent obstructions, for example, another valve or a wall. This dimension is, therefore, the minimum distance from centerline to centerline of two similar mounting surfaces placed on a manifold block. The fixing holes are at equal distances to this dimension.

   The valve manufacturer's attention is drawn to the fact that no part of the width of the complete valve assembly is to exceed this dimension.

4. Blind hole in the mounting surface to accommodate the locating pin on the valve. The minimum depth is 8/.31.

5. The maximum limit of the working pressure for subplates and manifold blocks with this mounting surface will be supplied by the manufacturer.

ANSI/B93.7M

## Recommended uses of the ports for various types of pressure relief valves.

Examples of symbols for direct operated pressure relief valves with main ports of 6.3/.25 maximum diameter, refer to figure 25.

| Description | External drain | Internal drain |
|---|---|---|
| **Pressure relief valve** | | |
| **Pressure relief valve with bypass check valve** | | |

Examples of symbols for pilot operated pressure relief valves with main ports of 6.3/.25 maximum diameter, refer to figure 25.

| Description | External drain | Internal drain |
|---|---|---|
| Pressure relief valve | | |
| Pressure relief valve with bypass check valve | | |

1. Port for remote controlled pilot; may be blocked if not needed.

# ANSI/B93.7M

**Recommended uses of the ports for various types of pressure relief valves.**

Examples of symbols for pilot operated pressure relief valves with main ports of 14.7/.58, 23.4/.92, 32/1.25 maximum diameter, refer to figures 26, 27 and 28.

| Description | Internal drain |
|---|---|
| Pressure relief valve | 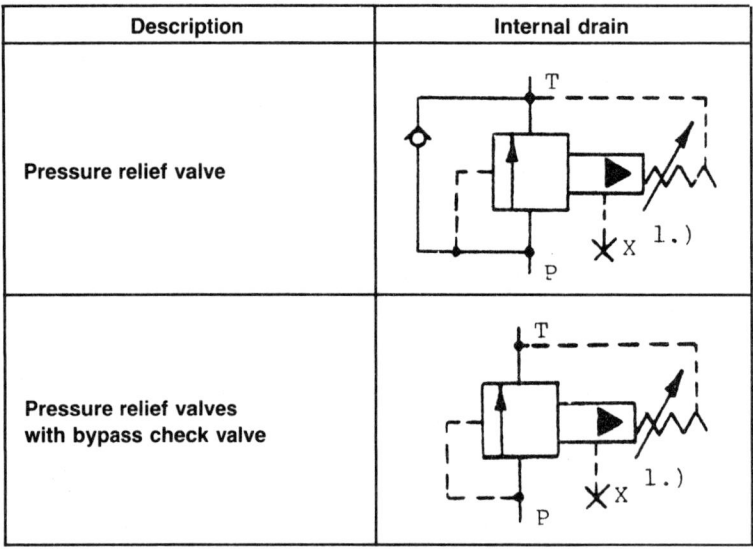 |
| Pressure relief valves with bypass check valve | |

1. Port for remote controlled pilot; may be blocked if not needed.

Examples of symbols for pilot operated pressure relief valves with main ports of 14.7/.58, 23.4/.92, 32/1.25 maximum diameter, refer to figures 29, 30 and 31.

| Description | External drain | Internal drain |
|---|---|---|
| Pressure relief valve | | |
| Pressure relief valve with bypass check valve | | |

1. Port for remote controlled pilot; may be blocked if not needed.

# ANSI/B93.7M

# APPENDIX
## to ANSI/B93.7M

This is not part of the ANSI/B93.7M Standard, but is included for information purposes only.

## 2 REFERENCES

NFPA/T3.5.34M, *National Fluid Power Association Recommended Standard Hydraulic fluid power - Code for identification of valve mounting surfaces.*

NFPA/T3.5.39M, *National Fluid Power Association Recommended Standard Hydraulic fluid power - Pressure Control valves (excluding pressure relief valves) sequence valves, throttle valves and check valves - Mounting surfaces.*

NFPA/T3.5.40M, *National Fluid Power Association Recommended Standard Hydraulic fluid power - Mounting surfaces for hydraulic flow compensated flow control valves.*

ISO 1302, *International Standard Technical drawings - Method of indicating surface texture on drawings.*

ISO 4401, *International Standard Hydraulic fluid power - Four-port directional control valves - Mounting surfaces.*

ISO 5598, *International Standard Fluid power systems and components - Vocabulary.*

ISO/DIS 5781[1], *International Standard Hydraulic fluid power - Pressure control valves (excluding pressure relief valves), sequence valves, throttle valves and check valves - Mounting surfaces.*

ISO 5783, *International Standard Hydraulic fluid power - Code for identification of valve mounting surfaces.*

ISO/DIS 6263[1], *International Standard hydraulic fluid power - Compensated flow control valves.*

ISO/DIS 6264[1], *International Standard Hydraulic fluid power - Pressure relief valves - Mounting surfaces.*

---

[1] Presently at the stage of draft.

ANSI/B93.9M-1969 (R1981)

## AN INDUSTRY STANDARD FOR FLUID POWER

American National Standard

Symbols for Marking

Electrical Leads and Ports

on Fluid Power Valves

(NFPA/T3.5.2M-1968)

Approved as an ANSI Standard
5 September 1969

published by

**NATIONAL FLUID POWER ASSOCIATION, INC.**

3333 N. Mayfair Road  /  Milwaukee, WI 53222  /  414-778-3344  /  TLX 26898

# FOREWORD

(This Foreword is not a part of American National Standard Symbols for Marking Electrical Leads and Ports on Fluid Power Valves, ANSI B93.9-1969.)

A firm foundation for valve port marking was laid late in 1963. At that time a survey on random sampling of current valves disclosed both harmony and discord in the way their ports were marked. Sometimes the actual component markings differed from those in the catalog description. The survey disclosed that valve manufacturers favored

- ... uniform port markings for both pneumatic and hydraulic valves

- ... one identification system rather than a preferred system with alternates

- ... identifying solenoids and leads in terms of the ports they control.

The results of the survey formed the basis for active committee work. Proposals from the valve component section of the NFPA Technical Board were submitted and given project status in January, 1965.

In January, 1966, the Symbols and Drafting Standards Coordinating Committee of the Technical Board was assigned the task of developing port markings that would be uniform for all fluid power components. Under the chairmanship of George Goepfrich, Skinner Precision Industries, Inc. and edited by J. L. Fisher, Jr., Bellows-Valvair, the first draft of this document was reviewed by the project group at its meeting in February, 1967. The second draft was reviewed and edited by the project group at its meeting in March, 1967.

The third draft was reviewed, edited, and labeled "Participant Consensus Draft" by the project group at its meetings in May and October, 1967. The document reached "Participant Consensus Draft" status on October 2, 1967.

After distribution for written comments, and their review at two meetings, a Summary Consensus Draft was reached on 15 July 1968. The Scope of the Summary Consensus Draft was narrowed to exclude actuators such as levers and pedals thereby satisfying all existing objections. Approval to ballot was granted on 24 July 1968.

Favorable ballot of both the Hydraulic Valve and Pneumatic Valve Sections of the NFPA resulted in the Proposed Standard being approved as a NFPA Recommended Standard by the Board of Directors of the Association on 10 November 1968.

ANSI/B93.9

## MEMBERS OF THE NFPA PROJECT GROUP RESPONSIBLE FOR THE DEVELOPMENT OF THIS STANDARD

| | |
|---|---|
| Bower, Allen - Section Chairman | Fluid Controls, Inc. |
| Peterson, Robert - Section Chairman (After 7-1-68) | Racine Hydraulics, Inc. |
| Goepfrich, George - Project Chairman | Skinner Precision Industries, Inc. |
| Morgan, James I. - Secretary | NFPA |
| Aslan, Wilfred | Alkon Products Corp. |
| Beckett, William N. | Beckett-Harcum Co. |
| Bowman, Donald L. | H. P. M. Division |
| Brown, James O. | Barksdale Valves |
| Clark, Richard J. | Racine Hydraulics, Inc. |
| Dickey, R. R. | Ross Operating Valve Co. |
| Doig, George | Numatics, Inc. |
| Fisher, J. L., Jr. | Bellows-Valvair |
| Forster, W. R. | Westinghouse Air Brake Co. |
| Green, R. H. | Denison Division |
| Greenwood, Martin | Alkon Products Corp. |
| Harte, Robert J. | Vickers Division |
| Hoffman, R. K. | C. A. Norgren Co. |
| Hunley, William L. | Tektro Fluid Power |
| Huntington, A. B. | Ross Operating Valve Co. |
| Kreider, W. | H. P. M. Division |
| Larsen, Borge E. | Allied Control Co., Inc. |
| Lear, Oliver G. | Sarasota Precision Products, Inc. |
| Mills, Nathan | Vickers Division |
| Olson, Paul | Westinghouse Air Brake Co. |
| Russell, J. G. | Parker Hannifin Corp. |
| Schilling, Wm. | Allied Control Co., Inc. |
| Sweeney, D. R. | Vickers Division |
| Spielman, C. R. | C. A. Norgren Co. |
| Wilkes, Roy | Miller Fluid Power Division |
| Whitmore, C. | Parker Hannifin Corp. |
| Winquist, R. | Parker Hannifin Corp. |

On 6 January 1969 the NFPA Recommended Standard was submitted, for promulgation as a USA Standard, to USASI Committee B93, Fluid Power Systems and Components. The NFPA and SAE are co-administrative sponsors of B93. Approval of Committee B93 was obtained on 3 March 1969.

On 3 March 1969, Standards Committee B93 was comprised of the following:

J. J. Pippenger, Chairman; Otto J. Maha, Vice Chairman; James I. Morgan, Co-Secretary; John G. Lippert, Co-Secretary.

| ORGANIZATION | REPRESENTATIVE | ALTERNATE |
|---|---|---|
| American Society of Agricultural Engineers | E. H. Fletcher | |

ANSI/B93.9

| | | |
|---|---|---|
| American Society of Lubrication Engineers | Karl G. Henrikson | |
| American Society of Mechanical Engineers | Henry Parsons | |
| American Society for Testing & Materials | J. D. Lykins | W. H. Millett |
| American Society of Tool & Mfg. Engineers | Kenneth E. Booher | |
| Automobile Manufacturers Association | Reino M. Mustonen | |
| Fluid Controls Institute | William Schilling | |
| Fluid Power Society | Prof. Russell Henke<br>Melvin Long<br>Frank L. Mackin<br>Frank Yeaple<br>Max Covert<br>Tobi Goldoftas<br>Dudley Pease<br>W. R. Smith<br>Lars G. Soderholm | |
| Industrial Truck Association | C. D. Gibson | R. T. McNeely |
| Material Handling Institute | Bernard Becker | Jack McPherson |
| Mechanical Packing Association | Guy Horvath | R. G. Singer |
| National Fluid Power Association | Otto J. Maha<br>J. L. Fisher, Jr.<br>W. R. Forster<br>W. J. Kudlaty<br>Z. J. Lansky<br>J. J. Pippenger | |
| National Hose Assemblies Mfrs. Association | Robert M. Byrne | |
| National Industrial Leather Association | E. R. Rath | |
| National Machine Tool Builders Association | D. G. Stewart | E. J. Koschella |
| Power Crane and Shovel Association | Wm. M. Shook | |

ANSI/B93.9

| | | |
|---|---|---|
| Rubber Manufacturers Association | L. Cranston | N. J. Cyphers |
| Society of Automotive Engineers, Inc. | John G. Lippert | |
| U. S. Coast Guard | LCDR J. W. Kime | |
| U. S. Dept. of Defense | C. A. Nazian | Jack T. Stevenson |

Suggestions for improvement gained in the use of this standard will be welcome. They should be sent to the American National Standards Institute, 1430 Broadway, New York, N. Y. 10018.

## REFERENCES

1. ANSI Standard Glossary of Terms for Fluid Power, B93.2

2. ANSI Standard Fluid Power Diagrams, Y14.17

3. ANSI Standard Graphic Symbols for Fluid Power Diagrams, Y32.10

ANSI/B93.9

1. INTRODUCTION

   Fluid power systems are those that transmit and control power through use of a pressurized fluid (liquid or gas) within an enclosed circuit. In such circuits, valves are devices which control fluid direction, pressure, or flow rate. The simplest valve usually has an inlet port and an outlet port. Sometimes one is the other, but often the inlet must be identified and distinguished from the outlet so that piping the valve incorrectly is avoided.

   In more complex valves, the importance of port identification is magnified. The valve may be multi-purpose and multi-ported. It may be special-purpose and special-ported. Its proper use will depend on the correct identification of all of its ports. Such identification is usually carried somewhere on the actual component adjacent to the port being named.

2. SCOPE

   This recommended standard offers symbols for the identification of valve ports, for the identification of valve pilot and solenoid actuators, and for the identification of valve solenoid leads.

3. PURPOSE

   Correct identification assists in providing:

   Safety to personnel
   Long life of the equipment
   Proper function of other system components
   Proper service and maintenance
   Universal understanding of fluid power valve function

4. TERMS AND DEFINITIONS

   For terms, abbreviations, and definitions applicable to fluid power, consult Ref. 1.

5. RULES

   5.1 Identify valve pilot and solenoid actuators with the same symbol as the port which is pressurized when the control is caused or permitted to actuate.

   5.2 Identify coil leads when the solenoid from which they originate cannot be determined visually.

ANSI/B93.9

      5.2.1 When solenoid coil leads are identified, use the same symbol as the port which is pressurized when the coil is energized.

  5.3 Identify common valve ports with the same symbol.

6. PORT SYMBOLS

Symbols in diagrams and on the actual component should be in agreement. See Ref. 2, Section 17-6.3.9.

  6.1 Arrow symbols are sufficient to indicate the direction of flow through valves containing two ports.

      6.1.1 A single arrow ( ⟶ ) is sufficient to indicate the direction of preferred flow.

      6.1.2 A single arrow ( ⟶ ) is sufficient to indicate the direction of free flow.

      6.1.3 A double headed arrow ( ⟵⟶ ) is sufficient to indicate flow when direction of flow is not significant.

  6.2 A and B are symbols identifying the working ports of valves when there are no more than two such ports.

      6.2.1 The symbol K identifies the working ports of valves which have three or more such ports. K shall always appear with a suffix digit symbol K1, K2, etc.

  6.3 The symbol D identifies a hydraulic <u>drain</u> port which usually is connected to an unrestricted line. The symbol D also identifies a pneumatic exhaust port which must not be subjected to restriction or pressure.

  6.4 The symbol E identifies the pneumatic <u>outlet</u> port that usually supplies a passage to atmosphere. (E on pneumatic valves corresponds to T on hydraulic valves.)

      6.4.1 If the valve so marked is used optionally with hydraulics the E port will correspond to T.

      6.4.2 E may appear with a suffix symbol. (EA, EB, E1) The suffix symbol takes the same identity

ANSI/B93.9

        as the working port it exhausts.

6.5   The symbol F identifies the controlled <u>flow</u> port of a flow control valve.

6.6   The symbol P identifies the <u>inlet</u> port of all valves (exception, see 6.1).

    6.6.1   P may appear with a suffix symbol (PA, PB, P1) The suffix symbol takes the same identity as the working port it pressurizes.

6.7   The symbol T identifies the <u>outlet</u> port usually connected to tank. The symbol T is used only with valves for hydraulic service.

    6.7.1   If the valve so marked is used optionally with pneumatics, the T port will correspond to E.

    6.7.2   T may appear with a suffix symbol (TA, TB, T1) The suffix symbol takes the same identity as the working port which it serves.

6.8   The symbol X:

    6.8.1   In pneumatic valves, the symbol X identifies an auxiliary port whose function may be one of the following: (Manufacturer's literature will designate specific use)

        6.8.1.1   A pneumatic pilot pressure supply port

        6.8.1.2   A pneumatic pilot control port (See option in par. 6.10)

    6.8.2   In hydraulic valves, the symbol X identifies an auxiliary port whose function may be one of the following: (Manufacturer's literature will designate specific use)

        6.8.2.1   A hydraulic pilot drain port

        6.8.2.2   A hydraulic pilot pressure supply port

        6.8.2.3   A hydraulic pilot control port

6.8.3 X may appear with a suffix symbol (XA, XB, X1) The suffix symbol takes the same identity as the working port pressurized when the pilot is caused or permitted to actuate.

6.9 The symbol Y:

6.9.1 In pneumatic valves, the symbol Y identifies an auxiliary port whose function may be one of the following: (Manufacturer's literature will designate specific use)

6.9.1.1 A pneumatic pilot pressure supply port

6.9.1.2 A pneumatic pilot control port (See option in par. 6.10)

6.9.2 In hydraulic valves, the symbol Y identifies an auxiliary port whose function may be one of the following: (Manufacturer's literature will designate specific use)

6.9.2.1 A hydraulic pilot drain port

6.9.2.2 A hydraulic pilot pressure supply port

6.9.2.3 A hydraulic pilot control port

6.9.3 Y may appear with a suffix symbol (YA, YB, Y1) The suffix symbol takes the same identity as the working port pressurized when the pilot is caused or permitted to actuate.

6.10 The symbol C plus a suffix symbol (CA, CB, C1) identifies a control port into which a pneumatic signal may be introduced to cause valve actuation. The suffix letter takes the same identity as the symbol for the working port which is pressurized when the control signal is given. (See also par. 6.8.1)

6.11 The symbol V identifies a control port which, when vented to a lower reference pressure, causes valve actuation.

6.11.1 V may appear with a suffix symbol (VA, VB, V1) The suffix symbol takes the same identity as the working port which is pressurized when V is vented.

**ANSI/B93.9**

6.12 The number symbols, 1, 2, 3, etc., identify the ports which are not otherwise described. (Manufacturer's literature will designate specific use)

7. EXAMPLES
(Symbols used will be found in Ref. 3)

7.1 Outlet to Atmosphere

7.1.1 Plain Orifice (Unthreaded)

7.1.2 Connectable Orifice (Threaded)

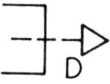 

7.1.3 Drain

7.2 Directional Control Valves

7.2.1 Check

7.2.2 Two-Way

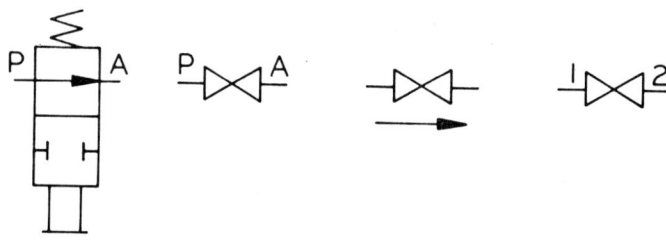

7.2.3 Three-Way

7.2.3.1 Normally Open

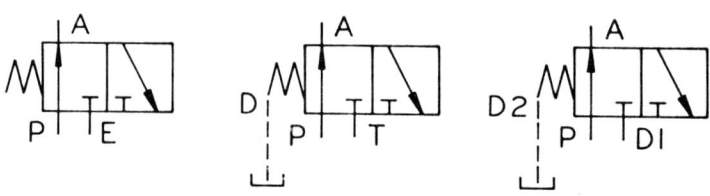

#### 7.2.3.2 Normally Closed

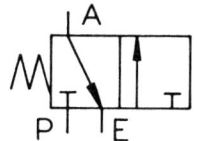 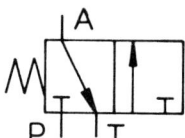 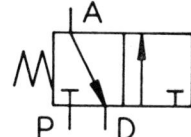

#### 7.2.3.3 Diverter

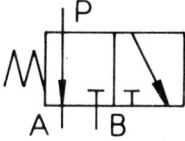

#### 7.2.3.4 Two-Pressure

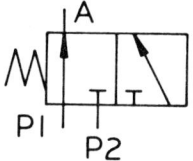

#### 7.2.3.5 Multipurpose

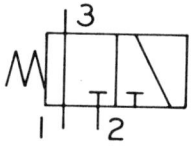

### 7.2.4 Four-Way

#### 7.2.4.1 Four Ports

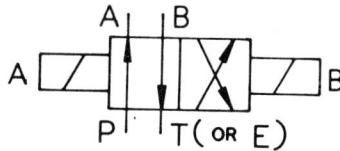 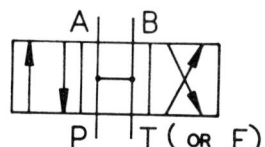

#### 7.2.4.2 Five Ports

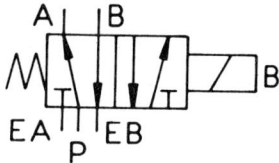 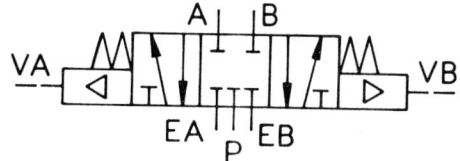

#### 7.2.4.3 Hydraulic Valve with Pneumatic Pilot (EX S & E)

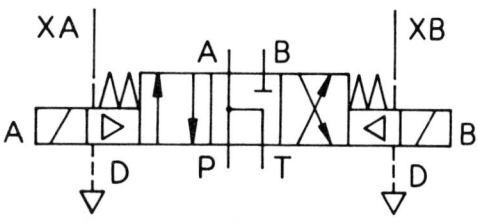

## 7.2.4.4 Pneumatic Valve with Pilot Supply and Control Supply

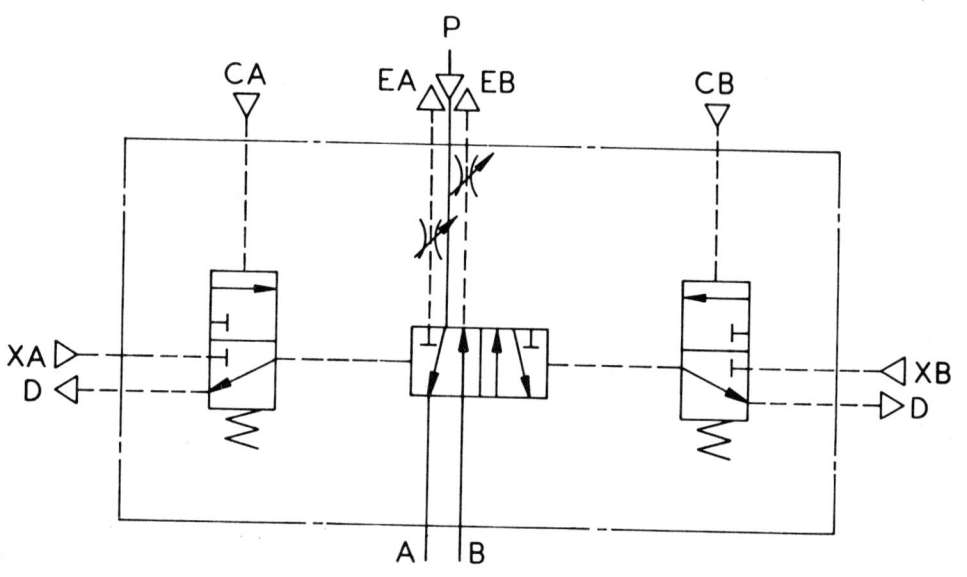

## 7.2.5 Multi Directional Multi Purpose

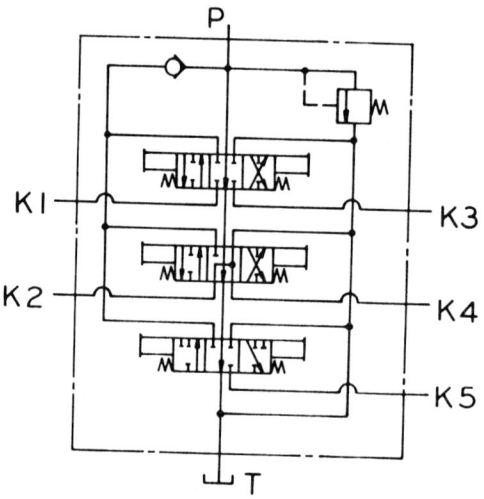

## 7.3 Flow Control Valves

### 7.3.1 Fixed Restriction

### 7.3.2 Adjustable Restriction

### 7.3.3 Adjustable Restriction with Bypass

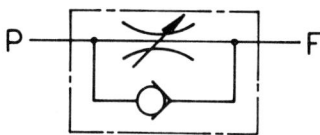

### 7.3.4 Adjustable, Pressure Compensated with Overload Relief

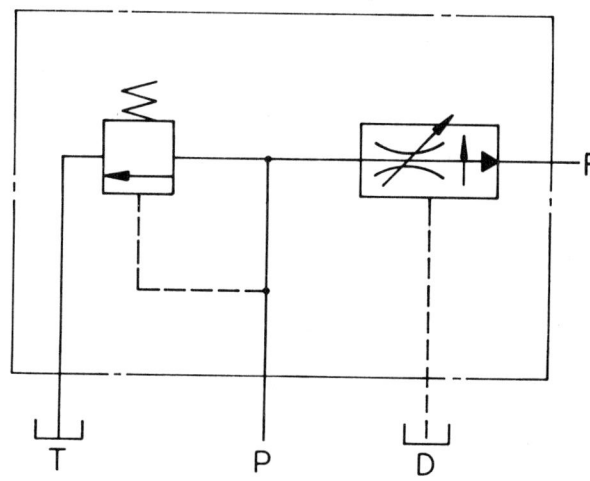

## 7.4 Pressure Control Valves

### 7.4.1 Relief

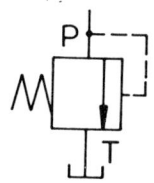 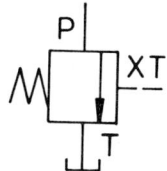

### 7.4.2 Reducing

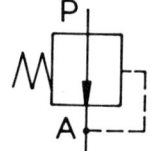 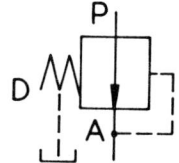

7.4.3  Reducing and Relieving (Adjustable)

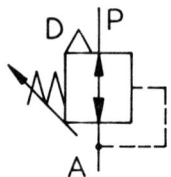

7.4.4  Relief, Balanced Type

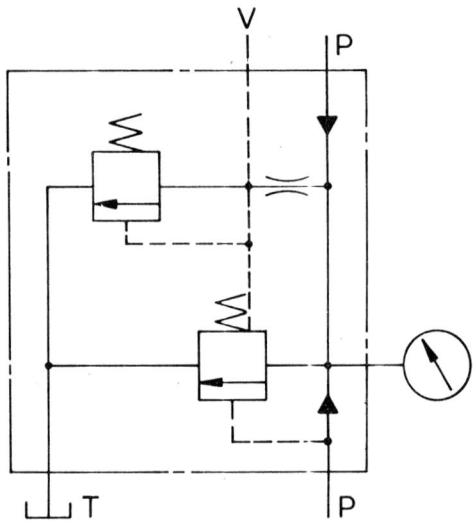

7.4.5  Sequence

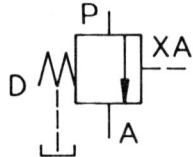

7.4.6  Counterbalance

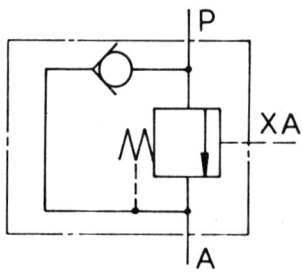

National **FLUID POWER** Association

ANSI/B93.40M-1976 (R1982)

## AN INDUSTRY STANDARD FOR FLUID POWER

American National Standard

Series of Mounting Interfaces

For 4567 Maximum psi (315 bar) Four-Port

Hydraulic Fluid Power Directional Valves

(NFPA/T3.5.9M-1973)

Approved as an ANSI Standard
28 May 1976

published by
**NATIONAL FLUID POWER ASSOCIATION, INC.**

3333 N. Mayfair Road  /  Milwaukee, WI 53222  /  414-778-3344  /  TLX 26898

## FOREWORD

This Foreword is not part of American National Standard Series of Mounting Interfaces for 4567 Maximum psi (315 bar) Four Port Hydraulic Fluid Power Directional Valves, ANSI/B93.40-1976.

This standard, which began as NFPA project T3.5.9, was approved by the NFPA Technical Board on 5 September 1969. It is somewhat similar to ANSI/B93.7-1968 (R1973) (NFPA/T3.5.1-1965), Mounting Surfaces of Sub-Plate Type Hydraulic Fluid Power Valves, except that it contains standard mounting surfaces for directional control valves used at the higher pressures of 4567 psi (315 bar).

A request to submit patterns was sent to the NFPA Valve Section members thru NFPA Headquarters on 5 January 1970. As a result of this survey, patterns were submitted by Denison, Vickers and Racine Hydraulics. A later submittal was received from Teledyne Republic Manufacturing on 9 December 1970.

Draft No. 1 was written on 30 January 1971. That draft was reviewed and revised at the Section's 16 March 1971 meeting. The Section reached consensus on 27 October 1971 and the General Review Draft was prepared on 2 March 1972.

The comments received in the General Review were resolved at the Section's 24 October 1972 meeting. The Ballot Draft was prepared on 19 December 1972. The balloting closed on 12 February 1973 and the Technical Board recommended approval of the document on 23 May 1973. The document was submitted to the NFPA Board of Directors and approved as NFPA Recommended Standard T3.5.9-1973 on 19 July 1973.

Members of the NFPA Project Group which developed this standard are listed on page 4.

On 16 August 1973, the NFPA Standard was submitted to ANSI Standards Committee B93 for promulgation as an ANSI Standard. Ballot was concluded on 26 October 1973, and resulted in one negative vote. Changes made to the standard caused the negative vote to be withdrawn on 1 October 1975. Approval by the ANSI Board of Standards Review was granted on 28 May 1976.

ANSI/B93.40

Members of the NFPA Project Group responsible for the development of this standard included:

| Name | Role | Organization |
|---|---|---|
| Anderson, Gerald | Project Chairman (1971 - present) | Applied Power Inc. |
| Clark, Thomas | Project Chairman (1970 - 1971) | Rexnord Inc. |
| Clark, Richard | Project Chairman (1970) | Rexnord Inc. |
| Olen, Robert | Section Chairman (1976 - present) | DeLaval Turbine Inc. |
| Mills, Nathan | Section Chairman (1971 - 1975) | Sperry Vickers |
| Peterson, Robert | Section Chairman (Prior to 1971) | Rexnord Inc. |
| Krehbiel, Rob | Section Vice Chairman (1971 to 1975) | Cessna Fluid Power Div. |
| Patel, Ratee | Secretary | U.S. Army - MERDC |
| Luecke, John R. | Director of National Technical Services | National Fluid Power Association |

| | | | | |
|---|---|---|---|---|
| Bowman, R. | Continental Hydraulics | | Martin, M. | Abex Corporation |
| Graber, H. | Cessna Fluid Power | | Mazur, J. | Linde Hydraulics |
| Grassl, R. | Galland Henning Nopak | | McKenna, L. | Bellows Int'l. |
| Hedge, J. | Abex Corporation | | Ostle, C. | Teledyne Republic |
| Jacoby, H. | Webster Electric Co. | | Smith, W. | Teledyne Republic |
| Keir, J. | Sperry Vickers | | Thomas, V. | Tomco, Inc. |
| Lindner, P. | HUSCO | | Zajac, T. | Parker Hannifin |

On 26 October 1973, ANSI Standards Committee B93 was composed of the following:
O. J. Maha, Chairman; James I. Morgan, Co-Secretary; R. Thomas Northrup, Co-Secretary.

AMERICAN SOCIETY OF AGRICULTURAL ENGINEERS
  E. H. Fletcher

AMERICAN SOCIETY OF LUBRICATION ENGINEERS
  H. Kaufman

AMERICAN SOCIETY OF MECHANICAL ENGINEERS
  H. Parsons
  T. R. Curran (Alternate)

AMERICAN SOCIETY FOR TESTING AND MATERIALS
  J. D. Lykins

AUTOMOBILE MANUFACTURERS ASSOCIATION
  R. Mustonen

CONSTRUCTION INDUSTRY MANUFACTURERS ASSOCIATION
  G. Stewart
  H. T. Larmore (Alternate)

FLUID CONTROLS INSTITUTE
  A. W. Churchill
  E. A. Bianchi (Alternate)

FLUID POWER SOCIETY
  T. Goldoftas
  Professor R. Henke
  M. E. Long
  F. L. Mackin
  W. R. Smith
  L. G. Soderholm
  F. Yeaple, Jr.

FLUID SEALING ASSOCIATION
  W. Krucke

INDUSTRIAL TRUCK ASSOCIATION
  C. D. Gibson
  R. T. McNeely (Alternate)

INSTRUMENT SOCIETY OF AMERICA
  A. I. Kutz

JOINT INDUSTRY COUNCIL
  R. Muhl

MATERIAL HANDLING INSTITUTE
  J. McPherson
  W. Chichester (Alternate)

NATIONAL FLUID POWER ASSOCIATION
  O. J. Maha
  J. L. Fisher, Jr.
  W. R. Forster
  W. J. Kudlaty
  Z. J. Lansky
  J. J. Pippenger

NATIONAL INDUSTRIAL LEATHER ASSOCIATION
  E. R. Rath

NATIONAL MACHINE TOOL BUILDERS ASSOCIATION
  J. I. Ehrhardt
  E. Loeffler (Alternate)

POWER CRANE AND SHOVEL ASSOCIATION
  W. M. Shook

RUBBER MANUFACTURERS ASSOCIATION
  W. J. Atwell
  N. J. Cyphers
  A. J. Jeffcott (Alternate)

SOCIETY OF AUTOMOTIVE ENGINEERS
  W. A. Hertel
  R. E. Lyons
  E. L. Falendysz
  J. E. Wieschel (Alternate)

SOCIETY OF MANUFACTURING ENGINEERS
  R. E. Willette

U. S. COAST GUARD
  LCDR G. A. Casimir

U. S. DEPARTMENT OF DEFENSE
  C. A. Nazian
  H. Y. Smith
  P. Hopler (Alternate)

INDIVIDUAL MEMBERS
  Professor E. C. Fitch, Jr.
  J. Johnson

## REFERENCES

1. American National Standard Glossary of Terms for Fluid Power, ANSI/B93.2-1971, and Supplements thereto. (ISO/TC 131/SC 1 (USA-2) 3)
2. SI units and recommendations for the use of their multiples and of certain other units, ISO 1000-1973.
3. American National Standard Dimensions for Mounting Surfaces of Subplate Type Hydraulic Fluid Power Valves, ANSI/B93.7-1968 (R1973).
4. American National Standard Decimal Inch, ANSI/B87.1-1965.

ANSI/B93.40

# SERIES OF MOUNTING INTERFACES FOR 4567 MAXIMUM PSI (315 BAR) FOUR PORT HYDRAULIC FLUID POWER DIRECTIONAL VALVES

## INTRODUCTION

In hydraulic fluid power systems, power is transmitted and controlled thru a liquid under pressure within an enclosed circuit. Typical components found in such systems are hydraulic valves. These devices control flow direction, pressure or flow rate of liquids in the enclosed circuit.

### 1. SCOPE

To include:

1.1 Subplates — 4567 maximum psi (315 bar) hydraulic service for directional control valves.

1.2 Dimensional Criteria

    1.2.1 Minimum surface dimensions.

    1.2.2 Sizes and locations of tapped holes for mounting bolts.

    1.2.3 Sizes and locations of ports.

    1.2.4 Sizes and locations of dowel or rest pins where required.

1.3 General Criteria

    1.3.1 Surface finish and flatness.

    1.3.2 Indication of tolerances where pertinent.

    1.3.3 Indication of appropriate corner breaks and radii.

    1.3.4 Use decimal inch system per Reference No. 4.

1.4 This Standard only applies to the dimensional criteria of products manufactured in conformance with this standard. It does not apply to their functional characteristics.

### 2. PURPOSE

To provide a document which specifies the dimensions of surfaces on which 4567 maximum psi (315 bar) directional hydraulic valves are mounted in order to ensure valve interchangeability on to which pipes are connected.

### 3. TERMS AND DEFINITIONS

For definitions of terms used, see Reference No. 1.

### 4. UNITS OF MEASUREMENT

4.1 Customary US units are used.

4.2 Approximate conversions to the International System of Units are given per Reference No. 2, and appear in parentheses following their US counterparts.

### 5. GENERAL

Ports are identified by numbers as a means of reference only.

## 6. SUBPLATE IDENTIFICATION CODING

6.1 Subplate identification code prefix is in accordance with Reference No. 3 as follows:

D — Directional Control Valves

6.2 Subplate identification code suffix is in accordance with Reference No. 3 as follows:

TABLE 1 — Subplate code suffixes

| Code Number | Valve Size |
|---|---|
| 02 | ¼ in |
| 06 | ¾ in |
| 10 | 1¼ in |

## 7. TOLERANCES — BREAKS — RADII

7.1 The tolerance on two place decimals is ± 0.010 in (± 0.254 mm).

7.2 The tolerance on three place decimals is ± 0.005 in (± 0.127 mm).

7.3 Break all sharp edges.

7.4 Maximum radii is 0.01 in (0.254 mm).

## 8. IDENTIFICATION STATEMENT

Use the following statement in catalogs and sales literature when electing to comply with this voluntary standard:

"Mounting surface dimensions conform to American National Standard, ANSI/B93.40-1976."

## 9. KEY WORDS

The following Key Words, useful in indexes and in information retrieval systems, are suggested for this recommended standard:

dimensions, directional control valve

dimensions, subplate mounting surfaces

fluid power

mounting surfaces, hydraulic directional control valve

315 bar hydraulic valve

valve, directional control

valve, hydraulic

valve, subplate type

ANSI/B93.40

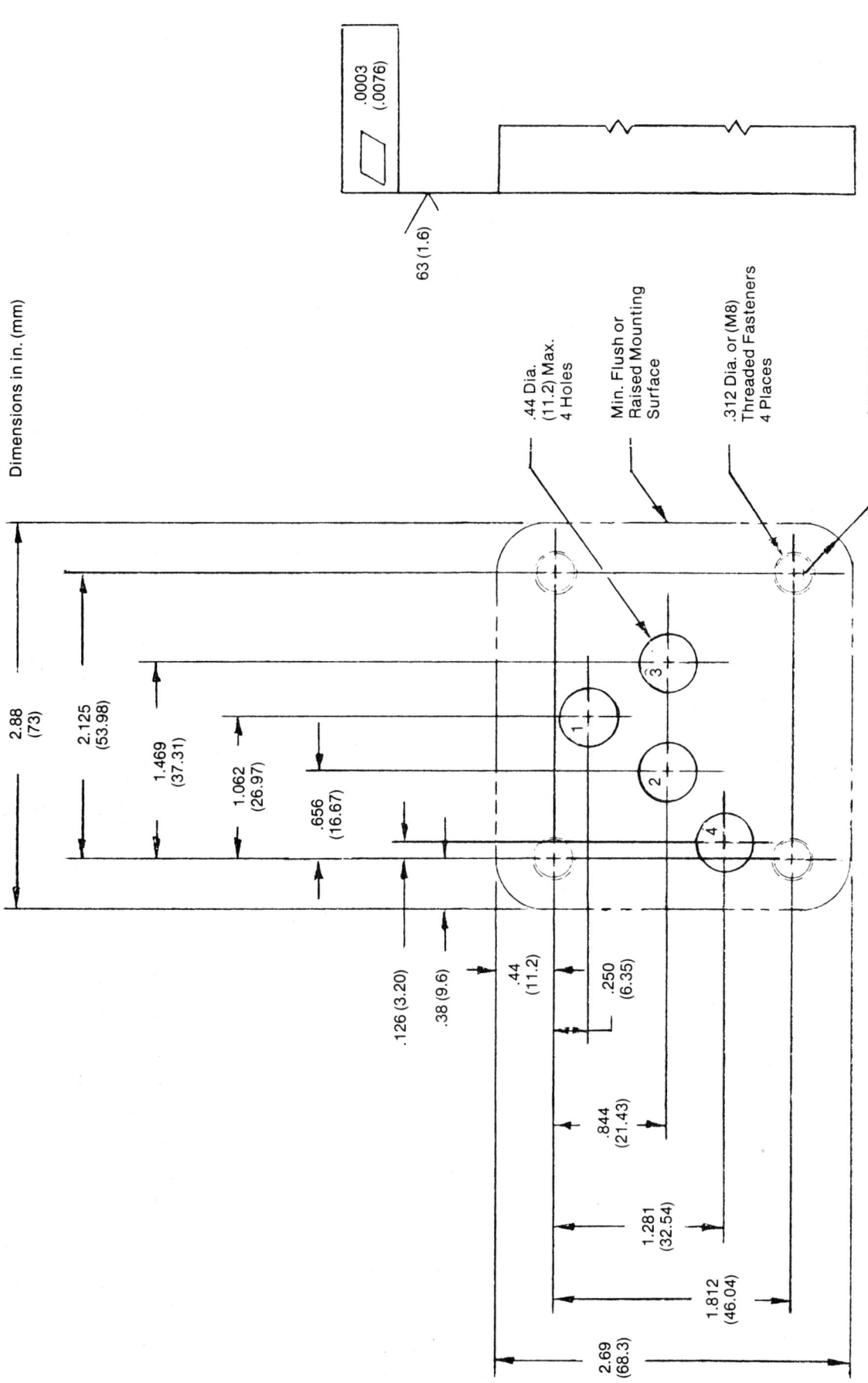

FIGURE 1 — 315 D 02 — ¼ inch directional control valve mounting surface

ANSI/B93.40

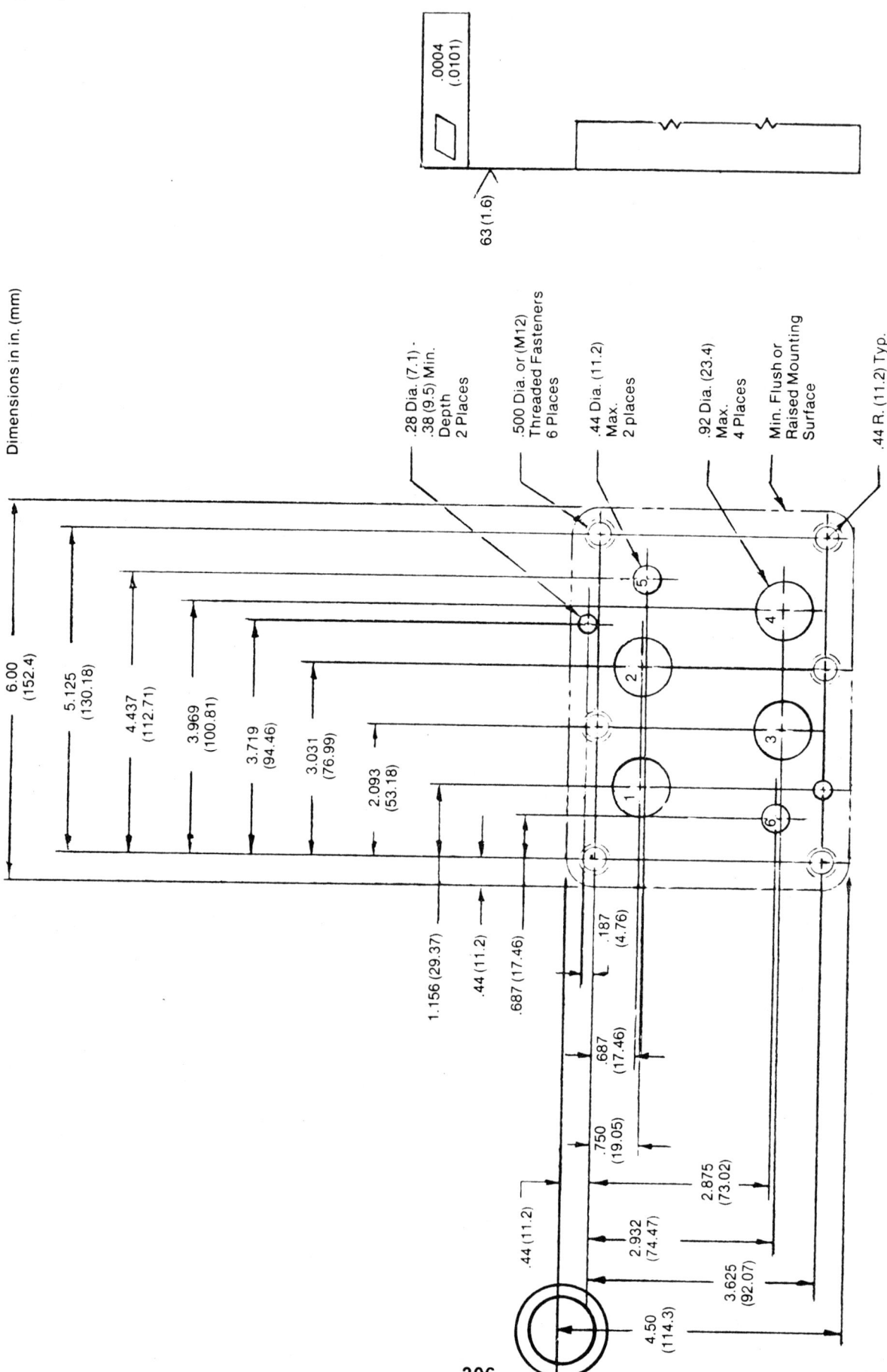

FIGURE 2 — 315 D 06 — ¾ inch directional control valve mounting surface

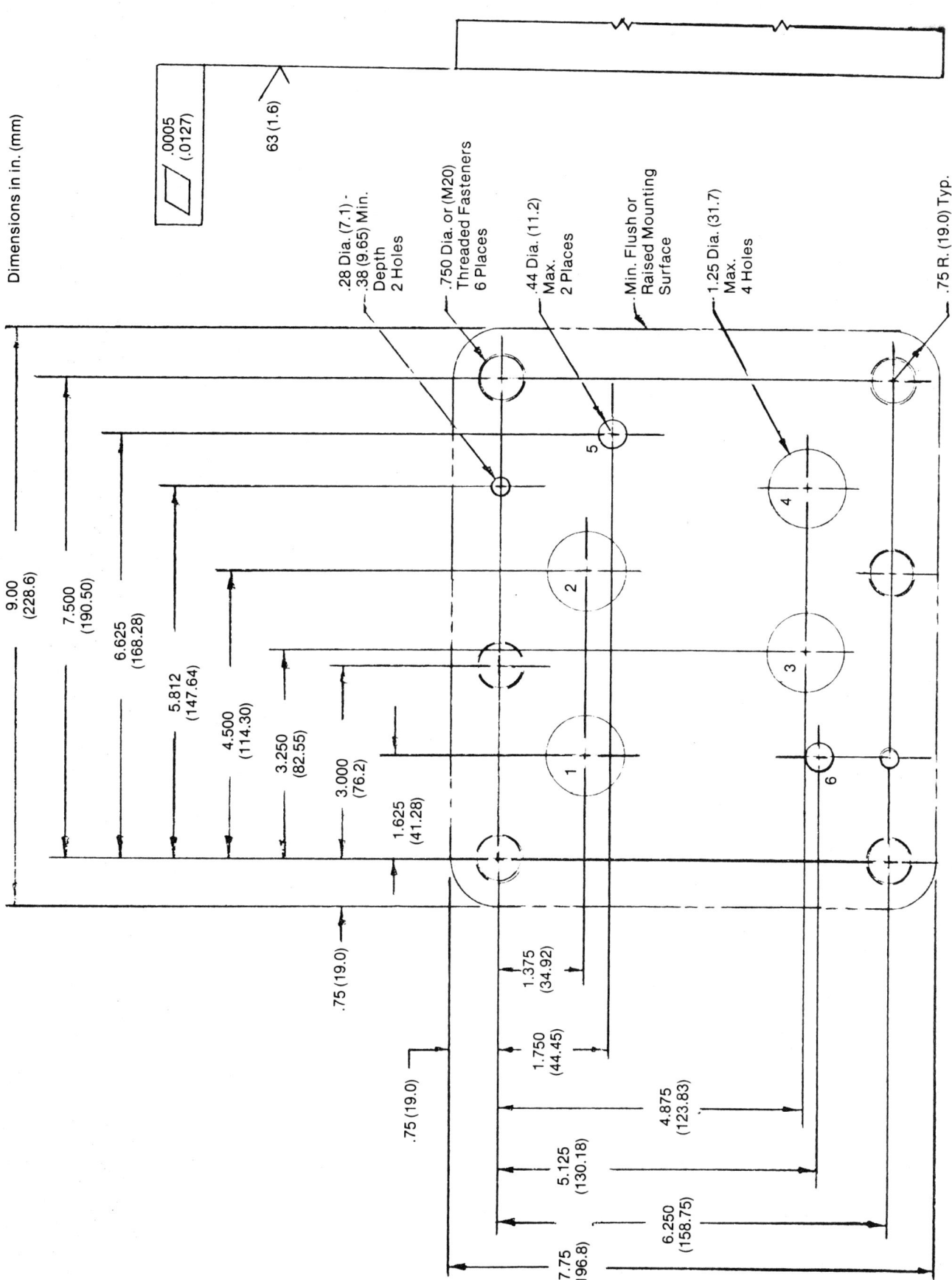

FIGURE 3 — 315 D 10 — 1¼ inch directional control valve mounting surface

NOTES

ANSI/B93.66M-1983

---

## AN INDUSTRY STANDARD FOR FLUID POWER

American National Standard

Hydraulic fluid power -

Directional control valve -

Method for determining the metering characteristics
(NFPA/T3.5.14M-1982)

Approved as an ANSI Standard
5 May 1983

**Descriptors:** control variable, displacement control variable; specified flow rate; flow rate; work port; specified pressure differential bar; pressure, load bar; pressure differential, measured bar; pressure, specified bar; pressure differential, tare bar; viscosity; aeration; filtration.

---

published by
## NATIONAL FLUID POWER ASSOCIATION, INC.

3333 N. Mayfair Road  /  Milwaukee, WI 53222  /  414-778-3344  /  TLX 26898

# ANSI/B93.66

## FOREWORD

This Foreword is not part of American National Standard Hydraulic fluid power - Directional control valve - Method for determining the metering characteristics, ANSI/B93.66M-1983. Control devices have always been an integral part of fluid power systems. The growth of the fluid power industry has precipitated many ways of defining and describing the control characteristics of fluid power valves.

As a result of the greatly expanded use of fluid power valves, it became apparent that there was a need for standard methods of testing and rating control devices which were accepted by the fluid power industry. In recognition of the need for standard test methods, a project group was established and assigned the Project Group number T3.5.14M.

In December 1971, at the request of NFPA, Mr. G. E. Maroney (Oklahoma State University) attended the NFPA Hydraulic Valve Section meeting. Mr. Maroney reported on the valve investigations being conducted at OSU for the US Army's Mobile Equipment Research and Development Command in consultation with a broadly based advisory group made up of valve users and manufacturers. Consistent with the Army's goal toward increased reliance on voluntary industrial standards, the results of the valve investigations were offered to NFPA for use in the development of voluntary industry standards. The offer was accepted.

Draft No. 1 of NFPA/T3.5.14M was prepared by OSU for review by the T3.5 Section. The Section reviewed Draft No. 1 on 7-8 December 1971. Comments made at that meeting were incorporated into the draft and Draft No. 2 was developed. The T3.5 Section reached consensus on Draft No. 2 on 22 March 1972.

A Section Review Draft was prepared on 26 June 1972 and the General Review Draft was completed on 27 July 1972. On the basis that the Headquarters Staff would incorporate the General Review Comments into the Ballot Draft, the Technical Board authorized ballot on 4 October 1973. The Ballot Draft was completed by Headquarters Staff on 23 January 1974.

The Ballot, which closed 6 March 1974, resulted in three negative votes. The comments accompanying two of the negative votes were reviewed, resolved, and incorporated into the document on 20 January 1974.

On 21 August 1975, the Technical Board recinded its recommendation for final approval, due to one unresolved negative ballot regarding SAE ARP 24 B. Headquarters Staff prepared the Second General Review on 5 June 1978.

The Second General Review closed 17 July 1978 and resulted with 10 comments. Project Group Chairman, Jack Curnow (Sperry Vickers) responded to Commentators. As a result, several of the comments were incorporated into the document.

The Technical Board granted approval to Ballot on 7 February 1979. T3.5.14M was forwarded to Headquarters. The NFPA Headquarters Staff prepared the Second Ballot on 15 February 1979.

As a result of the Second Ballot, 6 negative ballots were received and discussed at the 9 May 1981, T3.5 Section meeting.

Replies to the comments were sent out and 2 negative ballots remained unresolved. It was decided at the T3.5 Section meeting on 17 April 1981, the document because of changes in technical content, be resubmitted for ballot.

On 9 October 1981, the T3.5 Section decided no action be taken until the two negatives of the previous ballot be addressed.

**The negative ballots were finally resolved. At the 24 March 1981 T3.5 Section meeting, it was recommended the document be forwarded to the Technical Board for final approval. Because of technical changes as a result of the ballot process, Headquarters noted a third Ballot was required.**

The Technical Board granted approval for Ballot No. 3 on 13 May 1981.

The document was again forwarded to Headquarters. Headquarters Staff prepared NFPA/T3.5.14M for Ballot No. 3 on 13 November 1981.

Ballot No. 3 closed 14 December 1981 with 3 negative comments. The negative comments were reviewed and resolved.

Consequently, T3.5.14M was presented and granted approval by the Technical Board on 24 February 1982. The document was then submitted to the Board of Directors and granted final approval on 2 June 1982.

Project Group members who developed this standard:

**Jack Curnow**
Project Chairman
Sperry Vickers
(1974 - Present)

**Dave Prevallet**
Section Chairman
Gresen Manufacturing Co.

**Nathan Mills**
Section Chairman
(1970 - 1976)
Sperry Vickers

**Homer Graber**
Project Chairman
(1974)
Cessna Fluid Power Division

**Jack Kier**
Project Chairman
(1974)
Sperry Vickers

**Harold Jacoby**
Section Vice Chairman
Rexnord Inc.

**Wayland Tenkku**
Section Secretary
Fluid Controls Inc.

ANSI/B93.66

**Ed Saloum**
Technical Auditor
Snap-Tite, Inc.

**James C. White** *
Director of Technical Services
National Fluid Power Association

**G. Anderson**
Dynex/Rivett

**L. Coleman**
Continental Hydraulics

**T. Clark**
Rexnord, Inc.

**P. George**
Parker Hannifin

**R. Grassi**
Galland Henning Nopak Inc.

**J. Hedge**
Abex Corp.

**E. Maroney**
W. H. Nichols Co.

**J. Mazur**
Linde Hydraulics Corp.

**R. Olen**
Gresen Manufacturing Co.

**I. Sethi**
Abex Corp.

**C. Whitmore**
Parker Hannifin Corp.

**T. Zajac**
Parker Hannifin Corp.

On 22 October 1982, ANSI/B93.66M was submitted to ANSI Committee B93 for ballot. Balloting closed 3 December 1982 with unanimous approval.

ANSI/B93.66M was forwarded to ANSI Board of Standards Review on 5 April 1983 and granted approval on 5 May 1983.

The membership roster for the Standards Committee B93 at the time of ballot:

**Jack C. McPherson**
Chairman

**Daniel B. Shore**
Vice Chairman

**Allen E. Tucker**
Co-Secretary

**William G. Wagner**
Co-Secretary

**American Society of Agricultural Engineers**
Ed Fletcher

**American Society for Engineering Education**
William R. Smith

---

*Company affiliation has changed.

**American Society of Lubrication Engineers**
(to be named)

**Compressed Air & Gas Institute**
David E. Bonn
John Addington (alternate)

**Construction Industry Manufacturers Association**
Glenn Stewart

**Fluid Controls Institute**
Jude Pauli
Eric Bianchi (alternate)

**Fluid Power Distributor Association**
Thomas Neff

**Fluid Power Society**
Robert L. Firth
Carroll Grigsby
Robert W. Hanpeter
Marty King
Richard Read
Ronald Smith
Alan Tiedman

**Fluid Sealing Association**
John Scannell
Alex Pilecki (alternate)

**Instrument Society of America**
(to be named)

**Joint Industry Council**
Robert Muhl

**Material Handling Institute**
Jack C. McPherson
Willard Chichester (alternate)

**National Fluid Power Association**
Richard N. Bailey
John Bowin
Walter Forster
Z. J. Lansky
A. O. Roberts
Paul Schacht

**National Machine Tool Builders Association**
John B. Deam

**Rubber Manufacturers Association**
William A. Hertel
John R. Loder
E. J. McCarthy (alternate)

**SAE**
William A. Hertel
John T. Parrett
Henry Schultz
Daniel B. Shore
W. L. Snyder
Robert W. White

**U.S. Department of Defense**
William P. Coyne

**Company Member**
L. L. Schmaltz
Don McGeachy
John Welker (alternate)

**Individual Members**
Dr. E. C. Fitch, Jr.
Otto Maha
John J. Pippenger
Jack Walrad
Tom Wanke
Frank Yeaple

scs

# Hydraulic fluid power -
# Directional control valve -
# Method for determining the metering characteristics

## 0 INTRODUCTION

In hydraulic fluid power systems, power is transmitted and controlled through a liquid under pressure within an enclosed circuit. Some hydraulic valves are required to modulate flow or pressure with some specific relationship between the valve input and resultant output. The relationships between the valve input and the output flows for a given inlet pressure and output pressure are the metering characteristics of the product. The metering characteristics of a hydraulic direction control valve may be an important consideration when selecting a valve as part of a system.

## 1 SCOPE AND FIELD OF APPLICATION

This National Standard is intended to:

— Include the determination of the metering characteristics of a fluid power directional control valve;

— Provide a uniform procedure for obtaining and reporting the metering characteristics of a fluid power directional control valve.

## 2 REFERENCES

ANSI/B93.2-1971 and Supplement ANSI/B93.2A-1978, **American National Standard Glossary of Terms for Fluid Power.**

NFPA/T2.10.1M-1978, **National Fluid Power Association Recommended Standard Metric Units for Fluid Power Applications.**

ISO 1219-1976, **Fluid power systems and products - Graphic symbols.**

ANSI/Y14.17-1966 (R1980), **American National Standard Drafting Practices for Fluid Power Diagrams.**

ISO/DIS 4411, **Draft International Standard Hydraulic fluid power - Valves - Method of determining pressure differential flow characteristics.**

ASTM/D445-1972, **ASTM Determination of Kinematic Viscosity of Transparent and Opaque Liquids (and the Calculation of Dynamic Viscosity).**

## 3 TERMS AND DEFINITIONS

For definitions of terms not defined herein, see ANSI/B93.2 and ANSI/B93.2A.

**3.1 control variable(s):** The variable(s) which cause the controlled flow characteristic(s) of the valve to change.

**3.2 displacement control variable:** Mechanical displacement volume or electrical current used to control valve output.

**3.3 pressure, load:** Pressure which is measured at the work port when a circuit is working.

**3.4 specified flow rate $(Q_s)$:** Steady-state flow rate for the component.

**3.5 specified pressure $(P_s)$:** Rated steady-state operating pressure for the component.

**3.6 work port:** The port normally connected to the component(s) being controlled, e.g., cylinder, motor, etc.

## 4 UNITS OF MEASUREMENT

**4.1** Units of measurement are used in accordance with NFPA/T2.10.1M. This document is in agreement with ISO 1000.

**4.2** Approximate conversions to Customary US units are shown in parentheses after their metric counterparts and are made in accordance with NFPA/T2.10.1M.

## 5  GRAPHIC SYMBOLS

Graphic symbols are used in accordance with ISO 1219 and ANSI/Y14.17. When ISO 1219 and ANSI/Y14.17 are not in agreement, ISO 1219 governs.

## 6  LETTER SYMBOLS

The following letter symbols are used in this document:

|  |  |
|---|---|
| $\Delta P$ | pressure differential bar (psi) |
| $P_L$ | pressure, load bar (psi) |
| $\Delta P_m$ | pressure differential, measured bar (psi) |
| $P_s$ | pressure, specified bar (psi) |
| $\Delta P_t$ | pressure differential, tare bar (psi) |
| $Q$ | flow rate L/min (GPM) |
| $Q_s$ | specified flow rate L/min (GPM) |

## 7  GENERAL

**7.1** Set up and maintain apparatus in accordance with section 8 (Test Equipment).

**7.2** Run all tests in accordance with section 9 (Test Procedures).

**7.3** Present data from section 9 (Test Procedures) in accordance with section 11 (Data Presentation).

## 8  TEST EQUIPMENT

**8.1** Use a fluid power supply and circuitry as required similar to that in figure 1.

**8.2** Maintain the following fluid characteristics throughout all tests:

### 8.2.1  Viscosity

a) Use petroleum base fluid with nominal viscosity range of 21.8 - 26.4 mm/s (105 - 125 SUS) @ 50°C (122°F) with viscosity index (V.I.) of 95 ($\pm 10$). Maintain viscosity in accordance with table 1. Measure viscosity in accordance with ASTM/D445.

b) Use a Newtonian viscosity fluid; that is, one that does not contain polymeric materials used as thickeners or viscosity index improvers.

c) Select 46 mm/s (214 SUS), 68 mm/s (315 SUS), or 100 mm/s (464 SUS), with same variation permitted, if the 21.8 - 26.4 mm/s (105 - 125 SUS), nominal viscosity is not recommended by the valve manufacturer.

### 8.2.2  Aeration

Minimize fluid aeration by taking precautions such as proper system design and by adequate removal of air from the system before testing.

### 8.2.3  Filtration

Use a control filter which will limit the total number of particles greater than 10 micrometres (10 micron) less than 1 000 particles per millilitre (0.034 oz) in the fluid.

### 8.2.4  Pressure

Use of single pressure tappings may be employed. The center lines of all tappings should meet the pipe center line and be normal to it. No tapping should be situated on the lowest point of the pipe. The diameters of the pressure tappings shall be equal to or less than 0.1 times the pipe diameter but not less than 1 mm (0.039 in) nor greater than 6 mm (0.234 in). The pressure tapping hole shall have a length which is not less than twice its diameter.

## 9  TEST PROCEDURES

**9.1**  Test Temperatures

**9.1.1** Conduct all tests at one or both of the following fluid inlet temperatures:

a) 50°C (122°F)

b) 80°C (176°F). If 80°C exceeds the manufacturer's recommendations, use 63°C (149°F).

**9.2** Determine metering characteristics in desired flow path.

ANSI/B93.66

9.2.1 Inlet-Return;

9.2.2 Inlet-Work Port;

9.2.3 Work-Return Port.

9.3 Install in accordance with figure 1 by connecting appropriate ports.

9.4 Unused ports shall be blocked, drained to reservoir or appropriately pressurized depending on the circuitry and/or intended use of the component.

9.5 Adjust the test system to deliver specified flow rate at the specified temperature and at specified pressure.

9.6 Measure in accordance with 8.2.4 and record on a chart similar to figure 2 the pressure drop ($\Delta P_m$) between the desired ports and the flow rate (Q) in the desired path as a function of the displacement control variable for the entire range of the displacement control variable with the maximum work port pressures indicated in table 2.

9.7 Connect the lines used in 9.3 to complete the test circuit without the test valve as shown in figure 1 (schematic example of tare hookup).

9.8 Measure and record the tare pressure drop ($\Delta P_t$) at the same flow rate intervals and on the same chart used in 9.6. The flow through the work port loop may vary with the displacement control variable in some types of valves, resulting in a variable tare. A constant tare valve can be used for the "inlet-return" valve pressure differential ($\Delta P$) calculation which ignores the work port loop tare.

NOTE: If the tare pressure drop is small compared to the valve differential pressure, it can be ignored and the "Measured Pressure Drop" can be reported as the "Valve Pressure Differential."

9.9 Calculate the pressure differential ($\Delta P_t$) by subtracting the tare differential ($\Delta P_t$) from the measured pressure differential ($\Delta P_m$).

9.10 Special Metering Characteristics.

9.10.1 Tests may be run at additional increments of specified pressure, in any or all of the above mentioned (3) test modes, in order to adequately define a valve's special metering characteristics.

## 10 DATA ACCURACY

See Test conditions accuracy (table 1).

## 11 DATA PRESENTATION

11.1 Use metric units for all data presentation. If desired, also use "Customary US" units.

11.2 Plot the results of 9.9 on a graph similar to that shown in figure 3.

NOTE: Optional method - an X-Y plot of actual data can be used if desired.

11.2.1 Include the following information on the graph:

a) Label the graph, "Graph 1, Control (or Test) Number _____;

b) Indicate measured flow path including specified work ports where applicable;

c) Valve identification;

d) Date tested;

e) Test facility and location;

f) Instrument identifications and calibration dates;

g) Fluid viscosity;

h) Specified temperature;

i) Specified flow rate;

j) Specified pressure;

k) Maximum displacement control variable;

l) Curve labels - Tables similar to tables 3 and 4 may be used to identify the curves developed from the data collected.

## 12 SUMMARY OF DESIGNATED INFORMATION

The following designated information is needed when applying this recommended standard to a particular application or use:

12.1 Valve identification;

12.2 Specify flow paths required and identify indended ports;

12.3 Fluid viscosity;

12.4 Fluid temperature;

12.5 Flow rate;

12.5.1 Inlet to return;

12.5.2 Inlet to work port;

12.5.3 Work port to return.

12.6 Pressure

12.6.1 Supply;

12.6.2 Work port (load);

12.6.3 Return.

12.7 Displacement control variable

12.8 Special instructions

## 13 JUSTIFICATION STATEMENT

This document formalizes practices and equipment requirements which are based on research conducted at Oklahoma State University, and which are consistent industry practices.

## 14 TEST/PRODUCTION SIMILARITY

Utilize managerial controls necessary to maintain substantial similarity between test and production components or elements.

## 15 IDENTIFICATION STATEMENT

Use the following statement in catalogs and sales literature when electing to comply with this National Standard:

"Metering characteristics determined in accordance with ANSI/B93.66M-1983, **Hydraulic fluid power - Directional control valve - Method for determining the metering characteristics.**"

TABLE 1 - Test conditions accuracy

| Test Condition | Metric Unit | Customary U.S. Unit | Maintain within (±) of Actual |
|---|---|---|---|
| Electrical Current | A | A | 3% |
| Flow Rate | L/min | GPM | 2.5% |
| Force | N | lb (f) | 5% |
| Linear Mechanical Displacement (Distance) | mm | in | 3% |
| Mechanical Displacement (torque) | N·m | lb (f) - in | 5% |
| Pressure | bar | psi | 2.5% |
| Pressure Differential | bar | psi | 5% |
| Temperature | °C | °F | 4°C (6°F) |
| Viscosity | CSt or mm$^2$/s | SUS | 10% |
| Voltage | V | V | 3% |

ANSI/B93.66

TABLE 2 - Work port pressure values

| Desired Flow Path | Work Port Pressure |
|---|---|
| Inlet - Return | Work Ports Looped |
| Inlet - Work Port | 25%, 50%, 75% of Specified Pressure |
| Work - Return | 100%, 75% of Specified Pressure |

TABLE 3 - Inlet-return port graph - Curve labels

| CURVE | CURVE LABEL |
|---|---|
| $P (P_L = 0)$ @ $Q_s$ | A |

TABLE 4 - Inlet - work port graph - Curve labels

| Curve | Curve Label |
|---|---|
| $\Delta P(P_L = P_s/4)$ | B |
| Flow $(P_L = P_s/4)$ | C |
| $\Delta P(P_L = P_s/2)$ | D |
| Flow $(P_L = P_s/2)$ | E |
| $\Delta P(P_L = 3P_s/4)$ | F |
| Flow $(P_L = 3P_s/4)$ | G |

TABLE 5 - Work-return port graph - Curve labels

| Curve | Curve Label |
|---|---|
| $\Delta P(P_L = P_s)$ | H |
| Flow $(P_L = P_s)$ | J |
| $\Delta P(P_L = 3P_s/4)$ | K |
| Flow $(P_L = 3P_s/4)$ | L |

ANSI/B93.66

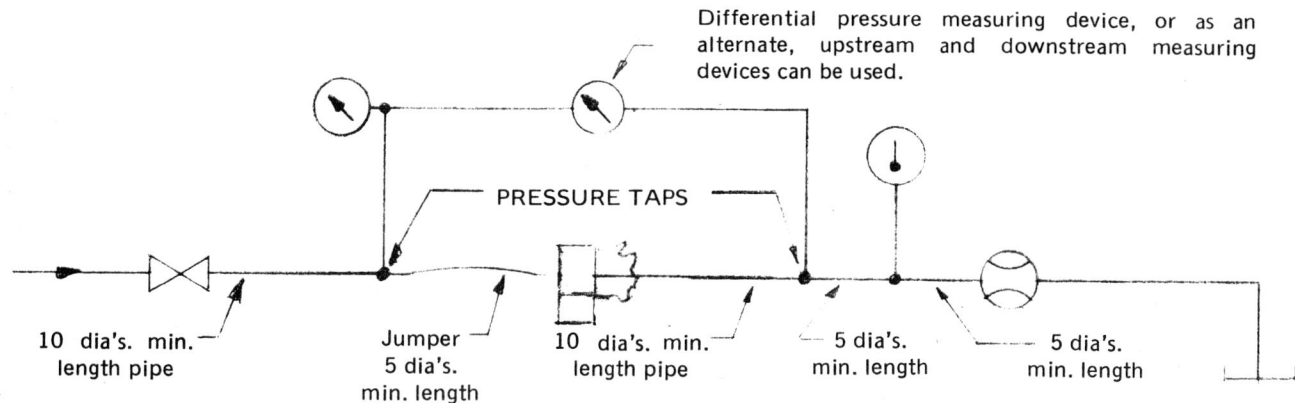

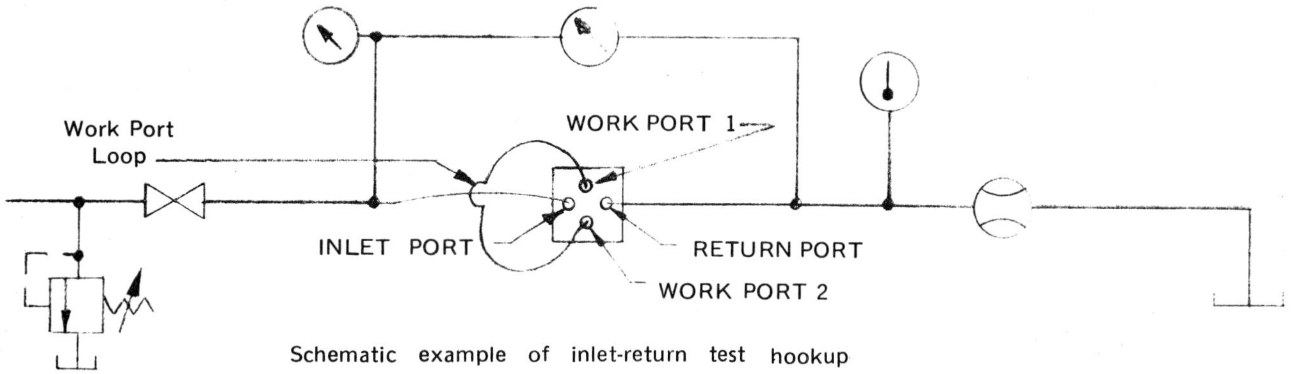

FIGURE 1 - **Test circuit diagram**

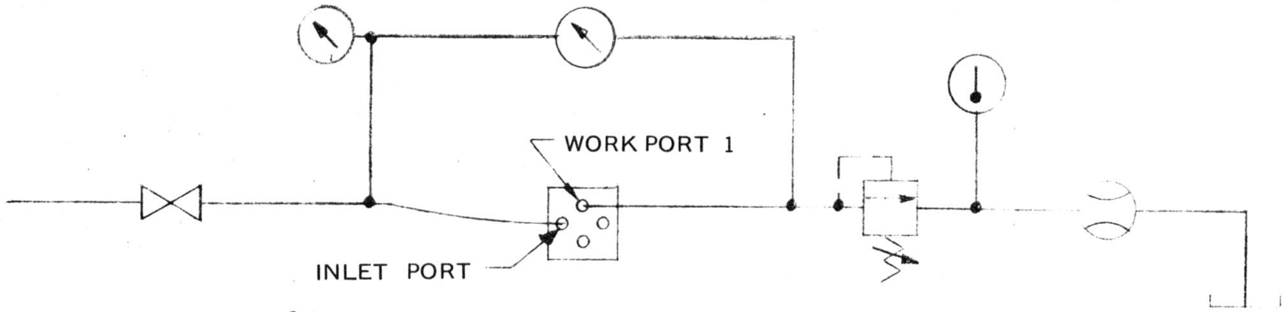

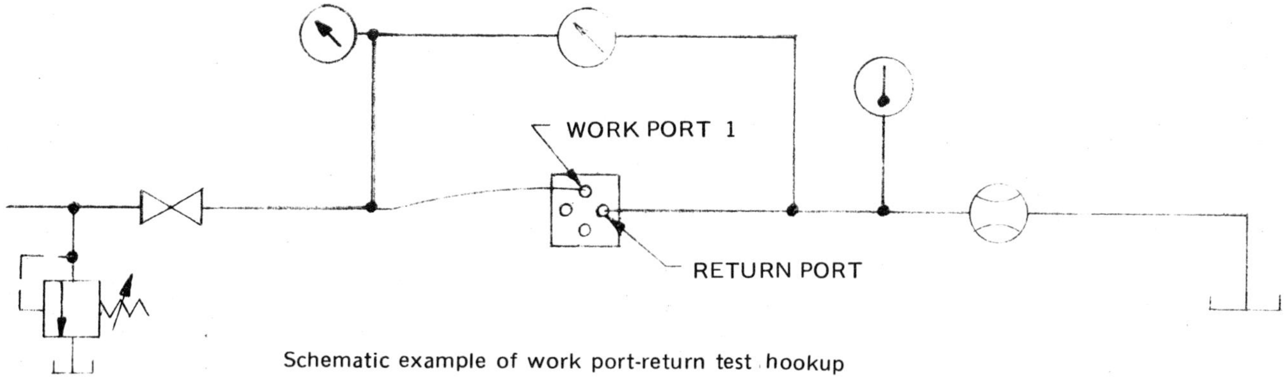

-217-

ANSI/B93.66

Date Tested: _____
Valve: _____
Specified Pressure: _____
Comments: _____

Test Facility and Location: _____
Specified Flow: _____
Temperature: _____
Fluid: _____
Viscosity: _____

| Graph Number | Displacement Control Variable cm (in) A L(in³) | Flow Rate L/min (GPM) (Q) | Measured Pressure Drop bar (psi) ($\Delta P_m$) | Tare Pressure Drop bar (psi) ($\Delta P_t$) | Valve Pressure Diff. ($\Delta P_M - \Delta P_T$) ($\Delta P$) |
|---|---|---|---|---|---|
| | | | | | |
| | | | | | |
| | | | | | |
| | | | | | |
| | | | | | |
| | | | | | |
| | | | | | |
| | | | | | |
| | | | | | |

FIGURE 2 - Typical metering test data summary

ANSI/B93.66

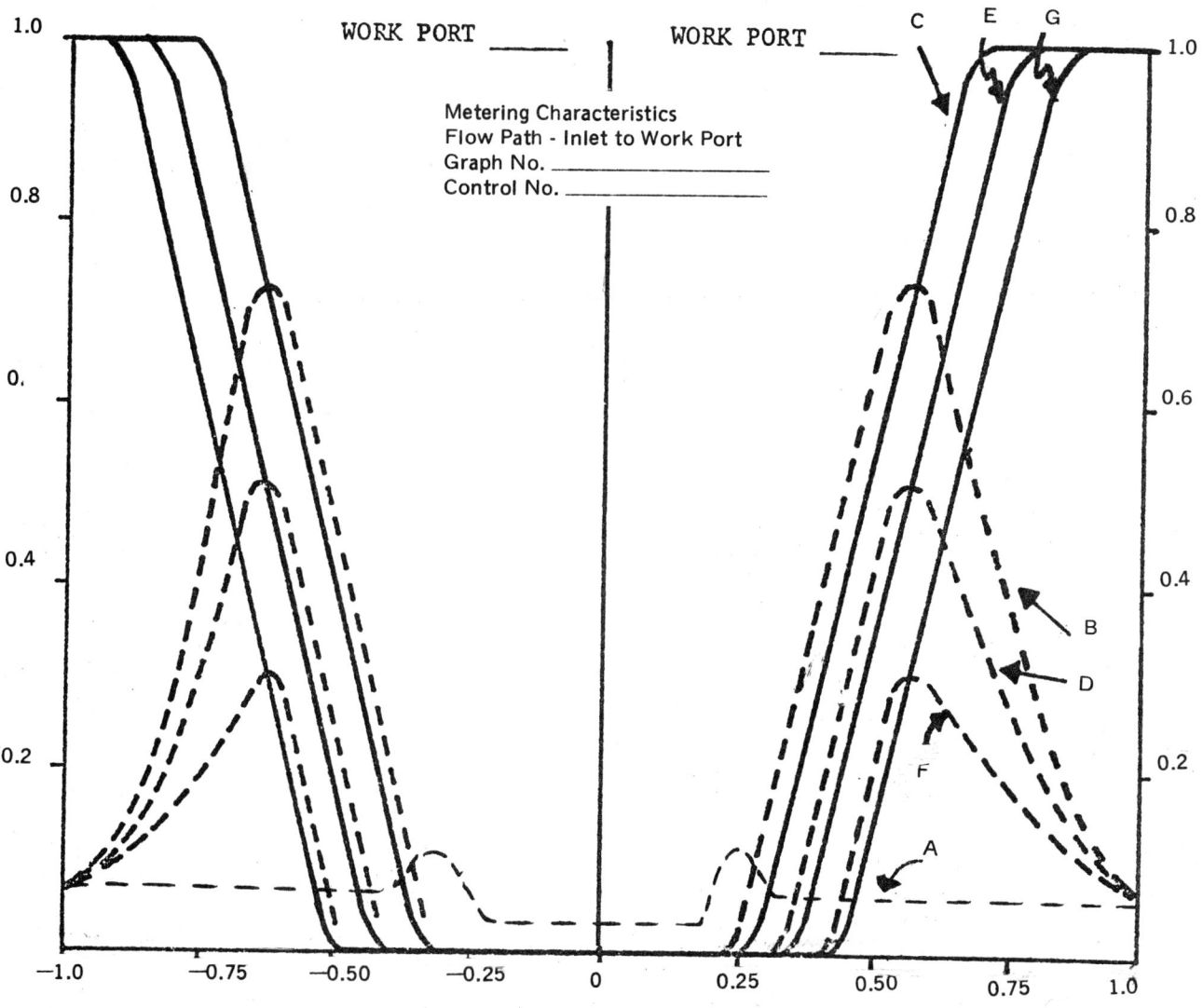

NOTE: Curves are shown for style of presentation only. No specific or related values are intended.

FIGURE 3 - **Typical plot of graph 1**

Appendix

to ANSI/B93.66M-1983

## 2 REFERENCES

ISO 5598[1], Fluid power systems and components - Vocabulary.

ISO 1000 -1981 SI Units and recommendations for the use of their multiples and of certain other units.

---

1) **Presently at the stage of draft.**

ANSI/B93.49M-1980

## AN INDUSTRY STANDARD FOR FLUID POWER

American National Standard

Hydraulic fluid power - Valves -

Pressure differential-flow characteristic -

Method of measuring and reporting
(NFPA/T3.5.28-1977)

Approved as an ANSI Standard
21 February 1980

**Descriptors:** flow characteristic, measuring; flow characteristic, reporting; fluid power; pressure differential, measuring; pressure differential, reporting; pressure drop; pressure loss; testing, flow characteristic; testing, pressure differential; testing, valve; valve, hydraulic fluid power; valve, port size.

published by
# NATIONAL FLUID POWER ASSOCIATION, INC.

3333 N. Mayfair Road / Milwaukee, WI 53222 / 414-778-3344 / TLX 26898

# ANSI/B93.49

## FOREWORD

This Foreword is not part of American National Standard Hydraulic fluid power - Valves - Pressure differential-flow characteristic - Method of measuring and reporting, ANSI/B93.49M-1980.

Control devices have always been an integral part of hydraulic fluid power systems. The rapid growth of the hydraulic control industry has precipitated a myriad of diverse ways of determining and reporting the pressure differential vs. flow characteristics of these control devices.

As a result of this diversity, it became apparent that there was a need for a standard method of testing and rating control devices which was acceptable to the fluid power industry. In recognition of this need, NFPA undertook project T3.5.7-19xx in March 1969. The original objective of this project was to establish a uniform way to determine the nominal and maximum flow capacity of a valve and to provide a means for applying "flow factors" to valves in such a way that the pressure drop experienced under various operating conditions could be readily calculated.

In June 1970, a program was instituted at Oklahoma State University by the U.S. Army Mobile Equipment Research and Development Command to investigate all phases of valve performance and to develop standard methods for determining the ability of a valve to meet its performance requirements. Consistent with the Army's goal toward increased reliance on voluntary industrial standards and at the invitation of NFPA, the results of this investigation were reported to NFPA Section T3.5 at its meeting in December 1971, and were offered to NFPA for use in developing voluntary standards. The offer was accepted.

The acceptance of the OSU work created 14 proposed NFPA recommended standards that pertained to valves. One of these was T3.5.12-19xx, which established a method for determining and reporting the pressure drop characteristics of any type of hydraulic fluid control device.

After extensively revising T3.5.7-19xx and circulating T3.5.12-19xx for General Review, the Section, at its meeting of 24 October 1972, again reviewed both documents and came to the following conclusions:

(1) That the establishment of a "flow factor" that can be applied to a hydraulic valve for calculating its pressure drop under various operating conditions is no longer a desirable objective.

(2) That it is more desirable to report the actual pressure differential vs. flow characteristic in the form of a curve than it is to establish a nominal and maximum flow capacity (or rating) for a control device.

(3) That any hydraulic standard sponsored by NFPA should apply to both mobile and industrial equipment.

(4) That given the above conditions, the two documents would become so similar that they should be merged into a single document.

The development of T3.5.28-1977, a merger of the two proposed standards, was assigned to a Project Group appointed by the Section on 24 October 1972. Draft No. 1 of the document was written on 15 December 1972, and was reviewed by the Project Group on 5 March 1973. The Final Working Draft resulted from this meeting. The NFPA Technical Staff prepared the General Review Draft on 19 June 1975.

General Review comments were answered at the 1 October 1975 Section meeting, and all comments were resolved by 2 April 1976. Technical Board approval to ballot was granted on 7 April 1976. The NFPA Technical Staff prepared the Ballot draft on 25 August 1976. Several negative comments received dealt with Section 8 of the document, and some of the requirements set forth in four clauses. These clauses were reworded and in Table 1 the flow, pressure, and pressure differential were changed. Along with the addition of SAE J 1117 as Reference No. 8, the negative comments were withdrawn.

On 2 November 1977, the Technical Board unanimously voted to recommend that this document be approved as an NFPA Recommended Standard. The NFPA Board of Directors granted final approval to NFPA/T3.5.28-1977 on 1 December 1977.

Project Group Members who developed this standard:

**Hunt, William**
Project Chairman (1976 - 1977)
Rexnord, Inc.*

**Gagnon, J. Robert**
Project Chairman (1975-1976)
HUSCO Division

**Olen, Robert**
Section Chairman (1976-1977)
Delaval Turbine Inc.*

**Krehbiel, Rob**
Project Chairman and Section Chairman (1975)
Cessna Fluid Power

**Mills, Nathan**
Section Chairman (1971-1975)
Sperry Vickers**

**Prevallet, David**
Section Secretary
Gresen Manufacturing Co.

**Hodgson, Robert**
Technical Auditor
Commercial Shearing, Inc.

**Luecke, John R.**
Director of National Technical Services
National Fluid Power Association*

**Becker, Lanson**
Hydreco

**Bethke, Donald**
HUSCO

**Bowman, Ron**
Continental Hydraulics Co.

* Company affiliation has changed since work on this Project Group.

** Retired.

In May of 1978, ANSI/B93.49M was submitted to ANSI Committee B93 for ballot.

# Hydraulic fluid power - Valves - Pressure differential-flow characteristic - Method of measuring and reporting

## 0 INTRODUCTION

In hydraulic fluid power systems, power is transmitted and controlled thru a liquid under pressure within an enclosed circuit. Hydraulic valves are required to modulate or direct the pressure and/or flow thru the system. These valves must operate with some desired efficiency criterion; the pressure differential-flow characteristic (commonly called pressure drop) is an important consideration in establishing a valve's ability to meet this criterion.

## 1 SCOPE

This National Standard includes:

— the determination of the pressure differential-flow characteristic of any fluid power valve;

— a uniform method of presenting the test data;

— standardized information for conducting comparative tests.

## 2 FIELD OF APPLICATION

This National Standard is intended to:

— provide a uniform laboratory procedure for measuring the pressure losses associated with any given flow path in a hydraulic valve;

— provide a standard means of reporting the pressure losses measured;

— establish uniform specified values for comparing the pressure losses of different valve designs.

## 3 REFERENCES

ISO 1000-1973, **SI units and recommendations for the use of their multiples and of certain other units.**

ISO 1219-1976, **International Standard Graphical Symbols for Hydraulic and Pneumatic Equipment and Accessories for Fluid Power Transmission.**

ASTM/D 445-1965, **American Society for Testing and Materials Standard Method of Test for Viscosity of Transparent and Opaque Liquids.**

ANSI/B93.9-1975, **American National Standard Symbols for Marking Electical Leads and Ports on Fluid Power Valves.**

## 4 TERMS AND DEFINITIONS

For definition of other terms used, see ANSI/B93.2-1971 and Supplement ANSI/B93.2A-1978, **American National Standard Glossary of Terms for Fluid Power.**

**4.1 Test Flow.** Any steady state flow rate required to conduct this test.

**4.2 Input Port.** Any port into which flow is directed for the purpose of this test.

**4.3 Output Port.** Any port from which flow exits for the purpose of this test.

**4.4 Tare Pressure.** The pressure losses between the pressure tapping points as generated by the test equipment exclusive of the test valve.

**4.5 Control.** Any adjustable feature integral with the test valve that varies the flow path and/or flow rate.

**4.6** An International Standard giving definitions of terms is in preparations.

## 5 UNITS

Units of measurement are used in accordance with ISO 1000.

## 6 GRAPHIC SYMBOLS

Graphic symbols used are in accordance with ISO 1219-1976 and ANSI/Y14-1974. Where ISO 1219 and ANSI/Y14 are not in agreement, ANSI/Y14 governs.

**11.3** Control the specified values to the required limits or to the standardized limits given in Section 14.

**11.4** Set the control for the flow path (or flow rate) required.

**11.5** Obtain enough data between zero flow and the specified maximum flow ($Q_m$) to produce a well-defined curve as illustrated in figure 3. If data is measured manually, record on a chart similar to figure 2.

**11.6** Read the upstream pressure ($P_1$) and the downstream pressure ($P_2$) at each flow selected in 11.5.

**11.7** Remove the test valve from the test circuit.

**11.8** Connect the lines used in 11.2 to complete the test circuit without the test valve.

**11.9** Record the upstream pressure ($P_1$) and the downstream pressure ($P_2$) at the same flow rate intervals and on the same chart used in 11.5.

**11.10** Calculate the measured pressure differential ($\Delta P_m$) by substracting the downstream pressure from the upstream pressure as read in 11.6.

**11.11** Calculate the tare pressure differential ($\Delta P_t$) by subtracting the downstream pressure from the upstream pressure recorded in 11.9.

**11.12** Calculate the pressure differential ($\Delta P$) by substracting the tare pressure differential ($\Delta P_t$) from the measured pressure differential ($\Delta P_m$).

## 12 DATA ACCURACY

TABLE 2 — **Data accuracy**

| Quantity | Metric Unit | Customary US Unit | Maintain within (±) of Actual |
|---|---|---|---|
| Flow | L/min | USGPM | 2.5% |
| Pressure | bar | psi | 2.5% |
| Pressure differential | bar | psi | 5.0% |
| Temperature | °C | °F | 4°C (6°F) |

## 13 DATA PRESENTATION

**13.1** Plot a curve of the pressure differential ($\Delta P$) vs. flow rate (Q) using the data obtained in 11.12.

**13.2** Use figure 3 as an example.

**13.3** Include the following information on the data plot:

    a) Valve description

    b) Description of fluid. (Identify fluid used and state specific gravity/mass density.)

    c) Viscosity of fluid (SUS and cSt)

    d) Fluid temperature

    e) Upstream port and size

    f) Downstream port and size

    g) Specific functional control position (s)

    h) Date of test

    i) Testing agency

**13.4** Identify ports in accordance with ANSI/B93.9.

## 14 SUMMARY OF DESIGNATED INFORMATION

**14.1** Use the following criteria when applying this recommended standard to a particular application or use:

    a) Description of valve including port size

    b) Description of fluid (Identify fluid used and state specific gravity/mass density.)

    c) Viscosity of fluid (SUS and cSt)

    d) Fluid temperature

    e) Maximum flow for this test ($Q_m$)

    f) Input port

    g) Output port

    h) Specific functional control position(s)

**14.2** Use the following standardized values in catalogs and sales literature:

**14.2.1** A fluid temperature of 50°C (122°F) and a viscosity range of 105-125 SUS (21.8 - 26.4 cSt).

ANSI/B93.49

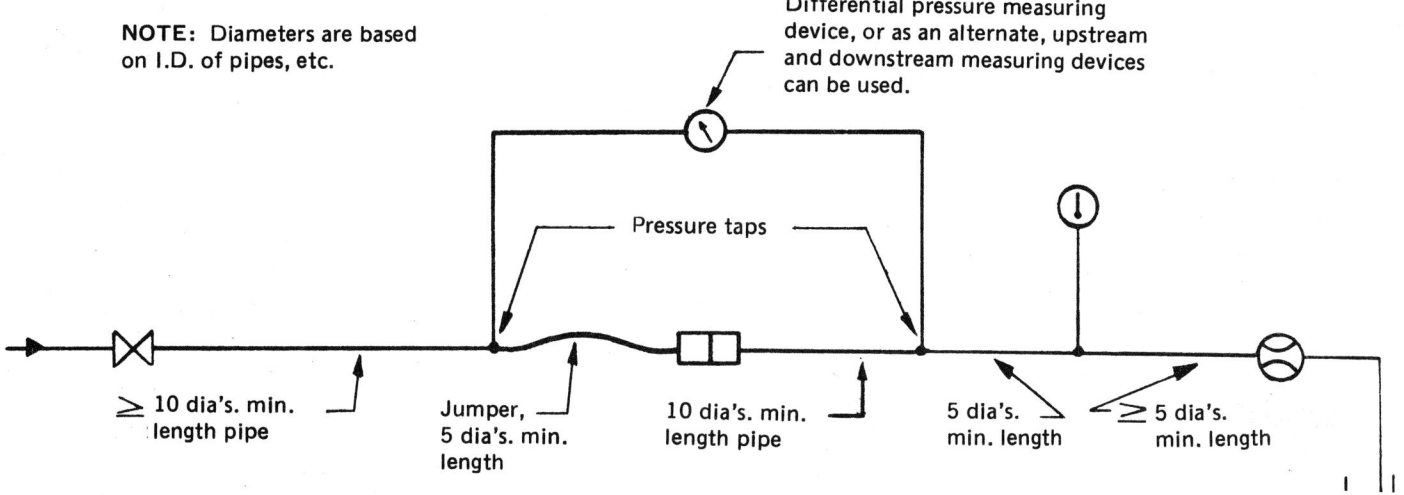

Schematic example of tare hook-up (see ISO/Dis 4411)

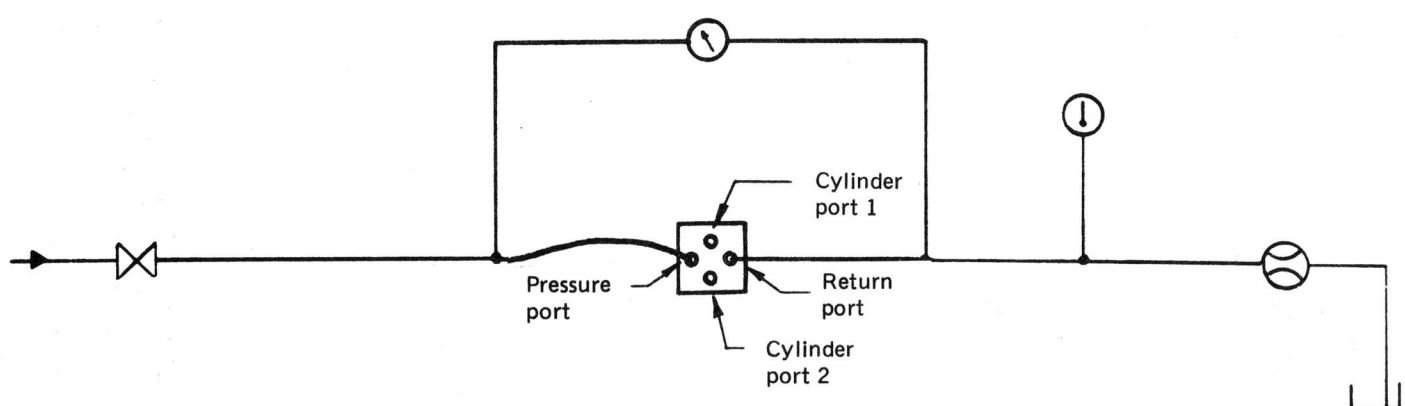

Schematic example of neutral flow hook-up

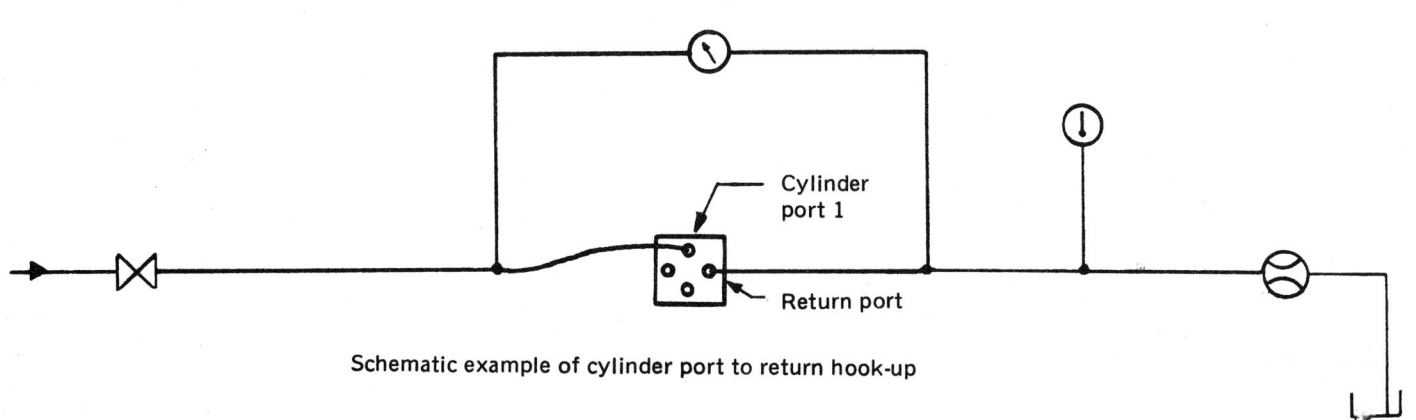

Schematic example of cylinder port to return hook-up

FIGURE 1 — **Schematic of a hydrualic valve test system**

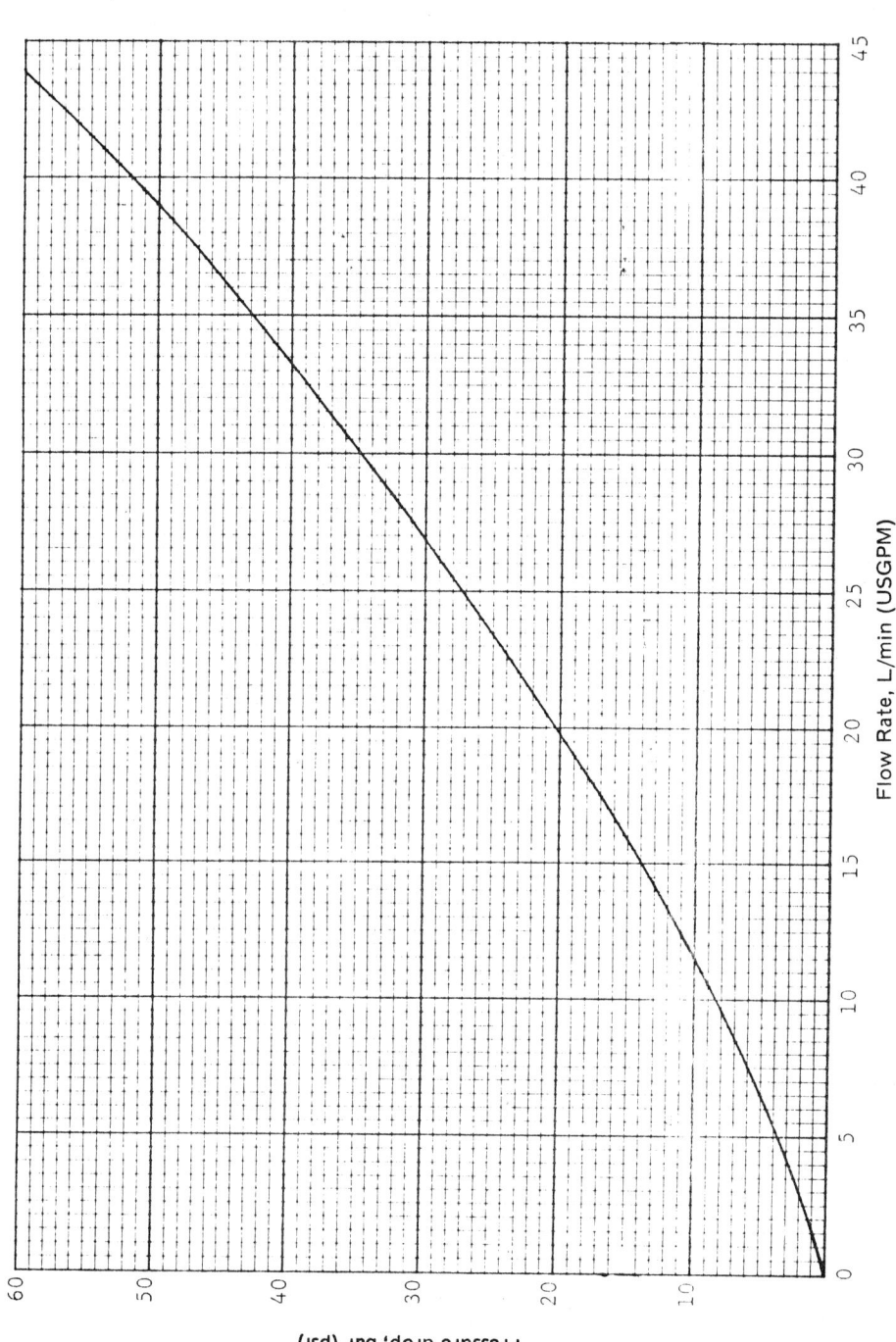

FIGURE 3 — Typical pressure differential vs. flow curve

ANSI/B93.55M-1981

## AN INDUSTRY STANDARD FOR FLUID POWER

American National Standard

Hydraulic fluid power -

Solenoid-piloted industrial valves -

Interface dimensions for electrical connectors
(NFPA/T3.5.29M-1980)

Approved as an ANSI Standard
24 August 1981

**Descriptors:** dimensions, electrical connector; electrical connector, valve; fluid power; requirements, electrical connector; specifications, electrical connector; valve, hydraulic fluid power.

published by
# NATIONAL FLUID POWER ASSOCIATION, INC.

3333 N. Mayfair Road / Milwaukee, WI 53222 / 414-778-3344 / TLX 26898

# ANSI/B93.55

**FOREWORD**

This Foreword is not part of American National Standard Hydraulic fluid power - Solenoid piloted industrial valves - Interface dimensions for electrical connectors, ANSI/B93.55M-1981.

This NFPA Recommended Standard originated at the 5 March 1973 meeting of NFPA Hydraulic Valves Section, when Mr. Werner Larsen, Sales Manager, Hydraulic Components Div., (Rexnord Inc.) expressed an urgent need for this standard. A Title, Scope and Purpose was prepared by Headquarters on 13 April 1973 which was approved by the Technical Board on 23 May 1973.

No significant work was done until the 20 April 1977 meeting at which time a Project Group was appointed with instructions to implement the Title, Scope and Purpose, which called for the formulation of a questionnaire to be circulated to interested parties.

The Project Group met on 20 September 1977 and drew up a questionnaire for submission to the Hydraulic Valve Section which met on 12 October 1977.

At the 12 October meeting the Hydraulic Valve Section voted to revise the Title, Scope, and Purpose to call for the immediate preparation of a draft standard.

This revision was approved by the Technical Board on 2 November 1977 with slight modification.

The Project Group met for the second time on 23 February 1978. The Group voted to adopt as a proposed standard, a connector that has been manufactured for several years, which led to the preparation of Draft No. 1.

The Project Group met on 1 November 1978 to critique the first draft. Several changes were made which were incorporated at the second draft. The Project Group also wished to have the following explanatory statement added to the Introduction: "The Project Group believes that the Three Pin Electrical Plug Connector in ISO Standard ISO/DIS 4400 will not be acceptable to large user groups in the U. S. For this reason it was decided to formulate a separate standard for use in the U. S."

The Hydraulic Valve Section, at its 7 December 1978 meeting, issued this document for General Review for itself (T3.5), as well as the Pneumatic Valve Section (T3.21), and the Pressure Switch Section (T3.29).

The Technical Board approved this document for General Review at its 7 February 1979 meeting. The General Review Draft was prepared on 17 April 1979 by Headquarters Technical Staff. All questions, comments and recommendations were resolved.

Approval to ballot was granted by the Technical Board on 16 November 1979, and Headquarters prepared a Ballot Draft on 26 November 1979.

Two negative ballots were received from the Ballot Draft. One commentor was concerned that no designation had been made for the polarity of the solenoid connection pins. The negative ballot was withdrawn with clarification that the standard related to A.C. solenoids only and not D.C. solenoids. The second negative comment stated that in 5.9 the wording "... with integral wire leads ... is ambiguous, and in 5.9.1 the wire size should not be specific. Mr. Clark, Project Chairman, clarified these comments by stating that the wording "with integral wire leads" is meant to describe the male half of the connector, and it was the intention of the document to specify only the connector, and the wires immediately attached to it. The negative comments were withdrawn.

On 7 May 1980, the Technical Board unanimously voted to recommend that this document be approved as an NFPA Recommended Standard. The NFPA Board of Directors granted final approval to NFPA/T3.5.29M-1980 on 4 June 1980.

Project Group Members who developed this standard:

**Clark, Thomas A.**
Project Chairman
Rexnord, Inc.

**Smith, Gary**
Project Chairman (1973 - 1975)
Rexnord, Inc.

**Prevallet, Dave**
Section Chairman
Gresen Mfg.

*****Olen, Robert**
Section Chairman (1975 - 1979)
DeLaval Turbine, Inc.

**Mills, Nathan**
Section Chairman (1973 - 1975)
Sperry Vickers

**Jacoby, Harold**
Section Secretary
Rexnord, Inc.

**Barthe, Henry**
Technical Auditor
Schroeder Brothers Corp.

**White, James**
Director of Technical Services
National Fluid Power Association

Anderson, G.
**Dynex/Rivett**

Coleman, L.
**Continental Hydraulics Division**

Curnow, J.
**Sperry Vickers**

Hedge, J.
**Abex Corporation**

Royston, D
**Sperry Vickers**

Stegner, J.
**Moog, Inc.**

Woodring, R.
**Parker Hannifin Corporation**

On 5 December 1980, B93.55M was submitted to ANSI Committee B93 for Ballot. The document received one negative ballot which was resolved with editorial changes to the document. ANSI/B93.55M was approved by ANSI's Board of Standards Review on 24 August 1981.

The membership roster of Standards Committt B93 at the time of ballot:

**Jack McPherson**
Chairman

**James C. White**
Co-Secretary

**Robert Uhl**
Co-Secretary

* Company affiliation has changed since preparation of document.

**American Society of Agricultural Engineers**
Ed Fletcher

**American Society of Lubrication Engineers**
William R. Smith

**American Society of Lubrication Engineers**
M. M. Gurgo

**American Society of Mechanical Engineers**
Robert Hildebrandt
Thomas R. Curran (alternate)
Frank Yeaple (alternate)

**Compressed Air & Gas Institute**
David E. Bonn
John Addington (alternate)

**Construction Industry Manufacturers Association**
Glenn Stewart

**Fluid Controls Institute**
Jude Paull
Eric Blanchi (alternate)

**Fluid Power Distributors Association**
Thomas Neff

**Fluid Power Society**
Edward Briggs
Ronald Brettnacher
Robert Firth
Carroll Grigsby
Robert Hanpeter
Marty King
Richard Read
Ronald Smith
Alan Tiedman

**Fluid Sealing Association**
John Scannell
Alex Pilecki (alternate)

**Instrument Society of America**
Aaron Kutz

**Joint Industry Council**
Robert Muhl

**Material Handling Institute**
Jack McPherson
Willard Chichester (alternate)

**National Fluid Power Association**
Richard Bailey
John Bowbin
James L. Fisher Jr.
Walter Forster
Z. J. Lansky
John Mueller
A. O. Roberts

**National Machine Tool Builders Association**
John Deam

**Rubber Manufacturers Association**
William Hertel
John Loder
E. J. McCarthy (alternate)

**Society of Automotive Engineers**
William Hertel
Eugene Falendysz
Henry Schultz
D. B. Shore
W. L. Snyder
Robert White

**Society of Manufacturing Engineers**
David Ashpole

**US Department of Defense**
Henry Schaefer
William Coyne (alternate)

**Company Members**
L. L. Schmaltz
Don McGeachy
John Welker (alternate)

**Individual Member**
Dr. E. C. Fitch Jr.
Tom Wanke
John J. Pippenger
Otto Maha

ss

ANSI/B93.55

## Hydraulic fluid power - Solenoid-piloted industrial valves - Interface dimensions for electrical connectors

### 0 INTRODUCTION

In hydraulic fluid power systems, power is transmitted and controlled thru a fluid under pressure within a closed circuit. A fluid power valve controls the fluid's direction, pressure or flow-rate.

When valves are solenoid operated, some designs use an electrical connector to connect the valve to the electrical control system. Users of solenoid operated valves benefit when valves from various manufacturers have a common electrical connector interface, and therefore can be interchanged when servicing or replacement is required.

### 1 SCOPE AND FIELD OF APPLICATION

**1.1** To include electrical plug type connector interface dimensions and configurations (not intended for current interruption) used with a single or double solenoid piloted hydraulic fluid power control valves used in industrial (in plant) applications.

**1.2** This National Standard is intended:

— to simplify variety, facilitate installation and servicing.

— to promote interchangeability and promote greater use of solenoid piloted hydraulic fluid power controls.

**1.3** This Standard only applies to the dimensional criteria of products manufactured in conformance with this standard. It does not apply to their functional characteristics.

### 2 REFERENCES

ANSI/B93.2M-1971 and Supplement ANSI/B93.2A-1978. **American National Standard Glossary of Terms for Fluid Power.**

NFPA/T2.10.1M-1978, **National Fluid Power Association Recommended Standard, Metric Units for Fluid Power Applications.**

ANSI/Y14.17-1974, **American National Standard, Drafting Practices for Fluid Power Diagrams.**

ANSI/B93.9M-1975, **American National Standard, Symbols for Marking Electrical Leads and Ports on Fluid Power Valves.**

UL-498-1974 (ANSI/C33.77-1973), **Underwriters' Laboratories Standard for Safety - Attachment Plugs and Receptacles.**

C22.2 No. 42-1959, **CSA Testing Laboratories Electrical Bulletin.**

FB10-1973, **National Electrical Manufacturers Association- General Standards for Plugs, Receptacles and Connectors of the Pin and Sleeve Type.**

UL-1063-1975 (ANSI/C33.124-1975), **Underwriters Laboratories Standard for Safety - Machine - Tool Wires and Cables.**

### 3 TERMS AND DEFINITIONS

For definitions of terms used, see ANSI/B93.2M and Supplement ANSI/B93.2A.

**3.1 electrical connector:** A two-piece instrument (plug and socket) which, when joined, provides electrical continuity.

### 4 UNITS OF MEASUREMENT

**4.1** Units of measurement are used in accordance with NFPA/T2.10.1M.

**4.2** Approximate conversions to Customary U.S. units are shown in parentheses after their metric couterparts and are made in accordance with NFPA/T2.10.1M.

### 5 ELECTRICAL CONNECTOR SPECIFICATIONS

**5.1** It is the purpose of this standard to specify the configuration of only the male half of the connector, which half is intended to mount on the valve. It is the duty of the manufacturer to provide the female half to mate properly both electrically and mechanically, so as to ensure electrical integrity and to seal out dust, dirt, and moisture and other common environmental matter.

**5.2** Use electrical connectors rated at 250 volts AC or DC.

**5.3** Use electrical connector rated at 8 amperes, steady state rating.

**5.4** Use $105°C$ ($221°F$) Class A insulation.

**5.5** Comply with regulatory specifications in accordance with UL-498 (ANSI/C33.77 and C22.2 No. 42).

**5.6 Terminals**

**5.6.1** Provide two or four live terminals and one ground terminal. Ground pin to make first break last by 1.5 mm (0.06 in).

**5.6.2** Use pins with diameters as shown in Figure 1.

**5.6.3** Seal connector pin ends to prevent fluid passage.

**5.7** Use a pin pattern as shown in Figures 1 and 2.

**5.8** Use an orientation key, as shown in Figures 1 and 2, to guard against improper assembly.

**5.9 Wire**

For connectors with integral wire leads, apply the following:

**5.9.1** Use 16 AWG copper wire.

**5.9.2** Provide wire insulation to meet requirements of UL-1063 (ANSI/C33.124) construction A.

**5.9.3** On five pin model, use red insulation on lead Nos. 1, 2, 4 and 5. Make lead No. 3 green. On the three pin model, use red for lead Nos. 2 and 3 and green for No. 1.

**5.9.4** Identify all red leads by number.

**5.10 Connector Orientation**

**5.10.1** Use the three pin connector for single solenoid valves. Pin No. 2 and No. 3 connect to the solenoid, and pin No. 1 connects to ground. (See Figure 2)

**5.10.2** Use the five pin connector for double solenoid valves. Pins No. 1 and 5 connect to solenoid "B", pins No. 2 and 4 connect to solenoid "A". Pin No. 3 connects to ground (See Figure 2).

**5.11 Material Used**

**5.11.2** Protect ferrous parts from corrosion equivalent to the protection provided by 0.005 (0.0002) electrodeposited zinc.

**5.11.3** Assure that plastic or elastomeric parts meet the requirements of UL-498 (ANSI/C33.77), C22.2 No. 42, and FB10, and are compatible with the hydraulic fluid with which they come in contact.

**5.12 Warning Statement**

If the voltage and current thru the connector is such that it may cause a hazard when interrupted under load, or is beyond the current interruption capacity of the connector, it is the manufacturer's responsibility to see that a statement is permanently attached to the valve indicating the hazard (See FB10).

**6  IDENTIFICATION STATEMENT**

Use the following statement in catalogs and sales literature when electing to comply with this voluntary standard:

"Electrical connector conforms to American National Standard, ANSI/B93.55M-1981, Hydraulic fluid power - Solenoid piloted industrial valves - Interface dimensions for electrical connectors."

ANSI/B93.55

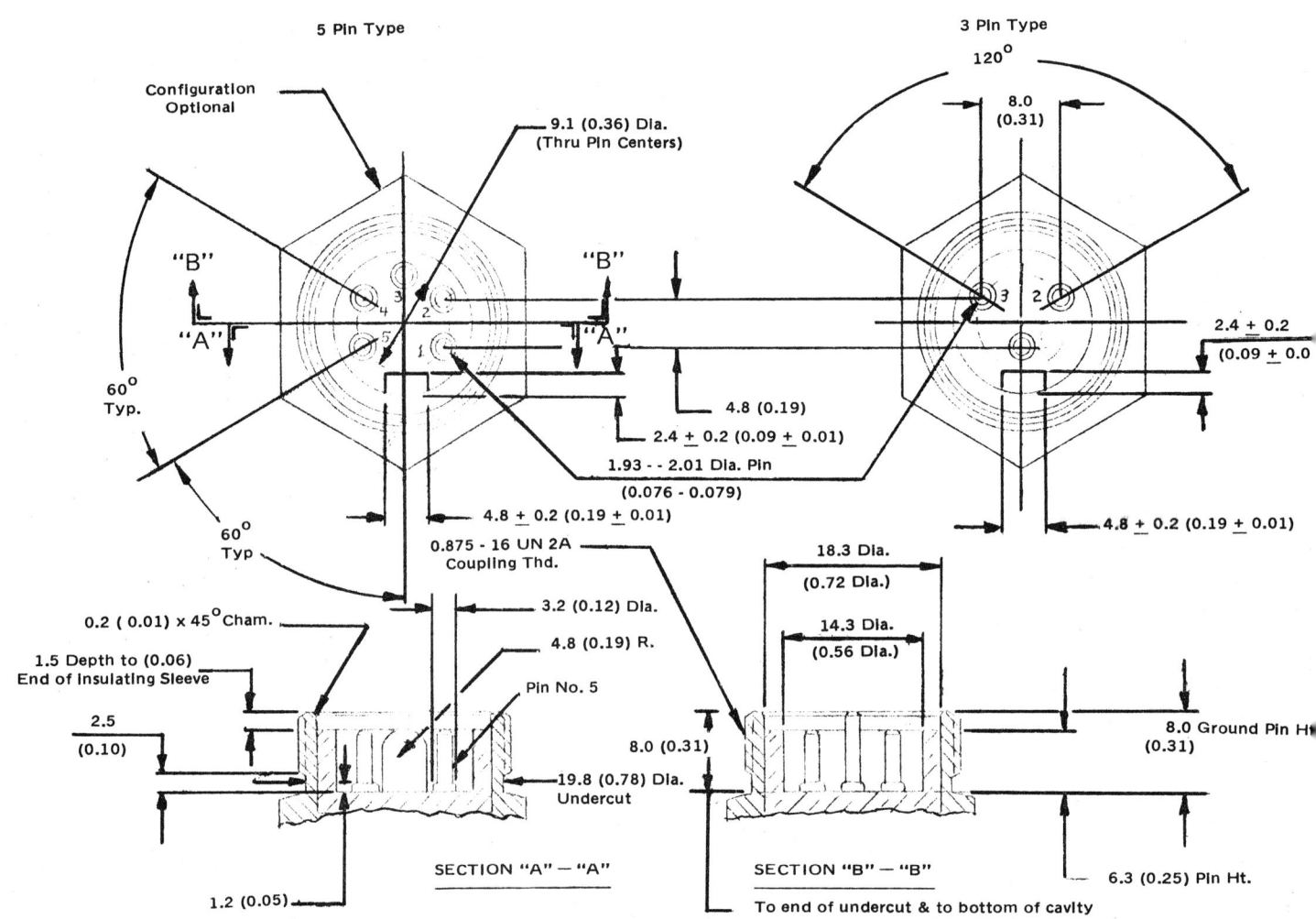

FIGURE 1 — Electrical connector, valve mounted half

ANSI/B93.55

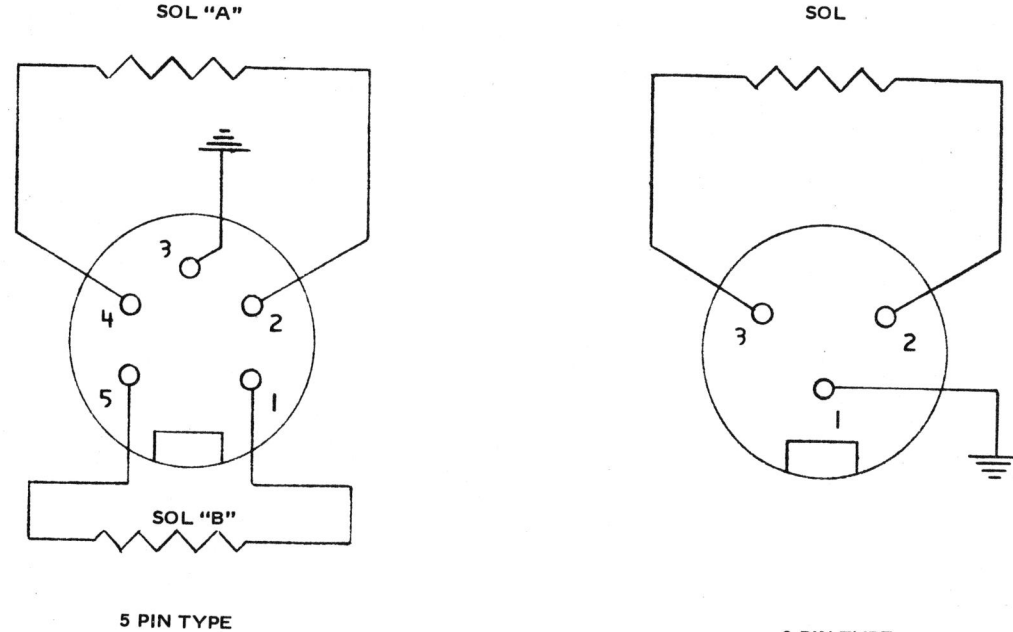

FIGURE 2 - Electrical Connector Schematic Diagram

## APPENDIX

## to ANSI/B93.55M-1981

## 2 REFERENCES

ISO 1000, SI units and recommendations for the use of their multiples and of certain other units.

ISO 5598[1], Fluid power systems and components - Vocabulary

---

[1] At present at the stage of draft.

NFPA/T3.5.33M-1985

---

**AN INDUSTRY STANDARD FOR FLUID POWER**

Hydraulic fluid power -

Cylinder actuator mounted valves -

Standard dimensions for mounting surfaces

Approved as an NFPA Recommended Standard
30 September 1985

---

Descriptors: dimensions, linear actuator mounted valve; fluid power; mounting surfaces; hydraulic fluid power vavles; valve; valve, control; valve, hydraulic; valve, linear actuator mounted.

---

published by
**NATIONAL FLUID POWER ASSOCIATION, INC.**

3333 N. Mayfair Road / Milwaukee, WI 53222 / 414-778-3344 / TLX 26898

## FOREWORD

This Foreword is not part of NFPA Recommended Standard -Hydraulic fluid power -Cylinder actuator mounted valves -Standard dimensions for mounting surfaces, NFPA/T3.5.33M-1985.

On 24 March 1981, the Hydraulic Valve Section approved the preparation of this standard as a Hydraulic Valve Section project. A Project Group was appointed and prepared a TSP.

The TSP was forwarded to and approved by the Technical Board on 24 February 1982, with modifications. "Cylinder" was added to the title and Scope.

Draft No. 1 was prepared and forwarded to NFPA Headquarters for review by the Project Group at their 29 September 1982 meeting.

The Hydraulic Valve Section met the afternoon of 29 September 1982 and discussed some of the changes that the Project Group recommended. The changes were incorporated into Draft No. 2. The NFPA Technical Stafff prepared Draft No. 2 on 3 December 1982 for review and comment by the Hydraulic Valve Section.

Draft No. 2 was reviewed by the Hydraulic Valve Section on 16 March 1983, and the Section recommended that the document be circulated for General Review. The General Review Draft was prepared by NFPA Headquarters Staff on 20 May 1983. Comments received as a result of the General Review Draft were reviewed and resolved.

Approval for ballot was granted on 9 February 1984 at the Technical Board meeting. NFPA Headquarters received the revised General Review Draft on 23 March 1984. The NFPA Technical Staff prepared a Ballot Draft on 13 April 1984.

Headquarters conducted a ballot on this proposed recommended standard on 13 April 1984, resulting in three negative comments.

The negative comments were withdrawn after Mr. Curnow, Project Chairman, clarified the intent of the document.

On 7 February 1985, the Technical Board voted unanimously to recommend to the Board of Directors that NFPA/T3.5.33M-19XX, be approved as an NFPA Recommended Standard and that after approval of this recommended standard it be submitted to ANSC B93 for promulgation as an ANSI standard.

\* **Company affiliation has changed.**

The Board of Directors concurred with the recommendations and granted final approval to NFPA/T3.5.33M-1985 on 30 September 1985.

Project group members who developed this document:

**Jack W. Curnow**
Project Group Chairman
Vickers, Incorporated

**David Prevallet**
Section Chairman
Dana Corp./Gresen Mfg. Co.

**Harold Jacoby**
Section Vice Chairman
Dana Corp./Racine Hydraulics Div.

**Wayland Tenkku** \*
Section Secretary
Fluid Controls Inc.

**Bruce McCord**
Technical Auditor
Aro Corporation

**Allen E. Tucker** \*
Director of Technical Services
National Fluid Power Association

**Charles G. Guthrie**
United Technologies/Fluid Power Systems

**Jack Johnson**
Milwaukee School of Engineering

**Robert Olen**
Dana Corp./Gresen Mfg.

mmr

# Hydraulic fluid power - Cylinder actuator mounted valves - Standard dimensions for mounting surfaces -

**NFPA/T3.5.33M**

## 0 INTRODUCTION

In hydraulic fluid power systems, power is transmitted and controlled thru a liquid under pressure within an enclosed circuit. Typical components found in such systems are hydraulic valves. These devices control fluid direction, pressure or flow rate of liquids in the enclosed circuit.

## 1 SCOPE AND FIELD OF APPLICATION

**1.1** This NFPA Recommended Standard includes:

a) a series of mounting surfaces which describe the working port interface between hydraulic fluid power directional, flow or pressure control valves and a hydraulic cylinder actuator for 350 bar maximum (5 040 psi) hydraulic service;

b) dimensional criteria;
   1) minimum surface dimensions,
   2) sizes and locations of tapped holes for mounting bolts,
   3) sizes and locations of ports,
   4) sizes and location of dowel or rest pins where required;

c) general criteria;
   1) surface finish and flatness,
   2) indication of tolerances where pertinent,
   3) this document will not attempt to define the physical orientation of the mounting surface with respect to actuator features.

**1.2** This NFPA Recommended Standard only applies to the dimensional criteria of products manufactured in conformance with this standard. It does not apply to their functional characteristics."

**1.3** This NFPA Recommended Standard achieves:

a) dimensional interchangeability;

b) simplification of variety.

## 2 REFERENCES

ANSI/B93.2-1971 and ANSI/B93.2A-1978, **American National Standard Glossary of Terms for Fluid Power.**

NFPA/T2.1.1 R1-1983, **National Fluid Power Association Recommended Standard Fluid power systems and products -Glossary**

NFPA/T2.10.1M-1978, **National Fluid Power Association Recommended Standard Metric Units for Fluid Power Applications.**

ANSI/B93.7-1968 (R1979), **American National Standard Dimensions for Mounting Surfaces of Subplate Type Hydraulic Fluid Power Valves.**

ANSI/B87.1-1965, **American National Decimal Inch.**

ANSI/B93.65M-1983, **American National Standard Hydraulic fluid power - Code for identification of valve mounting surfaces.**

## 3 TERMS AND DEFINITIONS

For definitions of terms used see ANSI/B93.2, ANSI/B93.2A and NFPA/T2.1.1 R1.

## 4 UNITS OF MEASUREMENT

**4.1** Units of measurement are used in accordance with NFPA/T2.10.1M. This document agrees with ISO 1000.

**4.2** Approximate conversions to Customary US units are shown in parentheses after their metric counterparts in the text, and in tables are shown under their metric counterparts separated by a line. Conversions are made in accordance with NFPA/T2.10.1M.

## 5 GENERAL

**5.1** Ports are identified by "A" and "B" as a means of reference only.

## 6 MOUNTING SURFACE DRAWING IDENTIFICATION CODING

**6.1** Identification code prefix is compatible with ANSI/B93.7 as follows:

A - Cylinder actuator mounted valves.

**6.2** Drawing identification code number is in accordance with NFPA/T3.5.34M and the following suffixes:

| Code Number | Nominal Port Size |
|---|---|
| 06 | 15.9 mm (.63 in.) |
| 07 | 19.0 mm (.75 in.) |
| 08 | 24.9 mm (.98 in.) |
| 09 | 31.4 mm (1.24 in.) |
| 10 | 38.0 mm (1.50 in.) |

# NFPA/T3.5.33M

## 7 TOLERANCES - BREAKS - RADII

**7.1** The tolerance of one place decimals is ±0.8 mm (± .031 in);

**7.2** The tolerance of two place decimals is ± 0.25 mm (± .010 in);

**7.3** Break all sharp edges.

**7.4** Maximum radii is 0.254 mm (.01 in).

**7.5** Surface Roughness:
N7 1.6 micrometre (63 microinch) AA.

**7.6** Surface Flatness:
.012 mm (.0005 in) over a distance of 25 mm (1.0 in), and .05 mm (.002 in) over the entire surface.

| ▱ | 0.05 (.002) |
|---|---|
|   | 0.012/25.0 (.0005/1.0) |

## 8 IDENTIFICATION STATEMENT

Use the following statement in catalogs and sales literature when electing to comply with this voluntary standard:

"Mounting surface dimensions conform to National Fluid Power Association Recommended Standard, NFPA/T3.5.33M -1985, **Hydraulic fluid power -Cylinder actuator mounted valves -Standard dimensions for mounting surfaces.**"

NFPA/T3.5.33M

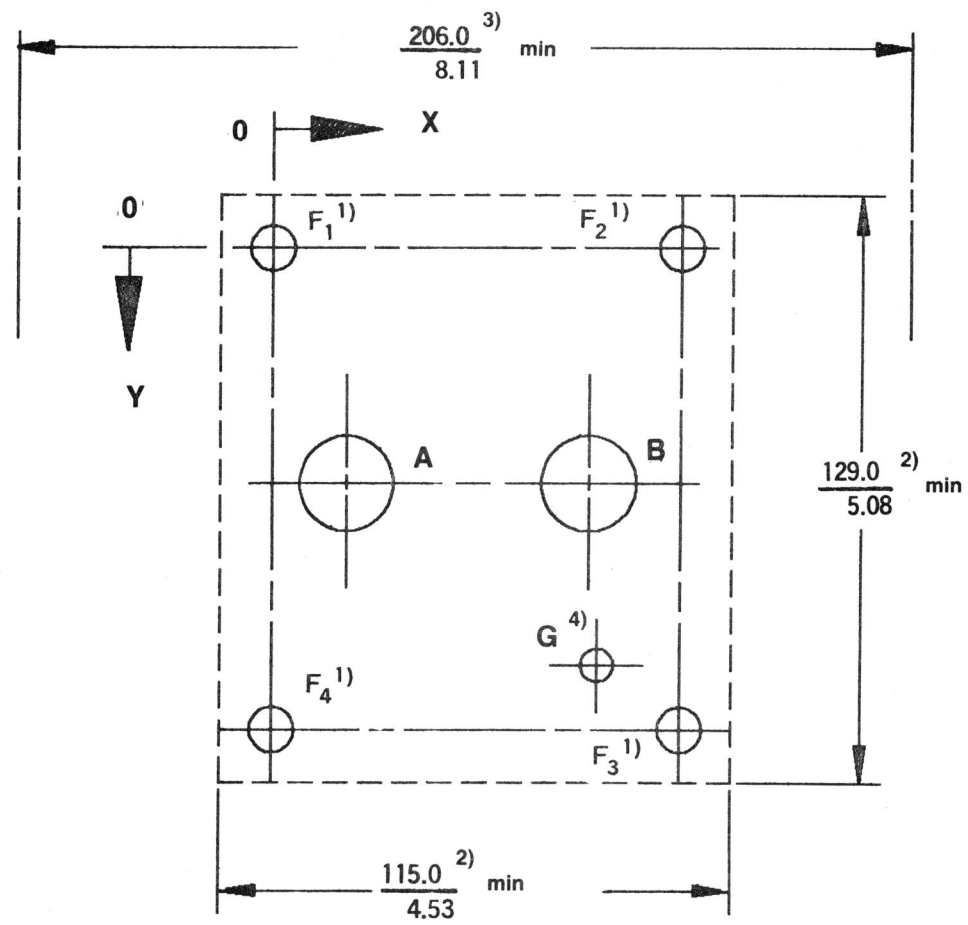

Figure 1 - A06 mounting surface for cylinder actuator mounted valve with 15.9 mm (.63 in) maximum port diameter. 5)

Table 1 - A06 mounting surface for cylinder actuator mounted valve with 15.9 mm (.63 in) maximum port diameter..

|   | A | B | G | $F_1$ | $F_2$ | $F_3$ | $F_4$ |
|---|---|---|---|---|---|---|---|
| X | 18.00 / .709 | 71.00 / 2.795 | 73.00 / 2.874 | 0 | 89.00 / 3.504 | 89.00 / 3.504 | 0 |
| Y | 51.50 / 2.028 | 51.50 / 2.028 | 91.50 / 3.602 | 0 | 0 | 103.00 / 4.055 | 103.00 / 4.055 |
| Ø | 15.90 / .63 | 15.90 / .63 | 9.00 / .354 | 10 / .375 | 10 / .375 | 10 / .375 | 10 / .375 |

) The minimum thread depth is 1.5 bolt diameter. The recommended engagement of fixing bolt thread for ferrous mounting is 1.25 bolt diameter.

) The dimensions specifying the area within the dotted lines are the minimum dimensions for the mounting surface. The corners of the rectangle may be diused to a maximum radius equal to the thread diameter of the fixing bolts. Along each axis the fixing holes are at equal distances to the mounting suface lges.

) This dimension gives the minimum spacing distance between the valve and the adjacent obstructions, for example a wall. The fixing holes are at equal stances to this dimension. The valve manufacturer's attention is drawn to the fact that no part width of the complete valve assembly is to exceed this mension.

) Blind hole in the mounting surface to accomodate the locating pin in the valve: the minimum depth is 8 mm (.31 in).

) The maximum limit of the working pressure for components with this mounting surface will be supplied by the manufacturer.

# NFPA/T3.5.33M

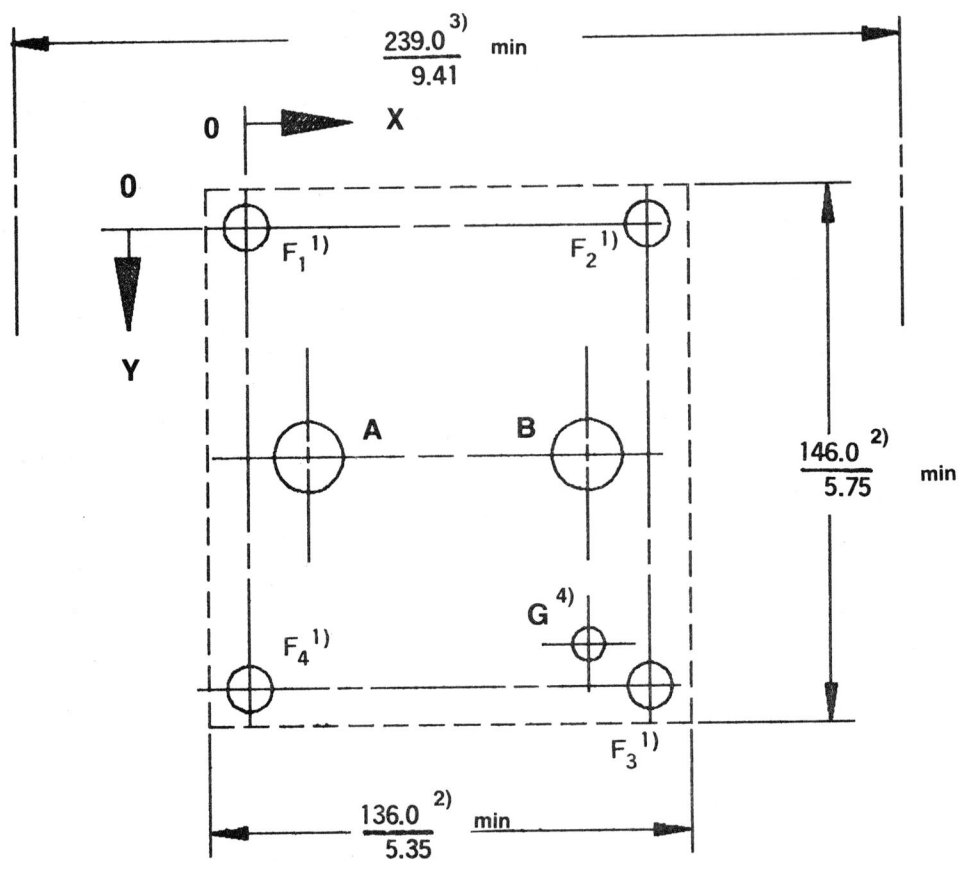

**Figure 2** - A07 mounting surface for cylinder actuator mounted valve with 19 mm (.75 in) in maximum port diameter. [5]

Table 2 - A07 mounting surface for cylinder actuator mounted valve with 19 mm (.75 in) maximum port diameter.

|   | A | B | G | $F_1$ | $F_2$ | $F_3$ | $F_4$ |
|---|---|---|---|---|---|---|---|
| X | 16.50 / .650 | 91.50 / 3.602 | 91.50 / 3.602 | 0 | 108.00 / 4.252 | 108.00 / 4.252 | 0 |
| Y | 59.00 / 3.323 | 59.00 / 3.323 | 107.00 / 4.213 | 0 | 0 | 118.00 / 4.646 | 118.00 / 4.646 |
| Ø | 19.00 / .75 | 19.00 / .75 | 9.00 / .354 | 12 / .500 | 12 / .500 | 12 / .500 | 12 / .500 |

1) The minimum thread depth is 1.5 bolt diameter.  The recommended engagement of fixing bolt thread for ferrous mounting is 1.25 bolt diameter.

2) The dimensions specifying the area within the dotted lines are the minimum dimensions for the mounting surface. The corners of the rectangle may radiused to a maximum radius equal to the thread diameter of the fixing bolts. Along each axis the fixing holes are at equal distances to the mounting surfa edges.

3) This dimension gives the minimum spacing distance between the valve and adjacent obstructions, for example a wall. The fixing holes are at equ distances to this dimension. The valve manufacturer's attention is drawn to the fact that no part width of the complete valve assembly is to exceed t dimension.

4) Blind hole in the mounting surface to accomodate the locating pin in the valve: the minimum depth is 8 mm (.31 in).

5) The maximum limit of the working pressure for components with this mounting surface will be supplied by the manufacturer.

NFPA/T3.5.33M

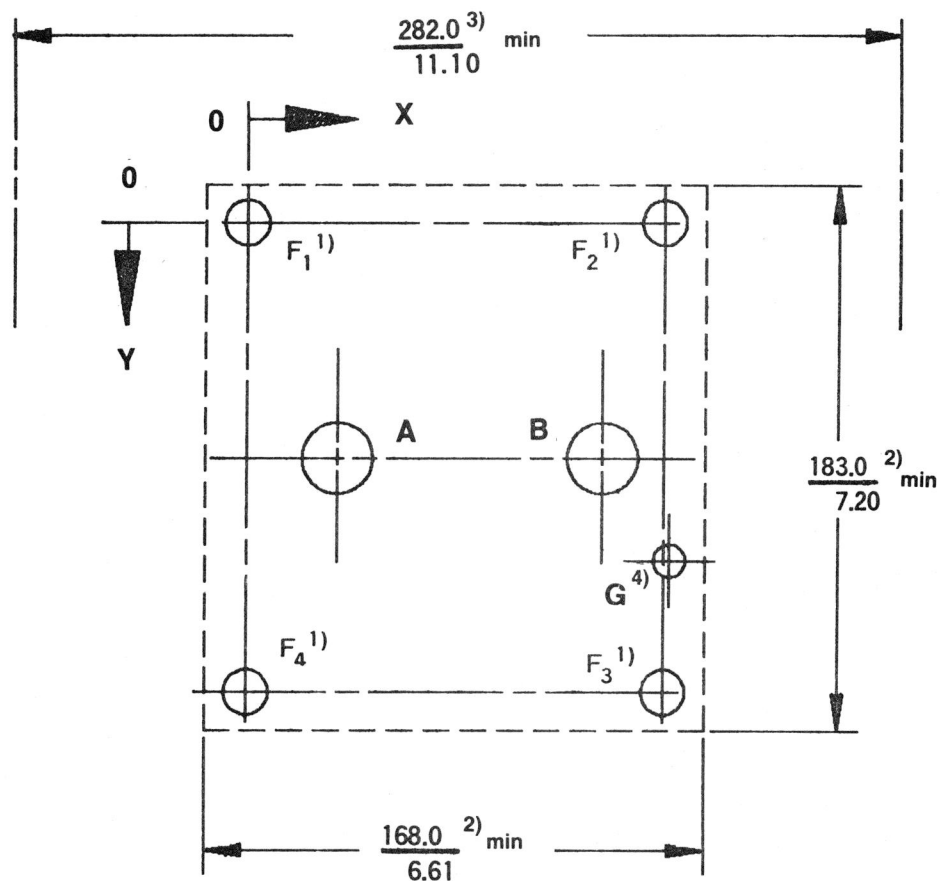

Figure 3 - A08 mounting surface for cylinder actuator mounted valve with 25 mm (.98 in) maximum port diameter. 5)

Table 3 - A08 mounting surface for cylinder actuator mounted valve with 25 mm (.98 in) maximum port diameter.

|   | A | B | G | $F_1$ | $F_2$ | $F_3$ | $F_4$ |
|---|---|---|---|---|---|---|---|
| X | 30.50 / 1.201 | 109.50 / 4.311 | 142.00 / 5.590 | 0 | 140.00 / 5.512 | 140.00 / 5.512 | 0 |
| Y | 77.50 / 3.051 | 77.50 / 3.051 | 111.50 / 4.390 | 0 | 0 | 155.00 / 6.102 | 155.00 / 6.102 |
| Ø | 25.00 / .98 | 25.00 / .98 | 9.00 / .354 | 12 / .500 | 12 / .500 | 12 / .500 | 12 / .500 |

1) The minimum thread depth is 1.5 bolt diameter. The recommended engagement of fixing bolt thread for ferrous mounting is 1.25 bolt diameter.

2) The dimensions specifying the area within the dotted lines are the minimum dimensions for the mounting surface. The corners of the rectangle may be adjused to a maximum radius equal to the thread diameter of the fixing bolts. Along each axis the fixing holes are at equal distances to the mounting surface edges.

3) This dimension gives the minimum spacing distance between the valve and adjacent obstructions, for example a wall. The fixing holes are at equal distances to this dimension. The valve manufacturer's attention is drawn to the fact that no part width of the complete valve assembly is to exceed this dimension.

4) Blind hole in the mounting surface to accomodate the locating pin in the valve: the minimum depth is 8 mm (.31 in).

5) The maximum limit of the working pressure for components with this mounting surface will be supplied by the manufacturer.

# NFPA/T3.5.33M

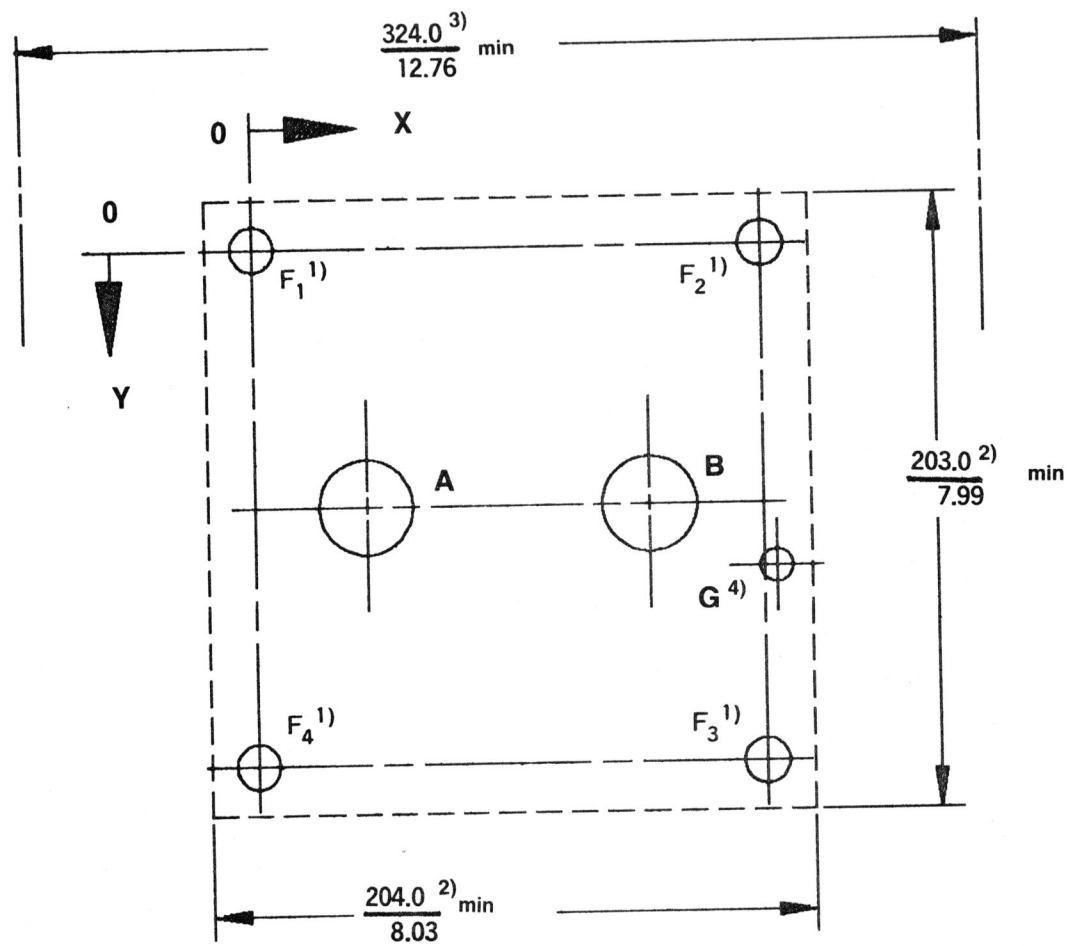

**Figure 4** - A09 mounting surface for cylinder actuator mounted valve with 31.9 mm (1.24 in) maximum port diameter. 5)

**Table 4** - A09 mounting surface for cylinder actuator mounted valve with 31.9 mm (1.24 in) maximum port diameter.

|   | A | B | G | $F_1$ | $F_2$ | $F_3$ | $F_4$ |
|---|---|---|---|---|---|---|---|
| X | 38.00 / 1.496 | 134.00 / 5.275 | 176.00 / 6.929 | 0 | 172.00 / 6.772 | 172.00 / 6.772 | 0 |
| Y | 85.50 / 3.366 | 85.50 / 3.366 | 106.50 / 4.193 | 0 | 0 | 171.00 / 6.732 | 171.00 / 6.732 |
| Ø | 31.4 / 1.24 | 31.4 / 1.24 | 9.00 / 3.54 | 16 / .625 | 16 / .625 | 16 / .625 | 16 / .625 |

1) The maximum thread depth is 1.5 bolt diameter. The recommended engagement of fixing bolt thread for ferrous mounting is 1.25 bolt diameter.

2) The dimensions specifying the area within the dotted lines are the minimum dimensions for the mounting surface. The corners of the rectangle may be radiused to a maximum radius equal to the thread diameter of the fixing bolts. Along each axis the fixing holes are at equal distances to the mounting surface edges.

3) This dimension gives the minimum spacing distance between the valve and adjacent obstructions, for example a wall. The fixing holes are at equal distances to this dimension. The valve manufacturer's attention is drawn to the fact that no part width of the complete valve assembly is to exceed this dimension.

4) Blind hole in the mounting surface to accomodate the locating pin in the valve: the minimum depth is 8 mm (.31 in).

5) The maximum limit of the working pressure for components with this mounting surface will be supplied by the manufacturer.

NFPA/T3.5.33M

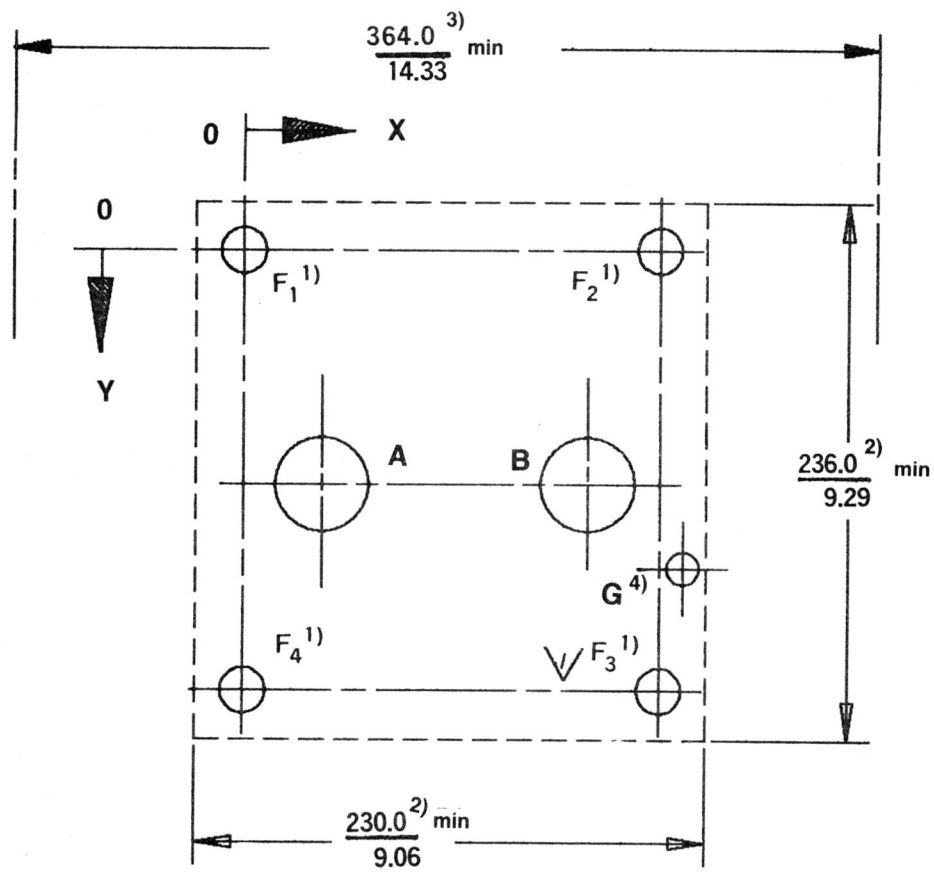

Figure 5 - A10 mounting surface for cylinder actuator mounted valve with 38.0 mm (1.50 in) maximum port diameter.[5]

Table 5 - A10 mounting surface for cylinder actuator valve with 38.0 mm (1.50 in.) maximum port diameter.

| A | B | G | $F_1$ | $F_2$ | $F_3$ | $F_4$ |
|---|---|---|---|---|---|---|
| 35.00 / 1.379 | 155.0 / 6.102 | 198.00 / 7.795 | 0 | 190.00 / 7.480 | 190.00 / 7.480 | 0 |
| 98.00 / 3.858 | 98.00 / 3.858 | 140.00 / 5.512 | 0 | 0 | 196.00 / 7.716 | 196.00 / 7.716 |
| 38.0 / 1.50 | 38.0 / 1.50 | 9.00 / .354 | 20 / .750 | 20 / .750 | 20 / .750 | 20 / .750 |

The minimum thread depth is 1.5 bolt diameter. The recommended engagement of fixing bolt thread for ferrous mounting is 1.25 bolt diameter.

The dimensions specifying the area within the dotted lines are the minimum dimension for the mounting surface. The corners of the rectangle may be used to a maximum radius equal to the thread diameter of the fixing bolts. Along each axis the fixing holes are at equal distances to the mounting surface es.

This dimension gives the minimum spacing distance between the valve and adjacent obstructions, for example a wall. The fixing holes are at equal ances for this dimenion. The valve manufacturer's attention is drawn to the fact that no part width of the complete valve assembly is to exceed this ension.

Blind hole in the mounting surface to accomodate the locating pin in the valve: the minimum depth is 8 mm (.31 in).

The maximum limit of the working pressure for components with this mounting surface will be supplied by the manufacturer.

## Annex to
## NFPA/T3.5.33M-1985

This Annex is not part of the NFPA/T3.5.33M-1985 standard, but is included for information purposes only.

**REFERENCES**

ISO 5598, *Fluid power systems and components - Vocabulary.*

ISO 1000, *SI units and recommendations for the use of their multiples and of certain other units.*

ANSI/Y14.5M, *Engineering Drawings and Related Documentation Practices Dimensioning and Tolerancing.*

ANSI/B93.65M-1983

---

**AN INDUSTRY STANDARD FOR FLUID POWER**

---

American National Standard

Hydraulic fluid power -

Code* for identification of valve mounting surfaces

(Technically identical to ISO 5783)
(Metric Only)
(NFPA/T3.5.34M-1982)

Approved as an ANSI Standard
13 July 1983

\* This standard defines an identification code that can be used in reports, catalogs and literature, but the standard specifically does not require marking hardware components.

Descriptors: hydraulic fluid power, fluid power; sub-plates; mounting surfaces; valves.

---

published by
**NATIONAL FLUID POWER ASSOCIATION, INC.**

3333 N. Mayfair Road / Milwaukee, WI 53222 / 414-778-3344 / TLX 26898

# ANSI/B93.65

## FOREWORD

This Foreword is not part of American National Standard Hydraulic fluid power - Code for identification of valve mounting surfaces, ANSI/B93.65M-1983 (technically identical to ISO 5783).

NFPA, thru its dual administration of the NFPA and ISO Technical Committees for Fluid Power, encouraged the participation of NFPA members in the development of ISO standards. While this work was underway at ISO, on ISO 5783 (Hydraulic fluid power - Code for identification of valve mounting surfaces), NFPA participants, in cooperation with NFPA Headquarters, developed a similar NFPA proposal. Thru the close liaison resulting from many of the same participants working at both the NFPA and ISO levels, identical documents were developed and hereby are being promulgated.

On 27 June 1980, T3.5.34M was sent out for a "Ballot of Consensus." Four negative ballots were received. Thru correspondence, Chairman Prevallet addressed the comments and they were subsequently withdrawn. On 11 February 1981, the Technical Board approved the document for ballot with the addition of the footnote which clarifies the intended use of this standard.

Headquarters Staff prepared T3.5.34M for Ballot on 15 May 1981. Ballot closed 11 August 1980 with one negative comment.

The document was forwarded to the Technical Board for approval. Approval was granted (24 February 1982) with the provision that the negative ballot be resolved.

Confirmation of withdrawal of the negative comment was received by Headquarters 4 March 1982. Headquarters Staff forwarded T3.5.34M to the Board of Directors for final approval. Final approval was granted 2 June 1982.

TAG/SC 5 and Project Group members who worked on this document:

**Dave Prevallet**
Section Chairman
Gresen Mfg. Co.

**Harold Jacoby**
Section Vice Chairman
Rexnord, Inc.

**Wayland Tenkku**
Section Secretary
Fluid Controls Inc.

**Henry Barthe**
Technical Auditor
Schroeder Brothers Corp.

**James C. White***
Director of Technical Services
National Fluid Power Association

**R. Bailey**
C. A. Norgren Co.

**J. Berninger (alternate)**
Parker Hannifin Corp.

**T. Clark**
Rexnord Inc.

**E. Cole (alternate)**
Schrader Bellows Div.

**J. Colter**
Skinner Valve Div.

**R. Culbertson**
Dynamco Inc.

**R. Falgerlie**
MAC Valves Inc.

**J. Fisher, Jr. (alternate)**
Schrader Bellows Div.

**C. Fletcher**
Fletch/Air

**E. Fletcher**
John Deere Product Engineering Center

**T. Frankenfield (alternate)**
Rexroth Corp.

**W. Forster***
WABCO Fluid Power Div.

**C. Grigsby**
Schrader Bellows Div.

**R. Hoffman (alternate)**
Parker Hannifin Corp.

**P. Huff (alternate)**
Dynamco Inc.

**E. Klein**
Rexroth/Mobile Hydraulics Div.

**J. Komendera**
Automatic Valve Corp.

**Z. Lansky**
Parker Hannifin Corp.

**J. Lytle**
Versa Prod.

**O. Maha**
Individual Member

**D. McGeachy**
Numatics Inc.

**R. Olen**
Gresen Mfg. Co.

**J. Pippenger**
Individual Member

**L. Schmaltz**
Ross Operating Valve

**J. Walrad**
Sperry Vickers

**J. Welker (alternate)**
Numatics Inc.

It is intended that this NFPA Recommended Standard after its approval, will be submitted to ANSI for promulgation as an American National Standard. This would accomplish total harmonization between identical ISO and ANSI Standards.

---

*Company affiliation has been changed.

The harmonization procedure will continue until identical NFPA and ISO documents are issued.

Proposed International Standard, ISO 5783 was drawn up by ISO/TC 131/SC 5 "Control components," the Secretariat of which is held by France: and by the ISO Technical Committee 131, "Fluid power systems and components," the Secretariat of which is held by the USA. This document has concluded final balloting in ISO and has been approved by the member nations. It is presently awaiting publication by ISO Central Secretariat. ISO 5783 was approved by the member bodies of the following countries:

| | |
|---|---|
| Australia | Italy |
| Austria | Japan |
| Belgium | Mexico |
| Brazil | Netherlands |
| Canada | Norway |
| Chile | Romania |
| Finland | South Africa, Rep. of |
| France | Spain |
| Germany | Sweden |
| Hungary | Switzerland |
| India | United Kingdom |
| Ireland | Yugoslavia |

Member bodies expressing disapproval of the document:

Czechoslovakia
USA

On 24 September 1982, ANSI/B93.65M was submitted to ANSI Committee B93 for ballot. Balloting closed 3 December 1982 with one negative vote. On 6 May 1983 ANSI B93 Committee was requested to review the unresolved negative and the ballots remained intact.

ANSI/B93.65M was forwarded to ANSI Board of Standards Review on 8 June 1983 and granted approval 13 July 1983.

The membership roster for the Standards Committee B93 at the time of ballot:

**Jack C. McPherson**
Chairman

**Daniel B. Shore**
Vice Chairman

**Allen E. Tucker**
Co-Secretary

**William G. Wagner**
Co-Secretary

**American Society of Agricultural Engineers**
Ed Fletcher

**American Society for Engineering Education**
William R. Smith

**American Society of Lubrication Engineers**
(To be named)

**Compressed Air & Gas Institute**
David E. Bonn
John Addington (alternate)

**Construction Industry Manufacturers Association**
Glenn Stewart

**Fluid Controls Institute**
Jude Paull
Eric Blanchi (alternate)

**Fluid Power Distributor Association**
Thomas Neff

**Fluid Power Society**
Robert L. Firth
Carroll Grigsby
Robert W. Hanpeter
Marty King
Richard Read
Ronald Smith
Alan Tiedman

**Fluid Sealing Association**
John Scannell
Alex Pilecki (alternate)

**Instrument Society of America**
(To be named)

**Joint Industry Council**
Robert Muhl

**Material Handling Institute**
Jack C. McPherson
Willard Chichester (alternate)

**National Fluid Power Association**
Richard N. Bailey
John Bowbin
James L. Fisher, Jr.
Walter Forster
Paul Schacht

**National Machine Tool Builders' Association**
John B. Deam

**Rubber Manufacturers Association**
William A. Hertel
John R. Loder
E. J. McCarthy (alternate)

**Society of Automotive Engineers**
William A. Hertel
Eugene Falendysz
John T. Parrett
Henry Schultz
Daniel B. Shore
William L. Snyder
Robert W. White

**U. S. Department of Defense**
William P. Coyne

**Company Member**
L. L. Schmaltz
Don McGeachy
John Welker (alternate)

**Individual Member**
E. C. Fitch, Jr.
Otto Maha
John J. Pippenger
Jack Walrad
Tom Wanke
Frank Yeaple

scs

**Hydraulic fluid power -
Code for identification of valve mounting surfaces**

## 0 INTRODUCTION

In hydraulic fluid power systems power is transmitted and controlled through a liquid under pressure within an enclosed circuit. The control and the regulation of the fluid are made by valves which can either be directly connected with pipes or be mounted on sub-plates.

## 1 SCOPE AND FIELD OF APPLICATION

**1.1** This National Standard defines an identification code for valve mounting surfaces, located on subplates defined in International Standards.

— Non-standard mounting surfaces cannot be identified by this code.

**1.2** This National Standard does not require that the hardware be marked with the identification code.

## 2 IDENTIFICATION CODE

Designate the mounting surfaces by the five groups of numbers or letters indicated below, written in the order given, and separated by dashes:

a) the reference "ISO" and the number of the International Standard in which the mounting surface is described;

b) two letters which specify all the necessary information to ensure interchangeability except main port sizes, locating pins and secondary ports (see note 1 in clause 3);

c) two numerals which specify the main port sizes in accordance with the following table;

**Table — Main port sizes according to their diameters**

| Size | Diameter of main ports mm |
|------|---------------------------|
| 00   | $0 < \phi \leq 1,6$       |
| 01   | $1,6 < \phi \leq 2,5$     |
| 02   | $2,5 < \phi \leq 4$       |
| 03   | $4 < \phi \leq 6,3$       |
| 04   | $6,3 < \phi \leq 10$      |
| 05   | $10 < \phi \leq 12,5$     |
| 06   | $12,5 < \phi \leq 16$     |
| 07   | $16 < \phi \leq 20$       |
| 08   | $20 < \phi \leq 25$       |
| 09   | $25 < \phi \leq 31,5$     |
| 10   | $31,5 < \phi \leq 40$     |
| 11   | $40 < \phi \leq 50$       |
| 12   | $50 < \phi \leq 63$       |
| 13   | $63 < \phi \leq 80$       |

d) one numeral which specifies the number of main ports;

e) one letter which indentifies the design variations (see note 2 in clause 3).

## 3 EXAMPLE OF USE OF THE CODE

Designate the mounting surface of a 4-port hydraulic directional control valve of maximum main port size 11,2 mm, such as described in ISO 4401[1]), as follows

ISO 4401 - AC - 05 - 4 - A

NOTES -

1 - The Secretariat of ISO/TC 131/SC5, (Association francaise de normalization (AFNOR), Tour Europe, Cedex 7, 92080 Paris la Defense, France), keeps the list of the two letters specified in 2 b) up to date and ensures that two mounting surfaces specified in two different International Standards but having the same geometrical impact are designated by the same two letters. They are assigned starting with AA in chronological order.

2 - By "design variations" is meant any addition or withdrawal of secondary ports or locating pins and, if necessary, any dimensional change for main ports in so far as such changes do not imply size modifications. Attention is drawn to the fact that a different letter can only be used if a new International Standard is adopted. The letter is assigned starting with A in the chronological order of modifications.

3 - If a new International Standard were adopted to modify the mounting surface mentioned in clause 3 and increased the maximum main port size to 12,5 mm, the new mounting surface would be designed as follows:

ISO ... - AC - 05 - 4 - B

If now it were decided in that International Standard to increase the maximum main port size to 14 mm, this mounting surface would be designated as follows:

ISO ... - AC - 06 - 4 - A

The letters AC would only be modified for another group of letters if, for example, fixing screw dimensions were changed or the port arrangement were different.

4 - This National Standard does not require that the hardware be marked with the identification code.

## 4 IDENTIFICATION STATEMENT

Use the following statement in test reports, catalogs and sales literature when electing to comply with this National Standard:

"Mounting surface identification code in accordance with ANSI/B93.65M-1983, **Hydraulic fluid power - Code for identification of valve mounting surfaces.**"

---

1) Similar to NFPA/T3.5.1 presently under revision.

# APPENDIX

## to ANSI/B93.65M-1983

**REFERENCES**

ANSI/B93.2-1971 and Supplement ANSI/B93.2A-1978, American National Standard, Glossary of Terms for Fluid Power.

NFPA/T2.10.1M-1978, National Fluid Power Association Recommended Standard, Metric Units for Fluid Power Applications.

ANSI/B93.7-1968 (R1979), American National Standard USA Standard Dimensions for Mounting Surfaces of Sub-plate Type Hydraulic Fluid Power Valves.

ISO 1000-1981, International Standard SI units and recommendations for the use of their multiples and of certain other units.

ISO 5598[1], Fluid power systems and components - Vocabulary.

---

1) At present at the stage of draft.

NFPA Recommended Standard
T3.7.2M-1968 (R1980)

## AN INDUSTRY STANDARD FOR FLUID POWER

Graphic Symbols

for

Fluidic Devices and Circuits

Approved as an NFPA Recommended Standard
1 February 1968

published by

# NATIONAL FLUID POWER ASSOCIATION, INC.

3333 N. Mayfair Road   /   Milwaukee, WI 53222   /   414-778-3344   /   TLX 26898

# FOREWORD

(This Foreword is not a part of NFPA Recommended Standard Graphic Symbols for Fluidic Devices and Circuits, T3.7.68.2)

On February 23, 1965, a general conference was called under the auspices of Standards Committee B93 of the United States of America Standards Institute. The 59 engineers and scientists assembled -- representing all segments of the industry, including manufacturers, government research and military agencies, and private companies doing research -- agreed that standards were needed, and voted unanimously to assign the responsibility of sponsoring these standards to the National Fluid Power Association.

The Board of Directors of the NFPA approved this action, and formed the Fluidic Devices Section under the jurisdiction of the NFPA Technical Board. The Section's organizational meeting was held May 25 and 26, 1965, in Cleveland, Ohio, at which time James I. Morgan, of Galland-Henning Mfg. Co., was appointed Chairman of the Section and Melvin E. Long, Penton Publishing Co., was appointed chairman of the Graphic Symbols Project Group. Meetings were held regularly on a bimonthly schedule.

Since January, 1967, the work has continued under Melvin E. Long, Chairman of the Fluidic Devices Section and Wayne Brown, Corning Glass Works, Chairman of the Graphic Symbols Project Group.

On January 18, 1967, 1500 copies of the first complete manuscript, "Progress Report on Formulation of Porposed NFPA Recommended Standards for Fluidic Devices" were distributed to all in the fluidics industry. Comments and criticisms resulting from the distribution of the progress report were resolved through the helpful assistance of the SAE A-6D Fluidics Panel chaired by Hans Stern of General Electric Co. and the Government Interagency Coordinating Committee for Fluidics chaired by R. Orin Gilbertson. A final draft was approved by the NFPA Project Group at its meeting in Cleveland on September 21, 1967. The draft was subsequently approved by the SAE during their meeting in Phoenix, Arizona, October 19, 1967, and by the chairman of the USGICC on October 19, 1967.

The NFPA Graphic Symbols and Drafting Standards Coordinating Committee, chaired by J. L. Fisher, Jr. of Bellows-Valvair, processed and approved the proposed standard as being compatible with all existing USA Standards.

Favorable ballot of the Fluidic Devices Section of the NFPA on 29 December 1967 resulted in the Proposed Standard being approved as a NFPA Recommended Standard by the Board of Directors of the Association on 1 February 1968. Editorial improvements, resulting from comments received with the ballots, were incorporated into this document at a project group meeting held on 27 March 1968.

NFPA/T3.7.2

## MEMBERS OF THE PROJECT GROUP RESPONSIBLE FOR THE
## DEVELOPMENT OF THIS STANDARD

| | |
|---|---|
| Long, Melvin E. -- Section Chairman | Penton Publishing Company |
| Brown, Wayne -- Project Chairman | Corning Glass Works |
| Streeter, David -- Vice Chairman | Norgren/Fluidics |
| Wilcox, Richard L. -- Secretary | Delavan Mfg. Co. |
| | |
| Allen, C. W. | Westinghouse Air Brake Co. |
| Auger, Raymond | Raymond N. Auger & Co. |
| Bart, Loren | The Modernair Corp. |
| Becker, Bruce H. | Imperial-Eastman Corp. |
| Belsterling, Charles A. | Conrac |
| Bracki, Kenneth A. | Parker Hannifin Corp. |
| Brown, James | Barksdale Valves |
| Bruinsma, Norman G. | Gleason Works |
| Carter, Howard C. | Barksdale Valves |
| Coleman, Richard R. | Univac |
| Curtiss, Howard A. | Norgren/Fluidics |
| Datwiler, Walter F. | Bendix Research Labs |
| Davis, R. K. | U. S. Naval Avionics Facility |
| Delmege, Arthur | Vickers Division |
| DiTirro, D. A. | Bellows-Valvair |
| Fisher, James L., Jr. | Bellows-Valvair |
| Gibbon, Vernon D. | NASA, Lewis Research Center |
| Gibson, Robert | Imperial-Eastman Corp. |
| Gieryn, Fred V. | Double A Products Co. |
| Gilbertson, R. Orin | U.S.N.O.T.S. -- China Lake, Calif. |
| Gluskin, Richard S. | Control Data |
| Gottron, Richard | Harry Diamond Labs |
| Griffin, William | NASA, Lewis Research Center |
| Hall, James | USAF Flight Dynamics Lab |
| Jacobs, Al | Milwaukee School of Engineering |
| Kautz, Will | The Modernair Corp. |
| Kepner, Hugh G. | Kepner Products Co. |
| Keto, Jorma R. | Harry Diamond Labs |
| Klessig, Ernest | Racine Hydraulics Inc. |
| Laakaniemi, R. N. | Johnson Service Co. |
| Lapinas, Z. J. | Aviation Electric |
| Lechner, T. J. | Johnson Service Co. |
| Luna, Sal F. | Bettis Atomic Power Laboratory |
| Maas, Margaret | Design News Magazine |
| Moffat, Allen J. | Beckett-Harcum Company |
| Morris, Allen | Hydraulics and Pneumatics Magazine |
| Morgan, James I. | National Fluid Power Association |
| Moynihan, Fred | Honeywell |
| Muehl, C. | Ames Research Center |
| Ott, Harold | U. S. Naval Air Engineering Center |
| Parsons, Alonzo R. | Honeywell |
| Pinkstaff, Carlos D. | Imperial-Eastman Corp. |
| Pollard, Fred | Grumman Aircraft |
| Rimek, K. O. | USAF Rocket Propulsion Lab |
| Sampson, Merritt B. | The S-P Manufacturing Corp. |

| | |
|---|---|
| Scudder, Kenneth R. | Harry Diamond Labs |
| Seleno, A. A. | Bendix Research Lab |
| Shinn, Jeffrey N. | General Electric Co. |
| Standen, G. W. | Aviation Electric |
| Stern, Hansjoerg | General Electric Co. |
| Wagner, R. E. | U. S. Army -- Watervliet Arsenal |
| Westveer, Robert C. | New York Air Brake Co. |

## REFERENCES

1. Glossary of Terms for Fluidic Devices and Circuits, NFPA T3.7.68.1
2. USA Standard Glossary of Terms for Fluid Power, B93.2
3. USA Standard Graphic Symbols for Fluid Power Diagrams Y32.10

NFPA/T3.7.2

## 1.0 INTRODUCTION

Fluid power includes the transmission and/or control of power thru a fluid (liquid or gas) under pressure within an enclosed circuit. Fluidics is one means of using fluid under pressure for control purposes. (For the standard definition of "fluidics", see Reference No. 1)

Circuits of interconnected fluidic devices can be depicted in diagrams made up of graphic symbols and lines to represent interconnected fluid conductors.

## 2.0 SCOPE

This document presents a system of graphic symbols for fluidic devices and circuits.

Elementary forms of symbols are: circles, squares, triangles, arcs, arrows, lines, dots, and crosses.

Symbols using words or their abbreviations are avoided. Symbols capable of crossing language barriers are presented herein.

Definitions of terms found in this standard are listed in Ref. 1.

## 3.0 PURPOSE

The purpose of this standard is to provide the circuit designer meaningful and specific graphic symbols for use in drawings and schematics which clearly define the function of the type of device employed to perform that function.

In preparing this document, the requirement for fluidic symbols to satisfy two basic needs was recognized. The first was the need of the system designer interested primarily in the function of the devices. The second was the circuit designer primarily interested in the operating principle. Functional symbols depict a fluidic function which may be performed by a single element or a combination of elements. Operating-principle symbols depict the fluid phenomena occurring in the interaction region of a fluidic element.

## 4.0 SYMBOL RULES

Functional representation is indicated by symbols enclosed within square envelopes. Operating-principle representation is indicated by symbols enclosed within circular envelopes. The difference in envelopes is deliberately intended to emphasize the difference in purpose of the symbols.

NFPA/T3.7.2

Functional Symbol        Operating-Principle Symbol

The cases where an operating principle symbol is not shown, for example, Schmitt Trigger, indicate that at present no single operating principle can perform the fluidic function, or that a combination of operating principles is required to perform the function.

4.1 The relative port locations for the symbols are arranged as follows:

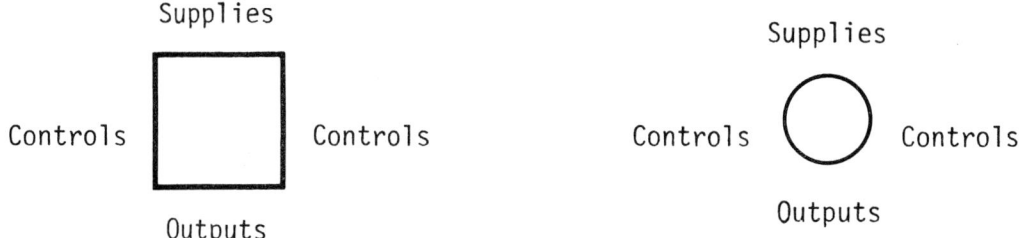

All symbols may be oriented in 90-deg increments from the position shown.

4.2 Specific ports are identified by the following nomenclature:

Supply Port - S;   Control port - C;   Output port - O.

The nomenclature shown on the graphic symbols need not be used on schematic diagrams. It is primarily intended to correlate the function of each port with the truth table.

4.3 Supply ports may be either active or passive. An inverted triangle, ▽ denotes a supply source connected to the supply port (active device)

Active   Devices

4.4 The fluid used is indicated by the inverted triangle.

A solid triangle, ▼ denotes liquid fluid.

An open triangle, ▽ denotes a gas fluid

4.5 An arrowhead on the control line inside the symbol envelope indicates continuous flow is required to maintain state (no memory, no hysteresis):

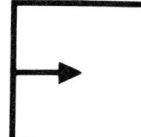 

Indicates no memory

4.6 Interconnecting fluid lines are shown with a dot at the point of interconnection:

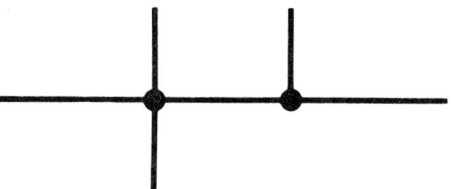

4.7 Crossing fluid lines are shown without dots:

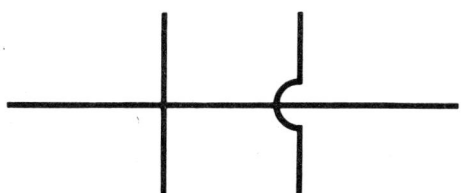

4.8 A small + on the output of a bistable device indicates initial or start-up flow condition:

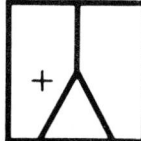 

NFPA/T3.7.2

## 5.0 GRAPHIC SYMBOLS

### 5.1 Bistable Fluidic Devices

#### 5.1.1 Flip Flop

##### 5.1.1.1 Functional Symbol

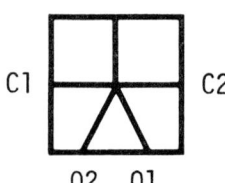

Truth Chart

| C1 | C2 | O1 | O2 |
|----|----|----|----|
| 1 | 0 | 1 | 0 |
| 0 | 0 | 1 | 0 |
| 0 | 1 | 0 | 1 |
| 0 | 0 | 0 | 1 |
| 1 | 1 | Not Valid | |

##### 5.1.1.2 Operating Principle Symbols

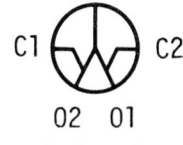

Wall Attachment    Induction    Edgetone

#### 5.1.2 Digital Amplifier

##### 5.1.2.1 Functional Symbol

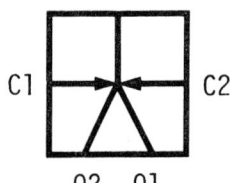

Truth Chart

| C1 | C2 | O1 | O2 |
|----|----|----|----|
| 1 | 0 | 1 | 0 |
| 0 | 1 | 0 | 1 |
| 0 | 0 | Undefined | |
| 1 | 1 | Undefined | |

##### 5.1.2.2 Operating Principle Symbols

Jet Interaction

#### 5.1.3 Binary Counter (SRT Flip-Flop)

##### 5.1.3.1 Functional Symbol

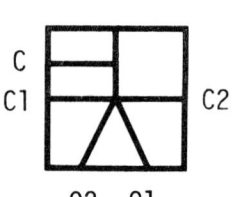

Truth Chart

| C | C1 | C2 | O1 | O2 |
|---|----|----|----|----|
| 0 | 1 | 0 | 1 | 0 |
| 0 | 0 | 0 | 1 | 0 |
| 0 | 0 | 1 | 0 | 1 |
| 0 | 0 | 0 | 0 | 1 |
| 1 | 0 | 0 | 1 | 0 |
| 0 | 0 | 0 | 1 | 0 |
| 1 | 0 | 0 | 0 | 1 |
| 0 | 0 | 0 | 0 | 1 |

- 258 -

NFPA/T3.7.2

### 5.1.4 Multivibrator

5.1.4.1 Functional Symbol

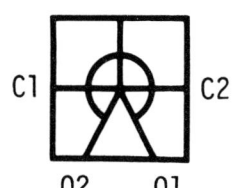

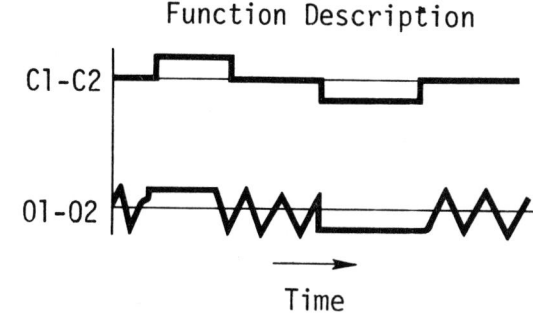

Function Description

5.1.4.2 Operating Principle Symbol

Wall Attachment

### 5.1.5 Oscillator

5.1.5.1 Functional Symbol

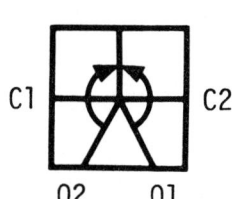

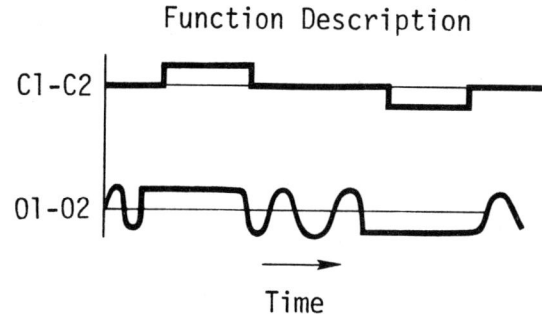

Function Description

5.1.5.2 Operating Principle Symbol

Jet Interaction

# NFPA/T3.7.2

## 5.2 Monostable Fluidic Devices

### 5.2.1 OR NOR  (Only one input defines a NOT element.)

#### 5.2.1.1 Functional Symbol

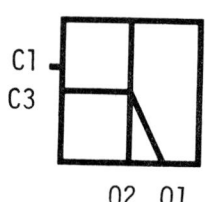

Truth Chart

| C1 | C3 | O1 | O2 |
|----|----|----|----|
| 0 | 0 | 0 | 1 |
| 1 | 0 | 1 | 0 |
| 0 | 1 | 1 | 0 |
| 1 | 1 | 1 | 0 |

#### 5.2.1.2 Operating Principle Symbols

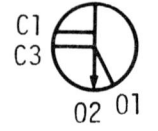

Wall Attachment    Wall Attachment (Internal Fluid Bias)    Turbulence

Vortex    Focused Jet    Geometric Bias

### 5.2.2 One Shot

#### 5.2.2.1 Functional Symbol

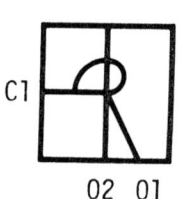

Function Description

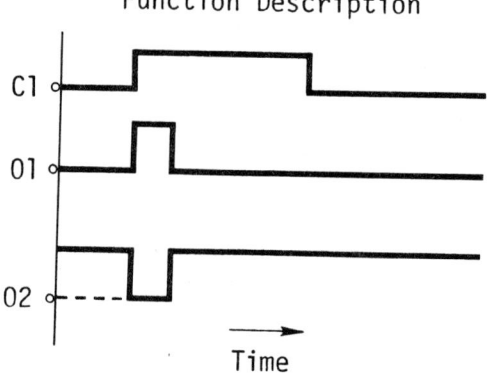

NFPA/T3.7.2

5.2.3 AND NAND

5.2.3.1 Functional Symbol

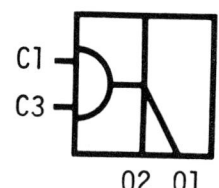

Truth Chart

| C1 | C3 | 01 | 02 |
|----|----|----|----|
| 0  | 0  | 0  | 1  |
| 1  | 0  | 0  | 1  |
| 0  | 1  | 0  | 1  |
| 1  | 1  | 1  | 0  |

5.2.4 Schmitt Trigger

5.2.4.1 Functional Symbol

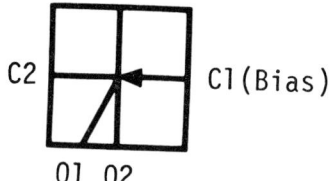

Truth Chart

|        |    | 01 | 02 |
|--------|----|----|----|
| C1 > C2 |   | 1  | 0  |
| C1 < C2 |   | 0  | 1  |
| C1 = C2 |   | Undefined | |

5.2.5 Exclusive OR (Active Only)

5.2.5.1 Functional Symbol

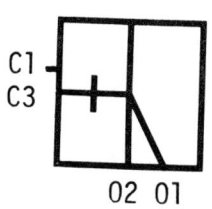

Truth Chart

| C1 | C3 | 01 | 02 |
|----|----|----|----|
| 0  | 0  | 0  | 1  |
| 1  | 0  | 1  | 0  |
| 0  | 1  | 1  | 0  |
| 1  | 1  | 0  | 1  |

5.2.5.2 Operating Principle Symbol

See "Passive Digital Device" Section.

## 5.3 Passive Digital Devices

### 5.3.1 2/3 And (Passive And - Exclusive OR)

5.3.1.1 Functional Symbol

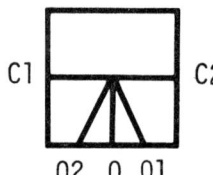

5.3.1.2 Operating Principle Symbol

Jet Interaction

Truth Chart

| C1 | C2 | O1 | O2 | O |
|----|----|----|----|---|
| 0  | 0  | 0  | 0  | 0 |
| 1  | 0  | 1  | 0  | 0 |
| 0  | 1  | 0  | 1  | 0 |
| 1  | 1  | 0  | 0  | 1 |

### 5.3.2 Exclusive OR

5.3.2.1 Functional Symbol

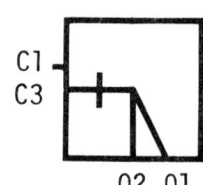

5.3.2.2 Operating Principle Symbol

Jet Interaction

Truth Chart

| C1 | C3 | O1 | O2 |
|----|----|----|----|
| 1  | 0  | 1  | 0  |
| 0  | 1  | 1  | 0  |
| 1  | 1  | 0  | 1  |
| 0  | 0  | 0  | 0  |

### 5.3.3 Passive OR

5.3.3.1 Functional Symbol

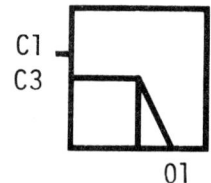

5.3.3.2 Operating Principle Symbol

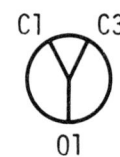

Jet Interaction (Passive)

Truth Chart

| C1 | C3 | O1 |
|----|----|----|
| 0  | 0  | 0  |
| 1  | 0  | 1  |
| 0  | 1  | 1  |
| 1  | 1  | 1  |

# 5.4 Proportional Fluidic Devices

## 5.4.1 Analog Amplifier

### 5.4.1.1 Functional Symbol

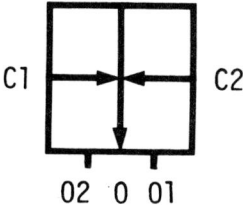

Function Descriptions

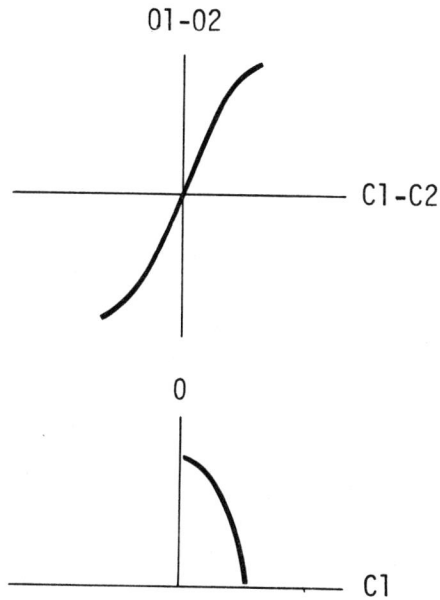

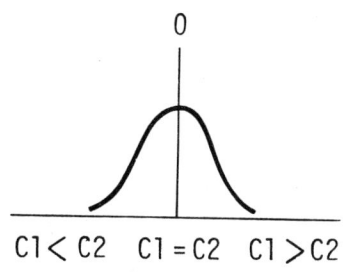

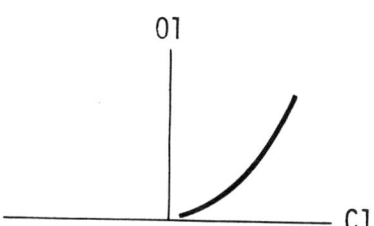

NFPA/T3.7.2

5.4.1.2 Operating Principle Symbols

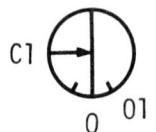
Jet Interaction
(Single Input)

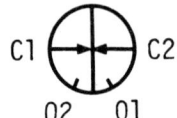
Jet Interaction
(Differential)

Separation Point
Control

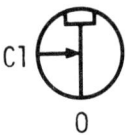
Impacting Jet
(transverse)

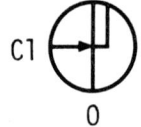
Impacting Jet
(direct)

Vortex

5.4.2 Pure Fluidic Throttling Valve

5.4.2.1 Functional Symbol

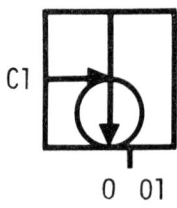

Function Description

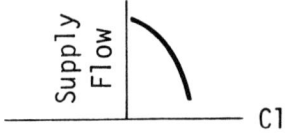

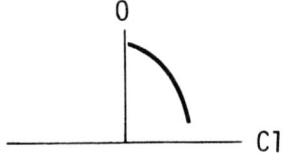

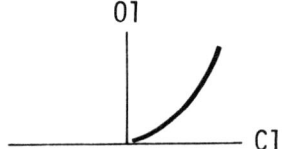

5.4.2.2 Operating Principle Symbol

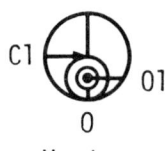
Vortex

NFPA/T3.7.2

5.4.3  Rate Sensor

    5.4.3.1  Functional Symbol

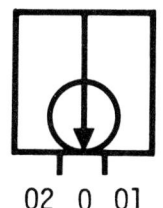

    5.4.3.2  Operating Principal Symbol

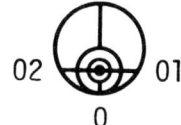

5.5  Impedance Symbols

    5.5.1  Resistance, general

    5.5.2  Resistance, Linear

    5.5.3  Resistance, non-linear

    5.5.4  Capacitance

    5.5.5  Inductance

    5.5.6  Diode  ( ←$\frac{free}{flow}$ )

    5.5.7  An inclined arrow, variable case.  , thru any of the impedance symbols denotes

- 265 -

# NOTES

ANSI/B93.14M-1971 (R1979)

## AN INDUSTRY STANDARD FOR FLUID POWER

American National Standard

Methods of Presenting Basic

Performance Data for Fluidic Devices

(NFPA/T3.7.3M-1970)

Approved as an ANSI Standard
14 July 1971

published by
## NATIONAL FLUID POWER ASSOCIATION, INC.

3333 N. Mayfair Road  /  Milwaukee, WI 53222  /  414-778-3344  /  TLX 26898

# FOREWORD

(This Foreword is not a part of American National Standard Method of Rating Performance of Fluidic Devices, ANSI/B93.14-1971.)

On February 23, 1965, a General Conference was called under the auspices of Standards Committee B93 of the American National Standards Institute. The fifty-nine engineers and scientists assembled -- representing all segments of the industry, including manufacturers, government research and military agencies, and private companies doing research -- agreed that standards were needed, and voted unanimously to assign the responsibility of sponsoring these standards to the National Fluid Power Association.

The Board of Directors of the NFPA approved this action, and formed the Fluidic Devices Section under the jurisdiction of the NFPA Technical Board. The Section's organizational meeting was held May 25 and 26, 1965, in Cleveland, Ohio, at which time James I. Morgan of Galland-Henning Mfg. Co., was appointed chairman of the Section. Shortly thereafter, A. H. Delmege of Vickers Division was appointed Chairman of the Performance Rating Project Group.

Since January, 1967, the work has continued under Melvin E. Long, Chairman of the Fluidic Devices Section. The final draft of the project group was completed on 18 January 1968. It was distributed for written comments on 18 July 1968. The comments received were reviewed at Project Group meetings on 20 November 1968 and 20 April 1969. On 30 April 1969 the Project Group agreed that all comments had been accommodated by minor revisions and that the resulting Summary Consensus Draft dated 9 June 1969 was ready for ballot.

Constructive comments accompanying the 100 percent favorable ballot which closed on 11 October 1969, were incorporated wherever possible. On 14 January 1970 the Technical Board added their approval. This was followed by Board of Directors approval on 15 January 1970 as NFPA Recommended Standard T3.7.3-1970.

Members of the NFPA Project Group preparing this standard are listed on page 4.

On 3 April 1970, the NFPA Recommended Standard was submitted to ANSI Standards Committee B93 for promulgation as an ANSI Standard. Favorable ballot was concluded on 4 May 1970. Approval by the ANSI Board of Standards Review was granted on 14 July 1971.

The membership roster for Standards Committee B93 at the time of approval is listed on page 4.

ANSI/B93.14

Members of the NFPA Project Group responsible for development of this standard include:

| | | |
|---|---|---|
| Delmege, Arthur | Project Chairman | Vickers Division |
| Shinn, J. N. | Project Vice Chairman | General Electric Co. |
| Long, Melvin | Section Chairman 1967-69 | Penton Publishing Co. |
| Brown, Wayne | Section Chairman 1969- | Corning Glass Works |
| Seleno, Andrew | Secretary | Bendix Research Labs |
| Morgan, James I. | NFPA, ANSI, and ISO Liaison | National Fluid Power Assn. |

| | |
|---|---|
| Bjornsen, B. | Johnson Service Co. |
| Curtiss, H. | Norgren/Fluidics |
| Datwyler, W. | Bendix Research Labs |
| Gluskin, R. | UNIVAC Division |
| Hayes, W. | Aviation Electric |
| Lapinas, Z. | Aviation Electric |
| Pinkstaff, C. | Imperial Eastman |
| Pollard, F. | Grumman Aircraft |
| Russell, G. | Norgren/Fluidics |
| Sampson, M. | The S-P Manufacturing Corp. |
| Standen, G. | Aviation Electric |
| Wilks, R. | Corning Glass Works |
| Young, A. | The S-P Manufacturing Corp. |

As of January 15, 1971, the Standards Committee B93 was comprised of the following:

John J. Pippenger, Chairman; Otto J. Maha, Vice Chairman; James I. Morgan, Co-Secretary; LeRoy Stoner, Co-Secretary.

AMERICAN SOCIETY OF AGRICULTURAL ENGINEERS
    E. H. Fletcher

AMERICAN SOCIETY OF LUBRICATION ENGINEERS
    Wm. Eismann

AMERICAN SOCIETY OF MECHANICAL ENGINEERS
    Henry Parsons

AMERICAN SOCIETY FOR TESTING AND MATERIALS
    J. D. Lykins
    W. H. Millett (alternate)

AUTOMOBILE MANUFACTURERS ASSOCIATION
    Reino Mustonen

CONSTRUCTION INDUSTRY MANUFACTURERS ASSOCIATION
    Glenn Stewart
    H. T. Larmore (alternate)

FLUID CONTROLS INSTITUTE
    A. W. Churchill

FLUID POWER SOCIETY
    Prof. Russell Henke
    Melvin E. Long
    Frank L. Mackin
    Tobi Goldoftas
    Wm. Smith
    Lars G. Soderholm
    Frank Yeaple

FLUID SEALING ASSOCIATION
    L. C. van Dyck

INDUSTRIAL TRUCK ASSOCIATION
    C. D. Gibson
    R. T. McNeely (alternate)

INSTRUMENT SOCIETY OF AMERICA
    Aaron I. Kutz
    T. R. Metz (alternate)

JOINT INDUSTRY COUNCIL
    Robert Muhl

MATERIAL HANDLING INSTITUTE
    Jack McPherson

NATIONAL FLUID POWER ASSOCIATION
    Otto J. Maha
    James L. Fisher, Jr.
    W. R. Forster
    W. J. Kudlaty
    Z. J. Lansky
    John J. Pippenger

NATIONAL INDUSTRIAL LEATHER ASSOCIATION
    E. R. Rath

NATIONAL MACHINE TOOL BUILDERS ASSOCIATION
    Joseph I. Ehrhardt
    Edward Loeffler (alternate)

POWER CRANE AND SHOVEL ASSOCIATION
    W. M. Shook

RUBBER MANUFACTURERS ASSOCIATION
    L. Cranston
    N. J. Cyphers (alternate)
    Frank Timmons (alternate)

SOCIETY OF AUTOMOTIVE ENGINEERS, INC.
    W. A. Hertel
    R. E. Lyons
    E. L. Falendysz

U. S. COAST GUARD
    Ens. John D. Richart

U. S. DEPARTMENT OF DEFENSE
    C. A. Nazian
    Hansel Y. Smith
    Paul Hopler (alternate)

## REFERENCES

1. American National Standard Glossary of Terms for Fluid Power, ANSI/B93.2-1971 (ISO/TC 131/SC 1 (USA-___)___)
2. International Standard Rules for the Use of Units of the International System of Units and a Selection of the Decimal Multiples and Sub-Multiples of SI Units, ISO/R 1000-1969
3. National Fluid Power Association Recommended Standard Graphic Symbols for Fluidic Devices and Circuits, NFPA/T3.7.2-1968
4. ISA Tentative Recommended Practice Dynamic Response Testing of Process Control Instrumentation, RP 26.1-1957 - Part 1 - General Recommendations
5. ISA Tentative Recommended Practice Dynamic Response Testing of Process Control Instrumentation, RP 26.2-1960 - Part 2 - Devices with Pneumatic Output Signals
6. International Standard Organization Recommendation for Standard Atmospheres for Conditioning and/or Testing Standard Reference Atmosphere Specifications, ISO/R 554-1967.

# Methods of Presenting Basic Performance Data for Fluidic Devices

## INTRODUCTION

In fluid power systems, power is transmitted and controlled thru a fluid (liquid or gas) under pressure within an enclosed circuit. Fluidics is one means of using fluid under pressure for control purposes. (For the standard definition of "fluidics", see Reference No. 1.

1. SCOPE

    To develop recommended standard methods of rating performance including:

    1.1 Descriptive data including information on function, type, configuration, and operating limits of the device.

    1.2 Operating and performance characteristics including both steady state and dynamic operating characteristics.

    1.3 Methods of performance measurement.

    1.4 Graphic means for uniformly presenting the data.

    1.5 Data of value for control system design.

2. PURPOSE

    2.1 To provide a recommended standard which contains the information necessary to permit uniform determination of a device's suitability as to impedance matching, output rating, sensitivity, etc., in a desired application.

    2.2 To recommend techniques thereby insuring uniformity in test data used in determining the performance characteristics.

3. TERMS

    For definition of terms not herein defined, see Reference No. 1.

ANSI/B93.14

4. UNITS

    4.1 The International System of Units (SI) is used in accordance with Reference No. 2.

    4.2 Approximate conversions to "customary US" units are given for information purposes. These appear in parentheses after their SI counterpart.

5. DESCRIPTIVE INFORMATION

Include the following information when describing the performance characteristics of a fluidic device:

    5.1 Descriptive Information

        5.1.1 Function. Descriptive name of basic function (e.g., OR device, proportional amplifier).

        5.1.2 Type. Descriptive name of operating principles of the device (e.g., wall attachment, turbulence amplifier).

        5.1.3 Configuration

            5.1.3.1 Schematic. Fluidic graphic symbol accompanied by the port markings.

            5.1.3.2 Actual

                5.1.3.2.1 Envelope Dimensions. Show significant dimensions on an outline drawing.

                5.1.3.2.2 Internal Cross-Section Dimensions. Power Nozzle, control nozzle and output receiver.

                5.1.3.2.3 Materials. List significant fabrication materials.

                5.1.3.2.4 External Connections. Furnish adequate information for connection to the device (e.g., tube size, type and size of fittings, manifold, etc.).

        5.1.4 Operating Limits

            5.1.4.1 Supply

5.1.4.1.1   Fluid type

5.1.4.1.2   Pressure range

5.1.4.1.3   Ripple and noise limitations

5.1.4.1.4   Moisture level

5.1.4.1.5   Micrometer filtration levels

5.1.4.1.6   Contaminant levels (other than solid particles, solvents, oil, etc.)

5.1.4.2   Environment

5.1.4.2.1   Operating temperature range - specify operating temperature range of the device.

5.1.4.2.2   Storage temperature

5.1.4.2.3   Shock and vibration

5.1.4.2.4   Radiation tolerance (magnetic or nuclear, or both)

5.1.4.2.5   Tolerance to acoustic energy

5.1.4.2.6   Environment contamination (e.g., dust, moisture, etc.)

5.2   Performance Characteristics, Steady State

5.2.1   Present the steady state characteristics by input, output, gain, switching and supply curves, and typical operating conditions.

5.2.2   Characteristic Curves. Use the devices for which characteristic curves are illustrated as examples. No attempt is made to include characteristics for all devices. Use the characteristic curves presented for guidelines for devices not included.

5.2.2.1   For active digital devices, furnish characteristic curves for supply, control, output, switching, and switching domain. Curve Figures 2, 5, 8, 9, 10, 11, and 12 apply to these devices.

ANSI/B93.14

5.2.2.2 For passive digital devices, furnish control, output, switching, and switching domain characteristics.

5.2.2.3 For proportional devices, furnish control, output, and gain characteristics. Curve Figures 1, 3, 4, 6, 7, and 12 apply to these devices.

5.2.2.4 Use the same general format as the illustrated examples for passive devices such as fluid resistors, diodes, etc.

TABLE 1 - Fluidic device performance characteristic curve index

| Device Type | Characteristic (Figure No.) | | | | | |
|---|---|---|---|---|---|---|
| | Supply | Control | Output | Gain | Switching | Switching domain |
| Digital, Active | 12 | 5 | 2 | NA | 8, 9 | 10, 11 |
| Digital, Passive | NA | X | X | NA | X | X |
| Analog | 12 | 3, 4 | 1 | 6, 7 | NA | NA |
| Others (Diode, Resistor) | X | X | X | X | X | X |

NOTES. - NA - Not Applicable
X - Format Optional

5.2.2.5 Select the proper characteristic curves that are applicable to a particular fluidic device type from Table 1.

5.2.2.6 Include the following qualifying statements when indicated at the bottom of the characteristic curve examples:

5.2.2.6.1 Use the curves as examples only and make no attempt to be all inclusive.

5.2.2.6.2 The pressures used in the characteristics are differential (gage) pressures from vent (ambient) pressure. Standard ambient conditions are per Reference No. 6 [(1.013 bar, abs (14.69 psia) and 20°C(68°F)]

5.2.2.6.3 The normalization of flows are with respect to supply flow and the normalization of pressures are with respect to supply pressure. For normalization of all data, use the flow as the average of the upper and lower limit of the flow band at the designated pressure.

5.2.2.6.4 The characteristics indicate the flow for one side only. Specify the null control pressure.

5.2.2.6.5 State the control values not shown for devices of more than two independent control inputs.

5.2.2.6.6 In those devices where the supply flow is a function of control flow or load flow, or both, correspond the supply characteristics to the flow at typical operating conditions.

5.2.2.6.7 Include the supply characteristic if a significant bias (not control) is employed in a fluidic device (e.g., impact modulators).

5.2.2.6.8 Indicate by the cross-hatched area the tolerance bands for the indicated operating conditions (i.e., supply pressure range, temperature range, etc.).

5.2.2.6.9 Specify the quiescent control flows or pressures.

5.2.2.7 Apply the descriptions in Tables 2 and 3 to the nomenclature used on the characteristic curves:

TABLE 2 - Symbols

| Symbol | Description | Units |
|---|---|---|
| P | Pressure | psig, |
| Q | Flow | scfm, |
| W | Flow | lb/sec, (mass) |
| $\Delta$ | Prefix to indicate incremental change from quiescent | -- |

TABLE 3 - Subscripts

| Subscripts | Description | Units |
|---|---|---|
| S, S1, S2,..., Sn | Indicating a supply parameter | --- |
| C, C1, C2,..., Cn | Indicating a control parameter | --- |
| O, O1, O2,..., On | Indicating an output parameter | --- |

5.2.2.8 For Output Characteristics, follow the examples provided by Figures 1 and 2.

5.2.2.9 For Input (control) Characteristics, follow the examples provided by Figures 3, 4, and 5.

5.2.2.10 For Gain Characteristics, follow the examples provided by Figures 6 and 7.

5.2.2.11 For Switching Characteristics, follow the examples provided by Figures 8, 9, 10, and 11.

5.2.2.12 For Supply Characteristics, follow the example provided by Figure 12.

5.3 Performance Characteristics, Dynamic

5.3.1 Present the dynamic performance characteristics by sinusoidal frequency response plots and step input transient response plots.

5.3.2 Consider the following factors when conducting dynamic response tests:

5.3.2.1 Divide into two categories the application of dynamic response test data.

5.3.2.1.1 Characterization of a control or measuring device.

5.3.2.1.2 Control system design.

5.3.2.2 Take and present the data per the appropriate recommended practice sections.

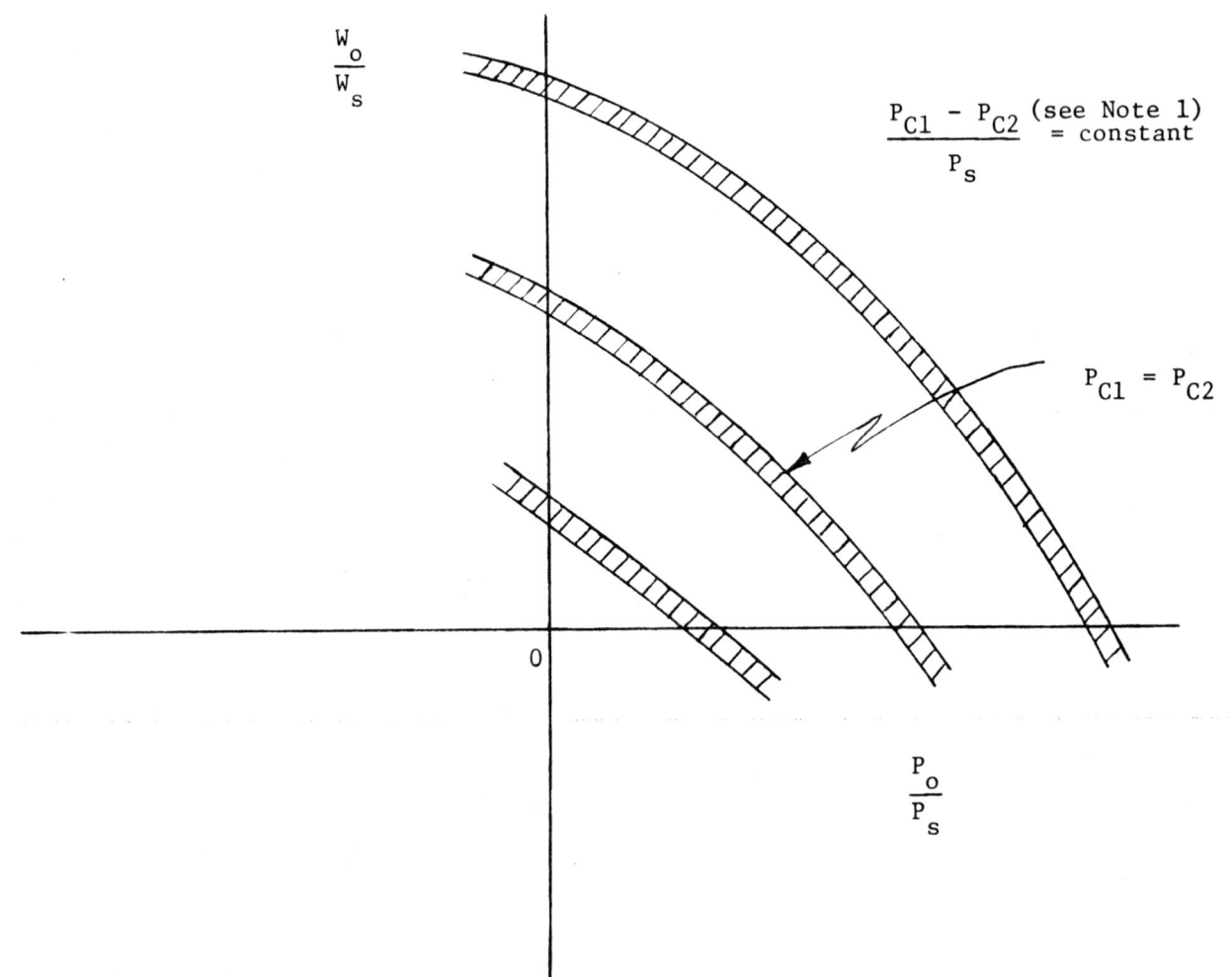

NOTES. -

1. For single-sided device $P_{C2} = 0$

2. Refer to qualifying statements 1, 2, 3, and 7.

FIGURE 1 - Output characteristic - analog device

ANSI/B93.14

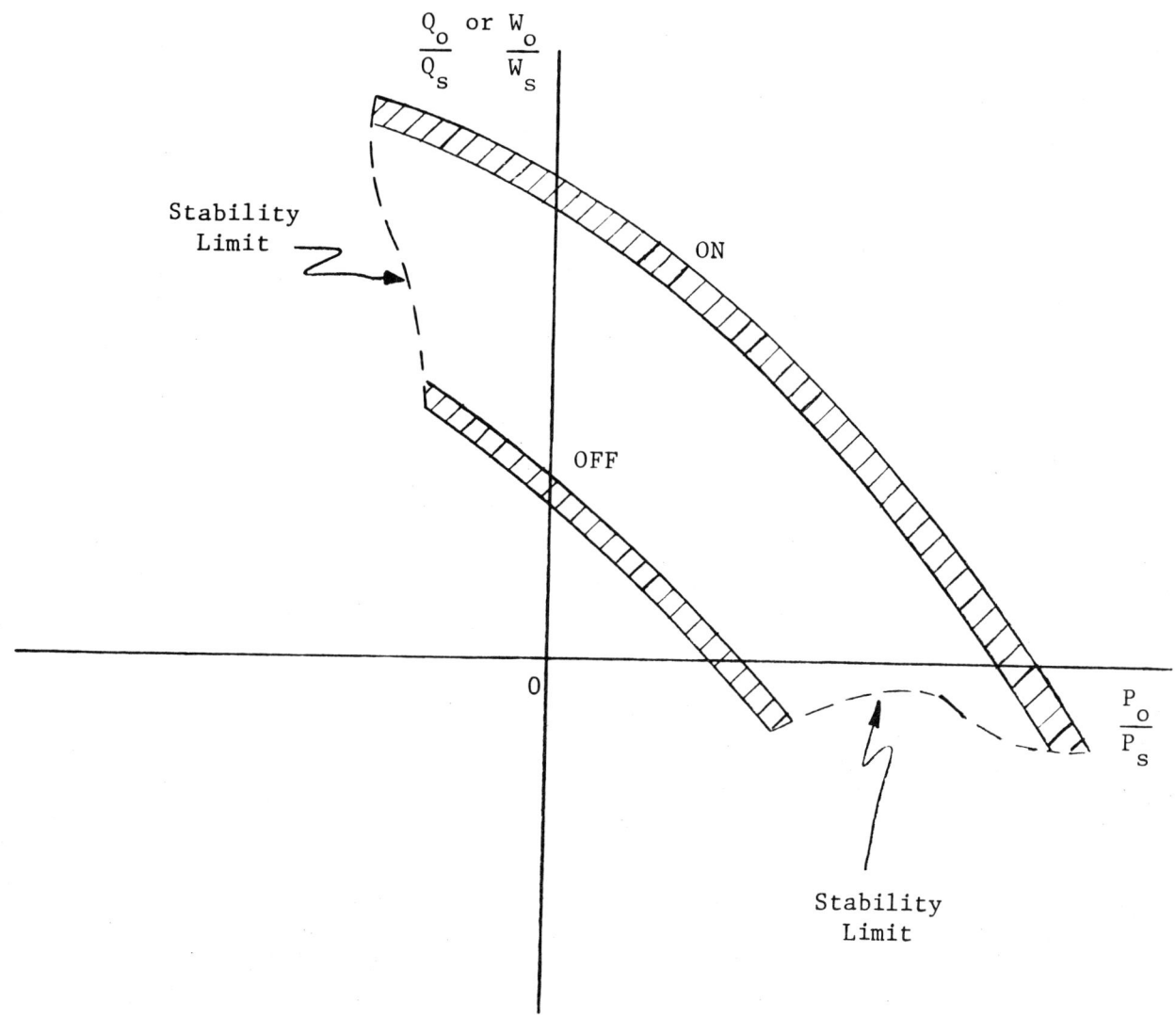

NOTE. - Refer to qualifying statements 1, 2, and 7.

FIGURE 2 - Output characteristic - digital device

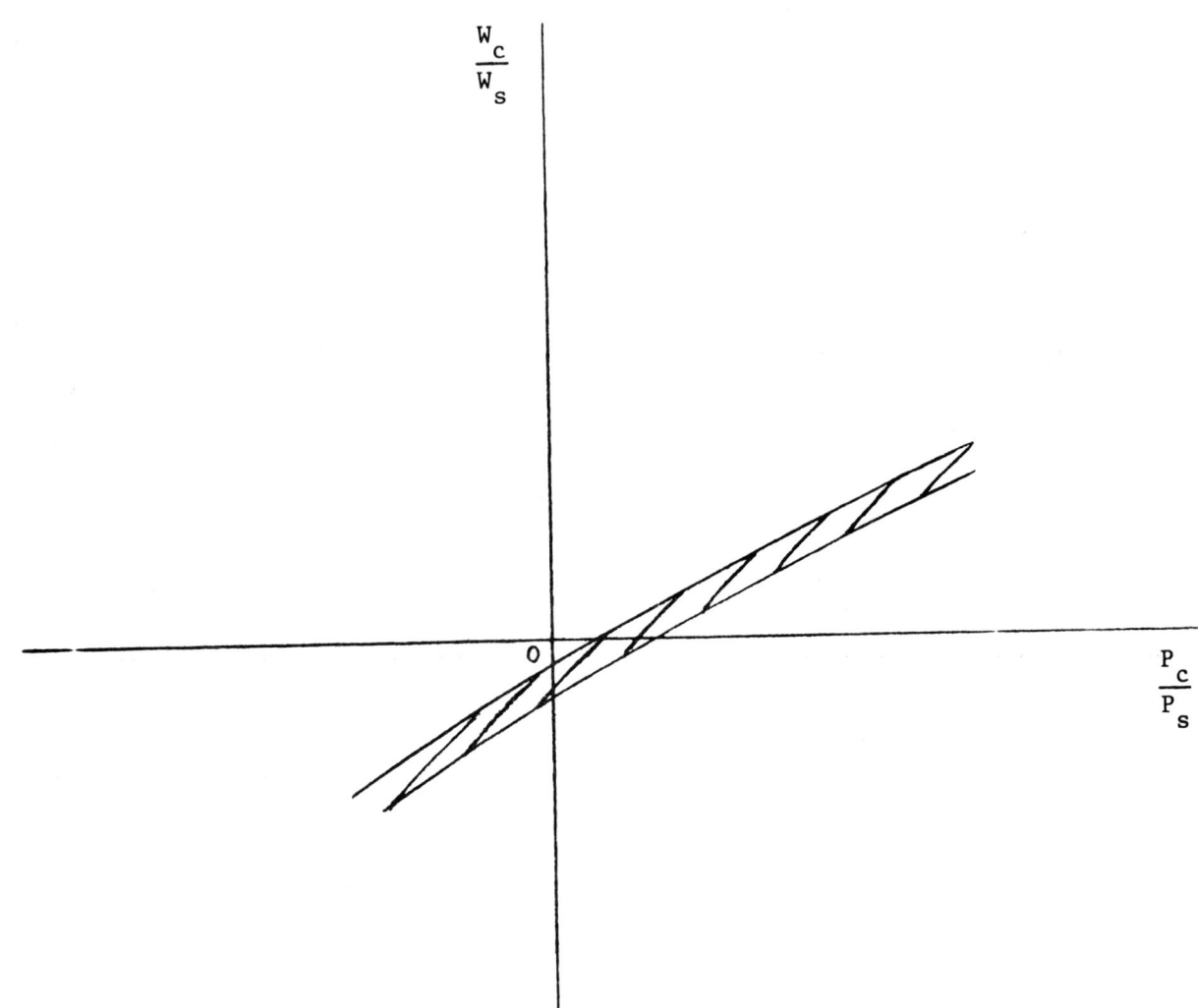

NOTE. - Refer to qualifying statements 1, 2, 3, and 7.

FIGURE 3 - Input (control) characteristic - single-ended analog device

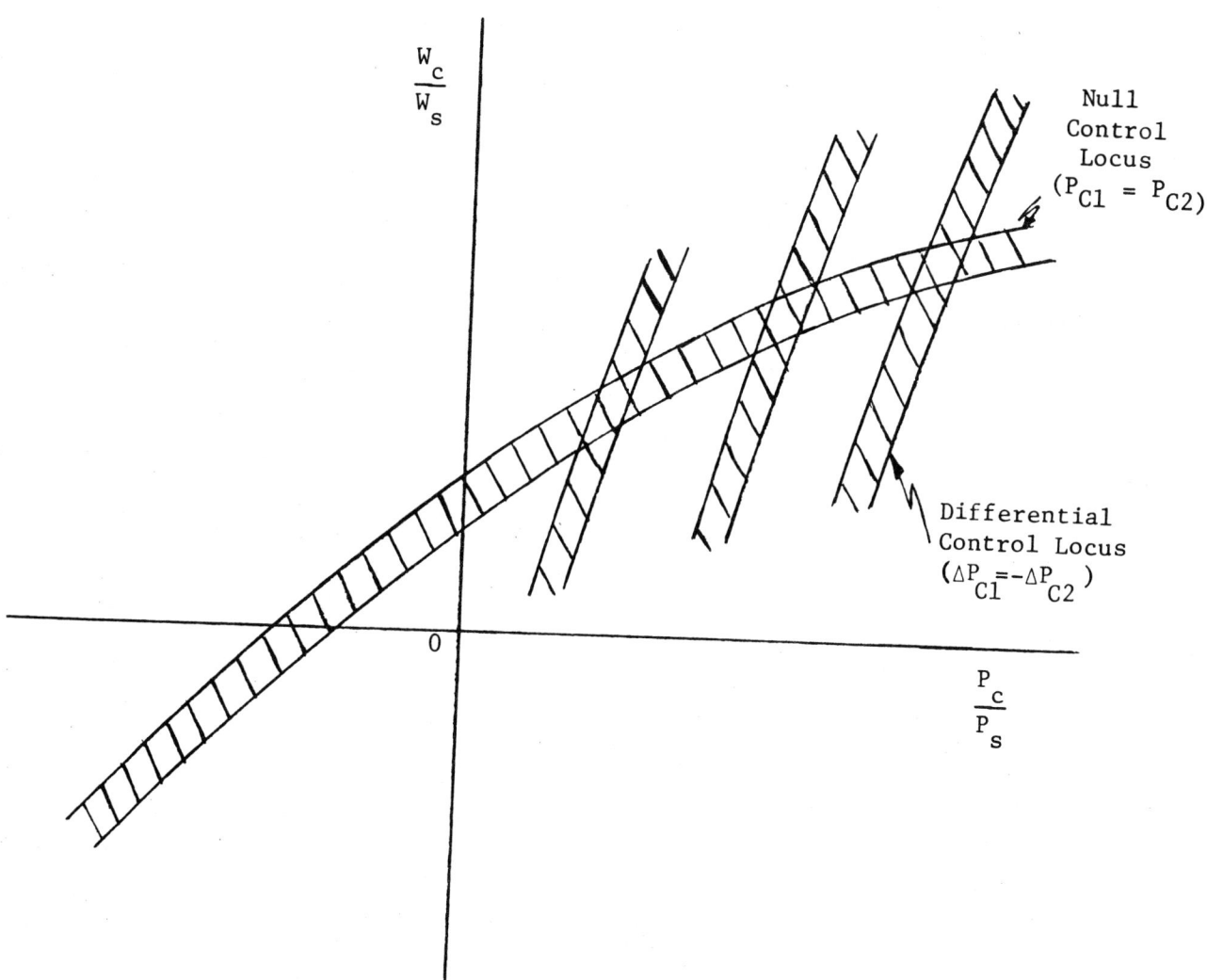

NOTE. - Refer to qualifying statements 1, 2, 3, and 7.

FIGURE 4 - Input (control) characteristic - differential analog device

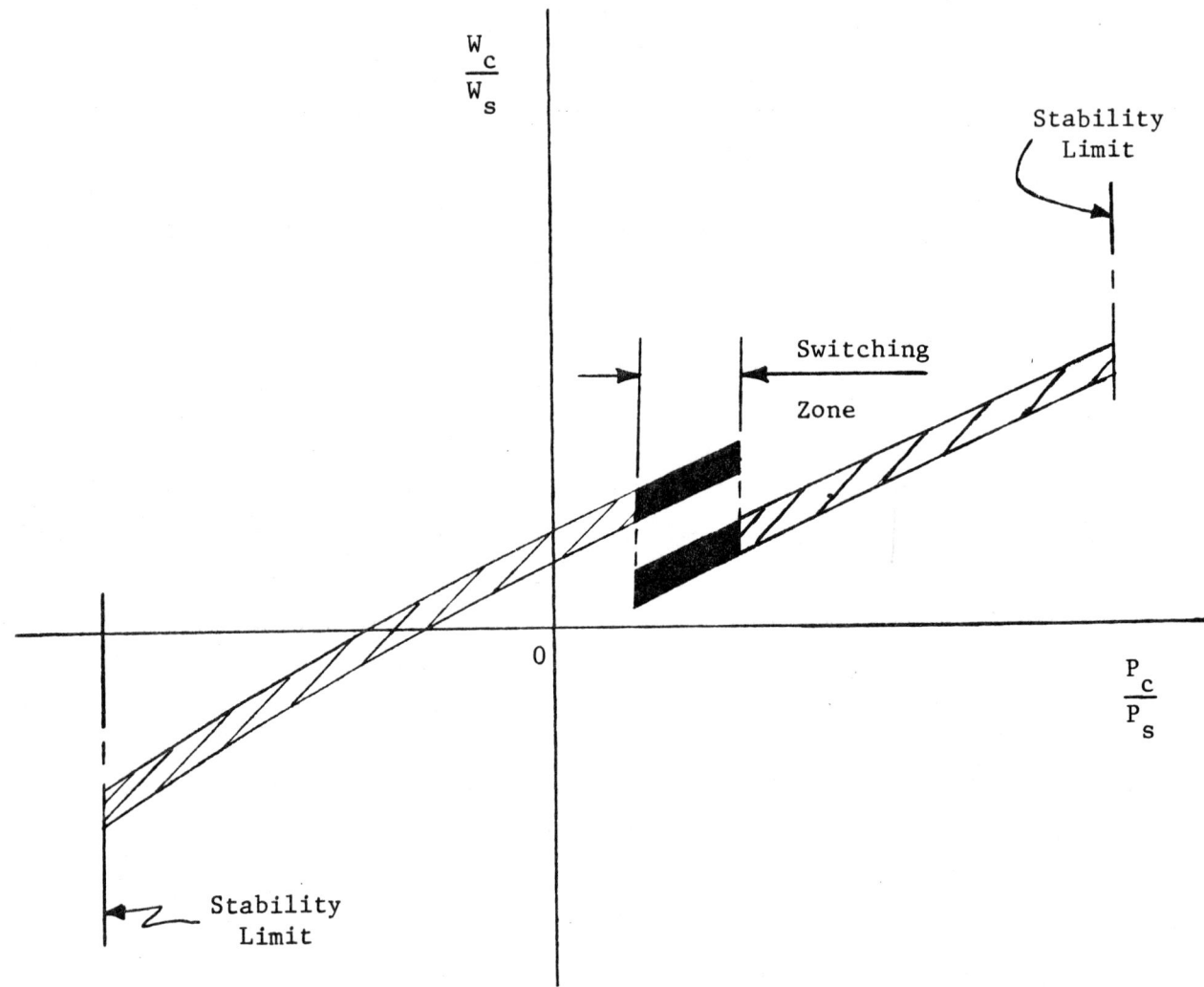

NOTE. - Refer to qualifying statements 1, 2, and 7.

FIGURE 5 - Input (control) characteristic - digital device

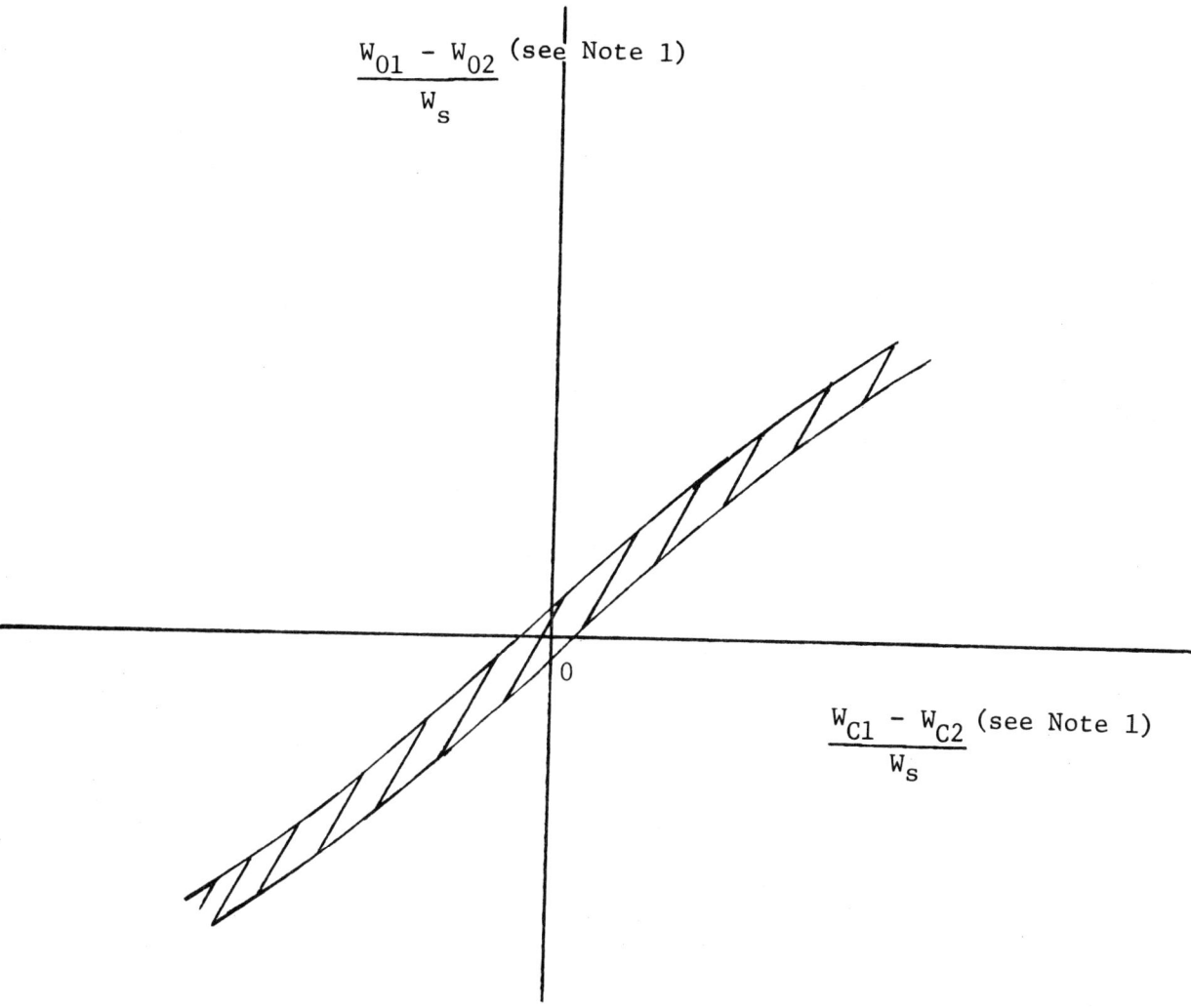

NOTES. -

1. For single-sided device $W_{02} = 0$
   $W_{C2} = 0$
2. Refer to qualifying statements 1, 2, 3, 7, and 8.

FIGURE 6 - Flow gain characteristic - analog device

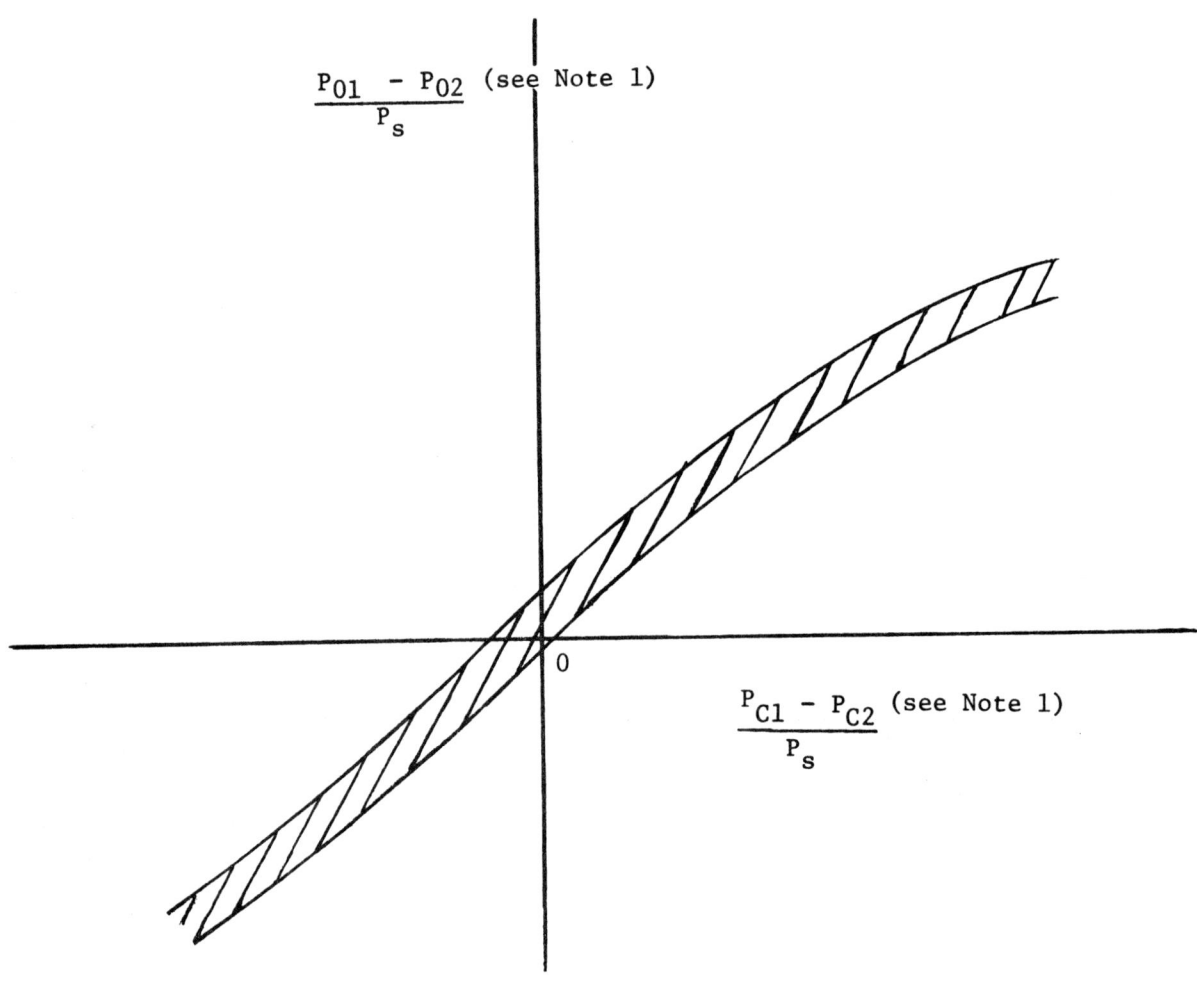

NOTES. -

1. For single-sided device $P_{02} = 0$
   $P_{C2} = 0$

2. Refer to qualifying statements 1, 2, 7, and 8.

FIGURE 7 - Pressure gain characteristic - analog device

ANSI/B93.14

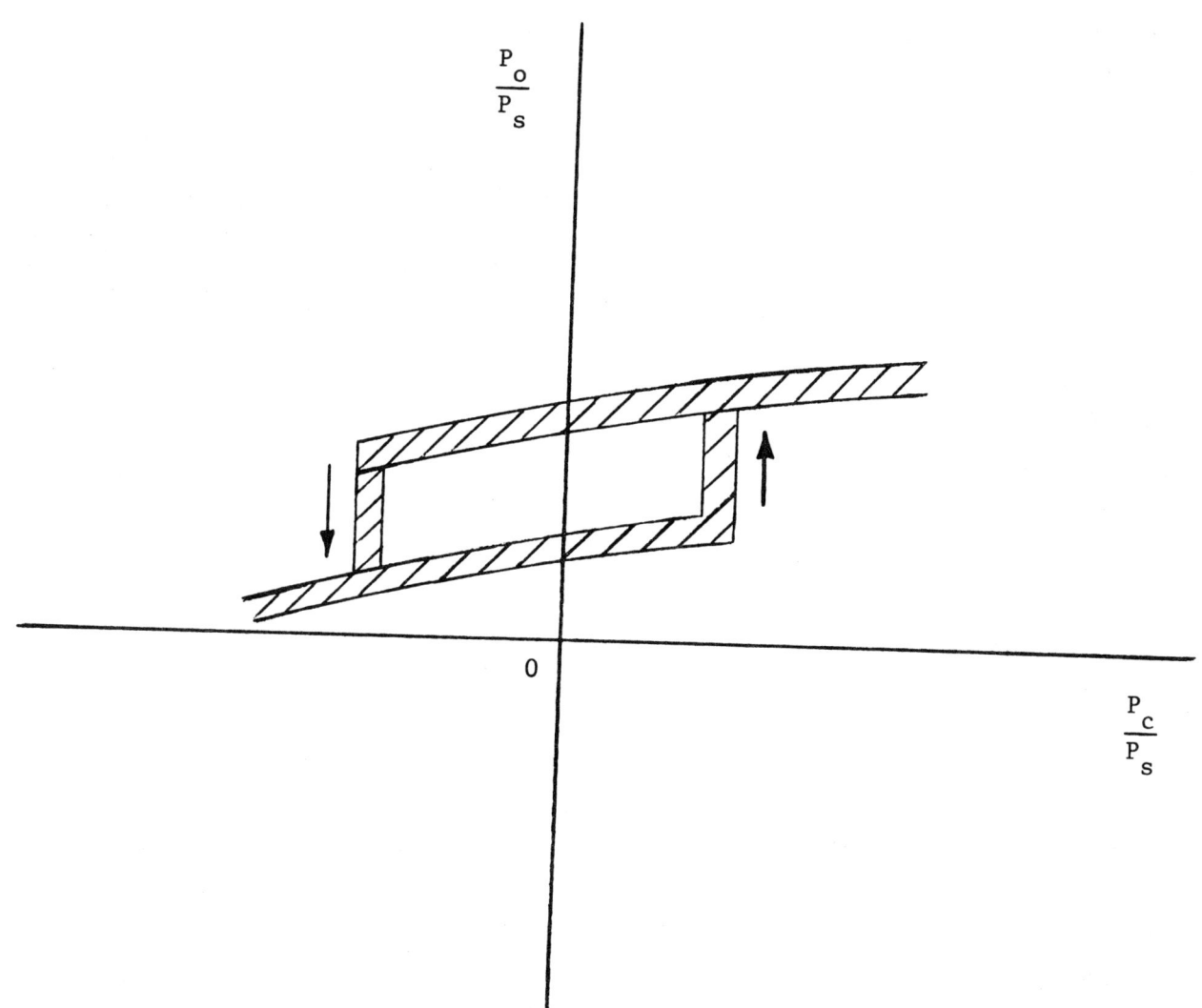

NOTE. - Refer to qualifying statements 1, 2, and 7.

FIGURE 8 - Pressure switching characteristic - digital device

ANSI/B93.14

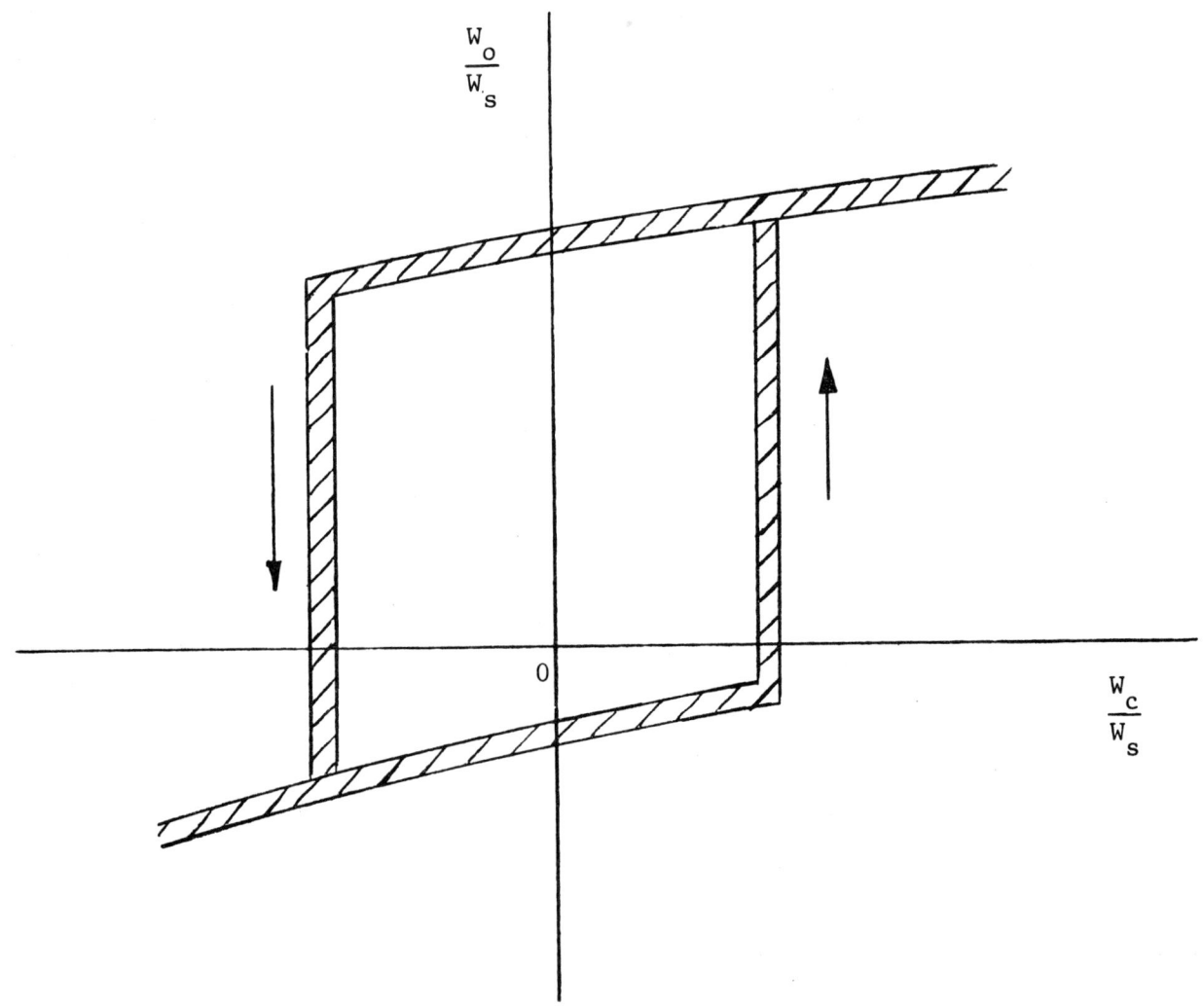

NOTE. - Refer to qualifying statements 1, 2, and 7.

FIGURE 9 - Flow switching characteristic - digital device

ANSI/B93.14

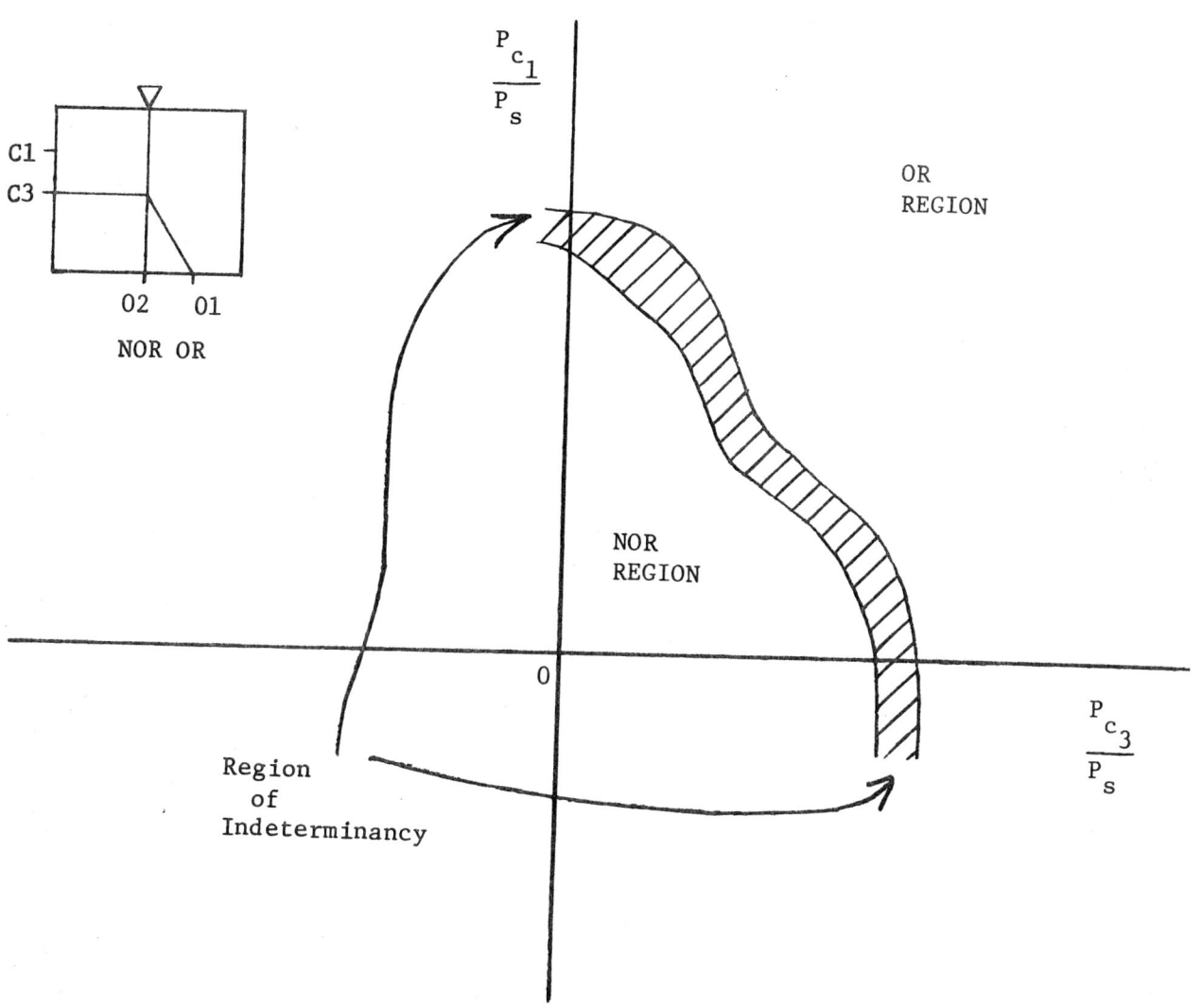

NOTE. - Refer to qualifying statements 1, 2, 4, and 7.

FIGURE 10 - Switching domain characteristic - digital device

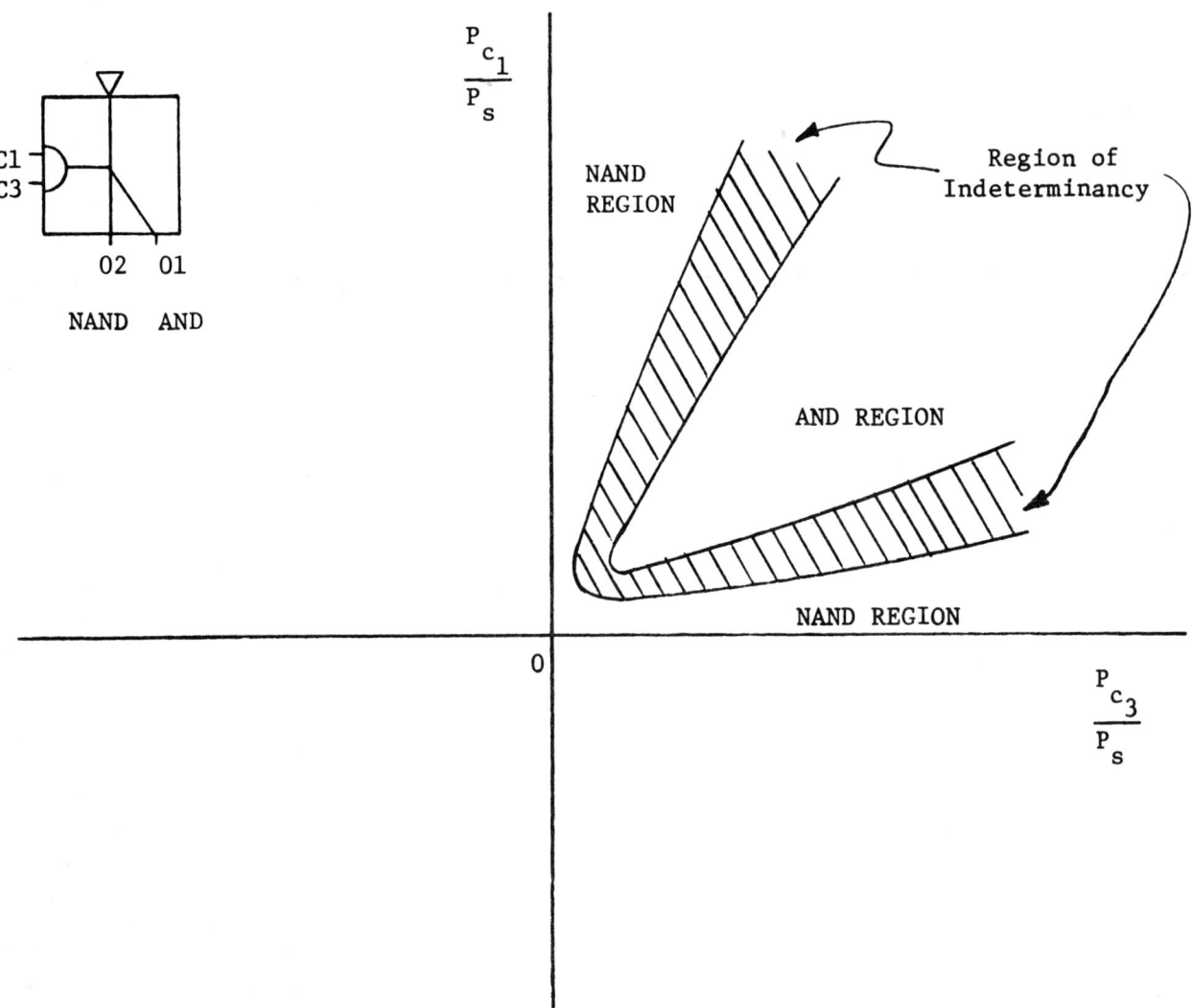

NOTE. - Refer to qualifying statements 1, 2, 4, and 7.

FIGURE 11 - Switching domain characteristic - digital device

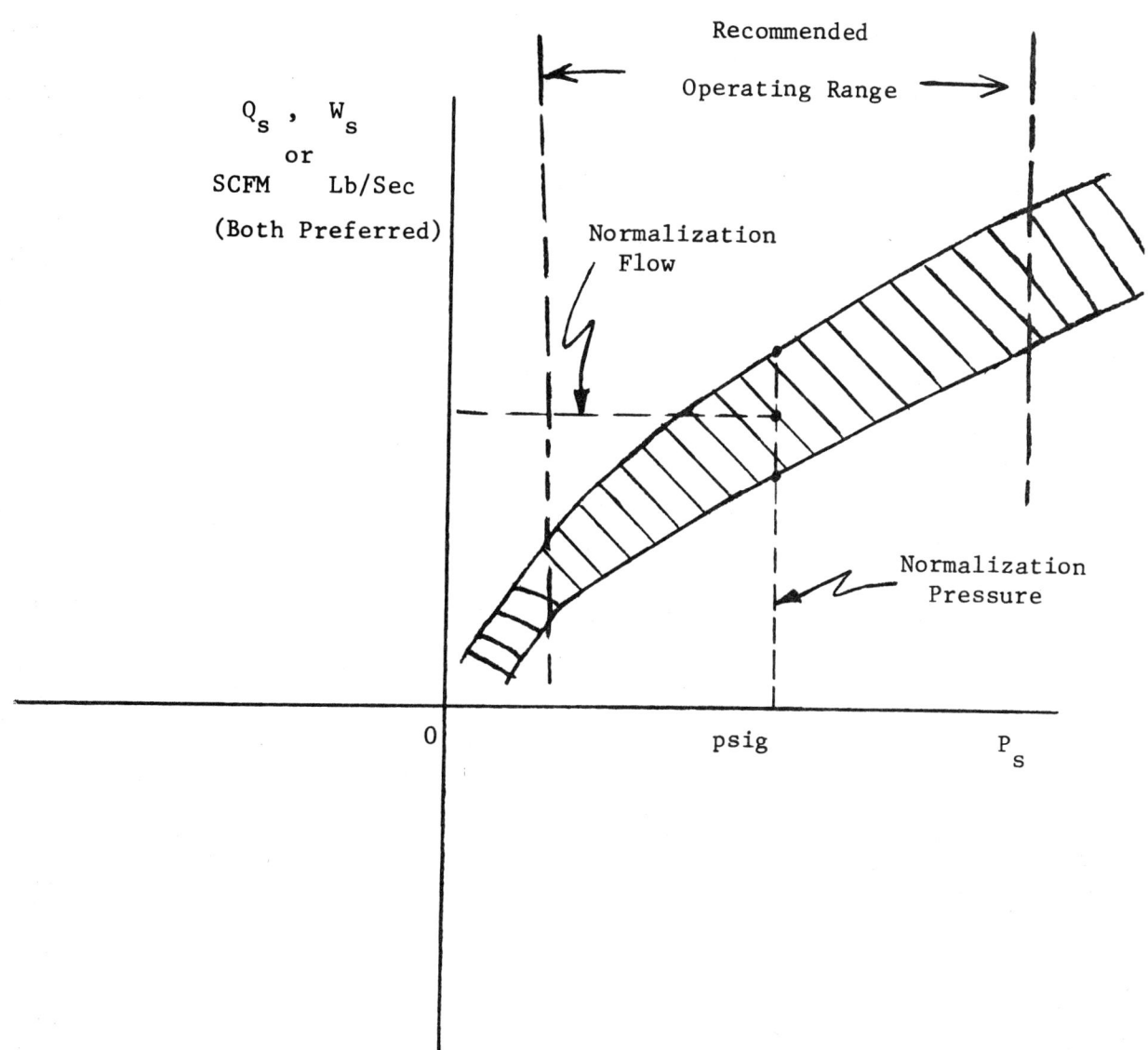

NOTE. - Refer to qualifying statements 1, 5, 6, and 7.

FIGURE 12 - Supply characteristics - analog or digital device

5.3.2.3 Interaction. When the relation between input and output of one component is affected by changing the characteristics of the other component.

5.3.2.4 Nonlinearity. Some common types are dead band, friction, hysteresis, velocity limiting components in a device, saturation, exponential measurements and valve characteristics. Not only causes distortion of the signal shape, but may also result in additional phase shift and attenuation.

5.3.2.5 Power Supply. Regulate the power supply to provide test pressures (flow) to within two percent of the designated levels.

5.3.3 Tests

5.3.3.1 Input Signals. Use sine wave and step input signals with designated standard output loads in clause 5.3.3.2.

5.3.3.1.1 Frequency Response Input Signals

5.3.3.1.1.1 Sine Wave Input Signals. Use the sine wave input signal for analog devices with a peak to peak magnitude of 10 percent of the actual input signal span as its center. The input span referred to here is that span that will drive the output signal through its full range.

The lowest test frequency is such that the magnitude of the output sine wave essentially equals that of the steady state output for the same input and there is essentially no phase shift between them (except if the device has pure integral or derivative functions). Increase the frequency until the magnitude of the output sine wave has been attenuated to 2% or less of the output magnitude at the lowest test frequency or the phase lag between the input and output signal is greater than $300°$. For integrator devices, the phase between the output and input is ideally a $90°$ lead with no attenuation. Thus, the highest test frequency corresponds to a $210°$ lag.

For differentiator devices, the phase between the output and input is ideally a 90° lag with no attenuation. Thus, the highest test frequency corresponds to a 390° lag.

5.3.3.1.1.2 Square Wave Input Signals. Use the square wave input signal for digital devices having a peak to peak magnitude of 100 percent of the actual input signal (i.e., from logic 0 to logic 1 level) span of the device being tested.

The lowest test frequency is such that the magnitude of the output square wave essentially equals that of the steady state output for the same input and there is essentially no phase shift between them. Increase the frequency until the magnitude of the output wave form is attenuated to 10 percent or less or until the output is intermittent.

5.3.3.1.2 Step Input Signals. Make the magnitudes of the step input signals to be (in percent of actual input span):

Analog

(1) 45 to 55
(2) 55 to 45
(3) 10 to 90
(4) 90 to 10

Digital

(1) From logic 0 to logic 1 level
(2) From logic 1 to logic 0 level

5.3.3.2 Loading. Unless otherwise specified, make the load equivalent to a fan-out of one (i.e., the load on the unit under test is equivalent to the input impedance of a similar device operating with the same power supply.

5.3.4 Test Equipment and Procedures

5.3.4.1 Data Required

5.3.4.1.1 Sine or Square Wave Tests

5.3.4.1.1.1 Signal frequency.

5.3.4.1.1.2 Amplitude of input and output signals at each frequency.

5.3.4.1.1.3 Phase relationship of input and output signals at each frequency.

5.3.4.1.1.4 Power supply variation (to device being tested) from lowest to highest frequency.

5.3.4.1.1.5 Condition of system (adjustment, etc.).

5.3.4.1.2 Step Tests

5.3.4.1.2.1 Input signal form from time zero to steady state.

5.3.4.1.2.2 Output signal form from time zero to steady state.

5.3.4.1.2.3 Power supply (to device being tested) variation during test.

5.3.4.1.2.4 Condition of system (adjustments, etc.).

5.3.4.1.3 Supporting Data. Comply with section 5.

5.3.4.2 Generalize Test Setup

5.3.4.2.1 Block Diagrams

5.3.4.2.1.1 Refer to Figure 13.

5.3.4.2.1.2 Use the transducers and amplifier-converters to convert the input and output signals to the form required for the display equipment.

5.3.4.2.2 Location of Output Signal Pickoff Point. Place the pickoff of the input and output signals as near to the ports of the device as physically practical.

5.3.4.3 Test Equipment

5.3.4.3.1 Signal Generation

5.3.4.3.1.1 Sine or Square Waves. Meet the following criteria for the sine or square wave generating device.

5.3.4.3.1.1.1 Require the frequency range to be consistent with the requirements of clause 5.3.3.1.1 and within the range of 1 to 1000 hertz.

5.3.4.3.1.1.2 Produce the sine or square wave generator to be capable of holding a given frequency and magnitude within two percent.

5.3.4.3.1.2 Step Tests. Produce the rise (fall) time of the step input signal to be less than 0.05 times the rise (fall) time of the output of the unit under test.

5.3.4.3.2 Data Evaluation

5.3.4.3.2.1 Sinusoidal Response. Measure the relative magnitudes and phase shifts in the following manner:

5.3.4.3.2.1.1 Use the root-mean-square values of the input and output for magnitude measurements when the output contains appreciable harmonic content.

5.3.4.3.2.1.2 Use the phase shift as the angle between the input and the fundamental of the output when harmonics are present in the output.

5.3.4.3.2.2 Step Response. Evaluate the time response from plot of the input and output signals as a function of time. Determine measurements or rise time, buildup time, etc. from these curves (see Figure 14).

5.3.4.3.3 Display Equipment

5.3.4.3.3.1 Use a multi-channel high speed recorder as one tool for dynamic test work. This recorder plots both input and output sine or step signals on the same time axis. Step test input and output can be read directly.

5.3.4.3.3.2 Use various combinations of oscilloscopes, single and dual beam, electronic switches, etc. to display input and output sine wave signals. Determine phase relationships and magnitude ratio with Lissajous figures or calibrated phase shift networks.

5.3.4.3.4 Transducers and Amplifier-Converters

5.3.4.3.4.1 Provide the transducer and amplifier-converter systems needed depending upon the type of signal needed to drive the display equipment as well as the nature of the signals used for the test (i.e., pneumatic, electrical, hydraulic, or mechanical).

5.3.4.3.4.2 Use strain gage or piezoelectric type transducers for converting pressures to electrical signals. Use linear variable differential transformers or multi-turn variable resistors for mechanical motion.

5.3.4.3.5 Test Equipment Performance Specifications. From the standpoint of results only, use the following recommended transducer-amplifier-converter display equipment system performance.

5.3.4.3.5.1 Signal Amplitude. Within 1 decibel (db) from 0.01 hertz to maximum frequency of test (at least to 1000 hertz).

5.3.4.3.5.2 Phase shift. Less than 5° from 0.01 hertz to maximum frequency of test (at least to 1000 hertz).

5.3.4.3.5.3 Good practice dictates the test equipment operate over a frequency range from one-tenth the lowest test frequency to 10 times the highest test frequency.

5.3.4.3.5.4 Drift. Less than five percent of the full span in four hours.

5.3.4.4 Testing Procedures. Use the following:

ANSI/B93.14

5.3.4.4.1   Blank Run. Whenever a new test setup is completed, make a "blank" run (with the device being tested by-passed) in such a manner that the phase shift be zero, and the attenuation zero db, as frequency increases.

5.3.4.4.2   Static Calibration. Statically calibrate the overall system with an independent standard (manometer, VTVM, etc.) on the actual signals.

5.3.4.4.3   Test Frequencies

   5.3.4.4.3.1   Select frequency points as the test progresses to provide small increments at the critical points or at abrupt changes in the phase or magnitude curves.

   5.3.4.4.3.2   Use at least twenty points for three decades of frequency.

   5.3.4.4.3.3   Select the lowest test frequency so that the magnitude of the output sine wave essentially equals that of the steady state output for the same input and there is essentially no phase shift between them (except if the device has pure integral or derivative functions) (see clause 5.3.3.1.1.1).

   5.3.4.4.3.4   Increase the frequency until the magnitude of the output sine wave is attenuated to two percent or less of the output magnitude at the lowest test frequency or the phase lag between the input and output signal is greater than $300°$.

   For integrator devices, the phase between the output and input is ideally a $90°$ lead with no attenuation. Thus, the highest test frequency corresponds to a $210°$ lag.

   For differentiator devices, the phase between the output and input is ideally a $90°$ lag with no attenuation. Thus, the highest test frequency corresponds to a $390°$ lag.

5.3.4.4.4 Correlating Results. Refer to clause 5.3.4.3.2 for interpretation of the display equipment information.

5.3.5 Data Presentation

5.3.5.1 Sinusoidal Response. Present the sinusoidal response as a plot of the output amplitude ratio in db and phase shift versus the input frequency in hertz.

5.3.5.1.1 Magnitude Ratio Plot

5.3.5.1.1.1 Use semi-log paper with the magnitude ratio plotted in db on the linear vertical axis.

5.3.5.1.1.2 Plot the frequency in hertz on the horizontal, logarithmic axis.

5.3.5.1.2 Phase Shift Plot

5.3.5.1.2.1 Plot the phase shift on semi-log paper with the phase shift in degrees on the line or vertical scale, and frequency in hertz on the horizontal logarithmic scale.

5.3.5.1.2.2 Identify the phase shift in degrees by a plus sign when the output leads the input.

5.3.5.1.2.3 Identify the phase shift in degrees by a minus sign when the output lags the input.

5.3.5.2 Step Test Data

5.3.5.2.1 Present step test data on linear coordinate paper, with the input and output signal magnitude in percent of its steady state change plotted on the vertical scale, (which may be suppressed to bring out detail) and elapsed time in appropriate units on the horizontal scale.

5.3.5.2.2 Figure 14, a typical time response of a device to a step input, defines the pertinent response characteristics of the device.

ANSI/B93.14

5.3.5.3 Supporting Information. Include the following information on each plot of dynamic response data:

    5.3.5.3.1 Dynamic Response Test

    5.3.5.3.2 Description of device or system tested, including model numbers, serial numbers, and sketch if appropriate.

    5.3.5.3.3 Curve identification as follows:

        5.3.5.3.3.1 Curve. Use identifying code.

        5.3.5.3.3.2 Input Signal. Use peak to peak magnitude and level for sine wave; direction and size for step test.

        5.3.5.3.3.3 Loading (including transmission). Describe or code to sketches on plot.

    5.3.5.3.4 Supply pressure or flow to device being tested.

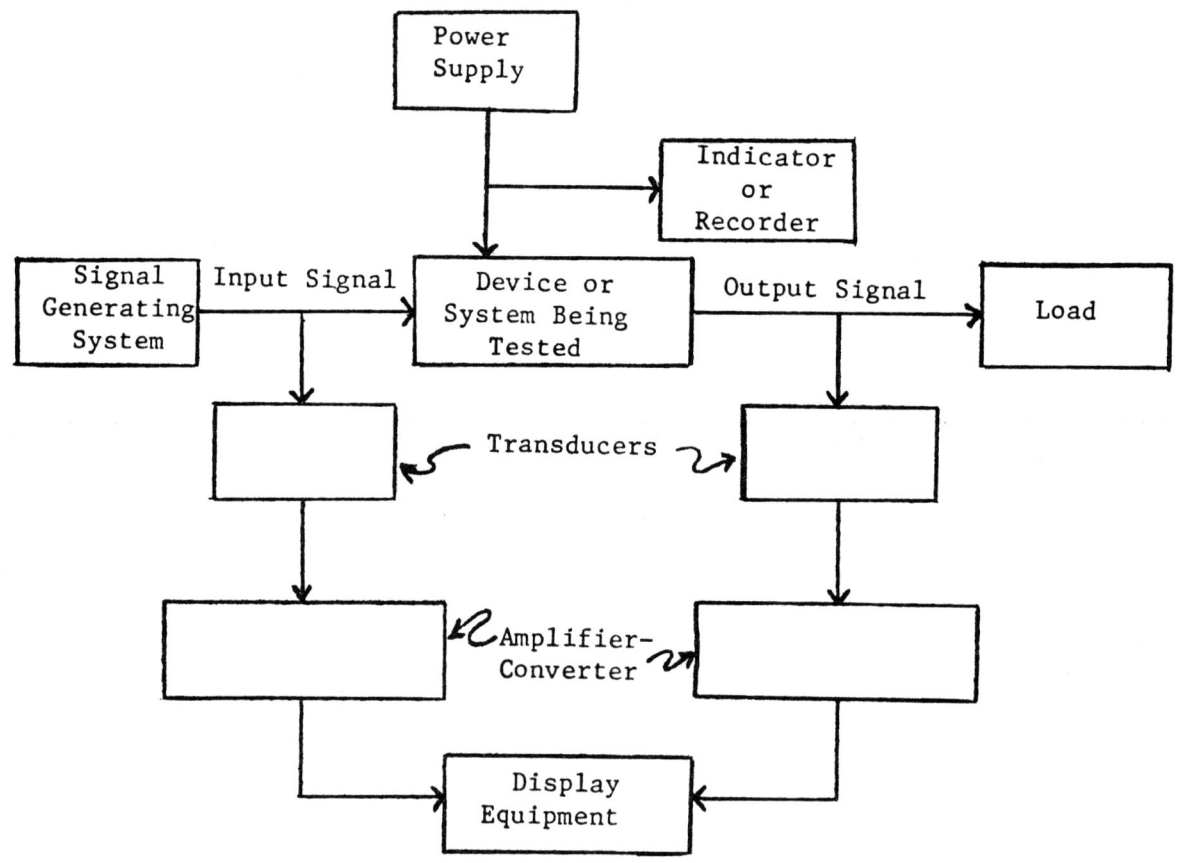

FIGURE 13 - Generalized dynamic test setup

ANSI/B93.14

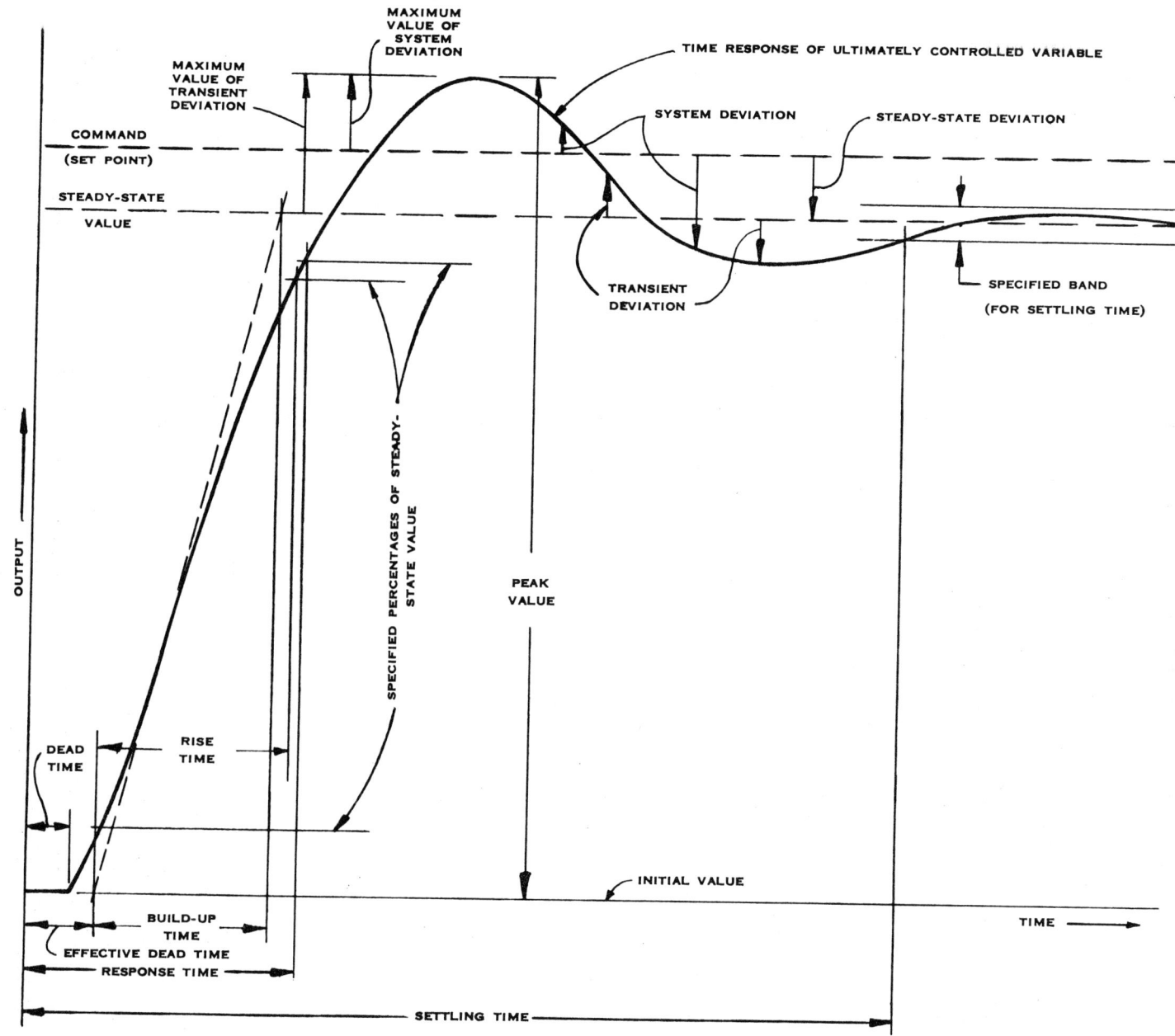

FIGURE 14 - Typical time response of a device to a step increase of input

- 299 -

# NOTES

ANSI/B93.39M-1978

## AN INDUSTRY STANDARD FOR FLUID POWER

American National Standard

Requirements for Presentation of Catalog Data,

Fluid Compatibility, Cleaning Media, Markings and

Dimensional Identification Codes and

Pressure Drop Characteristics for

Fluid Power Air Line Filters

(NFPA/T3.12.2M-1975)

Approved as an ANSI Standard
30 March 1978

published by
**NATIONAL FLUID POWER ASSOCIATION, INC.**

3333 N. Mayfair Road  /  Milwaukee, WI 53222  /  414-778-3344  /  TLX 26898

# FOREWORD

This Foreword is not part of American National Standard Requirements for Presentation of Catalog Data, Fluid Compatibility, Cleaning Media, Markings and Dimensional Identification Codes, and Pressure Drop Characteristics for Fluid Power Air Line Filters, ANSI/B93.39-1978.

In the past, tests of air line filters and the presentation of the measured performance data have not been done in a consistent manner, and therefore have provided only limited guidance to the user.

Recognizing this need, the NFPA FRL, Component Section initiated a program to develop a recommended standard for air line filters. The first draft of this document, a solid contaminant removal test, was written on 27 September 1969 and was reviewed by the Section on 10 December 1969. Subsequent drafts were reviewed on 16 June 1970 (Draft No. 2), 6 January 1971 (Draft No. 3), 19-20 March 1971 (Draft No. 4) and 5 January 1972 (Draft No. 5).

After reviewing Draft No. 5, the Section agreed to follow an international trend of dividing a proposed standard into sub-parts to expedite development. Draft No. 6 was prepared as eight individual proposals.

On 22 March 1972, all of the sub-parts were circulated for General Review. All comments resulting from the General Review were resolved by 6 September 1972. The proposed standard was balloted on 6 November 1972.

The first sub-part, Air Line Filter Terms (T3.12.2.1-1973), completed ballot on 19 February 1973. The one negative ballot on this sub-part was resolved on 23 May 1973. The NFPA Board of Directors approved NFPA/T3.12.1 on 11 November 1973; however, the document was never printed because its contents were incorporated in the second supplement to the ANSI Glossary of Terms for Fluid Power.

The second, third and fourth sub-parts, Catalog Data (T3.12.2.2), Fluid Compatibility (T3.12.2.3) and Port Identification and Dimensional Codes (T3.12.2.4) also completed ballot on 19 February 1973. On 2 January 1975, all negative ballots on these three sub-parts were resolved. The NFPA Technical Board recommended final approval with the condition that Headquarters editorially prepare the three sub-parts as one document before submitting it to the Board of Directors for final approval. This was completed by 5 March 1975, and was reviewed for accuracy and clarity at the Section meeting on 12 March 1975. The NFPA Board of Directors approved T3.12.2 as an NFPA Recommended Standard on 4 May 1975.

The fifth sub-part, Pressure Rating (T3.12.2.5-19xx), was replaced by two pressure rating projects based on NFPA's basic pressure rating standard, NFPA/T2.6.1-1974. These two projects are NFPA/T3.12.10-1976 and T3.12.11-19xx.

On 25 April 1975, the NFPA Technical Staff prepared the Ballot Draft of sub-part six, NFPA/T3.12.2.6-19xx. Replies to those who submitted negative ballots were formulated on 1 October 1975 at the T3.12 Section meeting. By 16 October 1975, all replies had been accepted by the recipients and negative ballots changed to the affirmative. The NFPA Technical Board recommended approval of T3.12.2.6-19xx on 5 November 1975. On 11 December 1975, the NFPA Board of Directors granted approval to NFPA/T3.12.2.6-1975.

The seventh and eighth sub-parts, Liquid and Solid Removal Characteristics (T3.12.2.7-19xx and T3.12.2.8-19xx), are in preparation.

ANSI/B93.39

On 22 January 1976, NFPA Headquarters editorially merged NFPA/T3.12.2-1975 and NFPA/T3.12.2.6-1975 to form the present document before submitting it to ANSI/B93–Fluid Power Systems and Components.

Members of the NFPA Project Group responsible for the development of this standard included:

| | | |
|---|---|---|
| Colter, John | Project Chairman (1973-1977) and Section Chairman (1973-1977) | Watts Fluid Power |
| Draxler, Walter | Project Chairman (1973) | RegO Division/ Golconda Corp.* |
| Ray, Walter | Project Secretary | Master Pneumatic Detroit, Inc. |
| Bailey, Richard | Section Chairman (1977-present) | C. A. Norgren Co. |
| Baker, R. C. | Section Chairman (1970-1973) | Ace Controls, Inc. |
| Mueller, John | Technical Auditor | The Weatherhead Co. |
| Luecke, John R. | Director of National Technical Services | National Fluid Power Association |

| | |
|---|---|
| Bailey, R. | C. A. Norgren Co. |
| Malinowski, L. | Schrader Fluid Power Div. |
| Sessoms, W. | Schrader Fluid Power Div. |
| Slater, R. | Fairchild Industries |
| Thrasher, G. | Master Pneumatic-Detroit, Inc. |
| VanderHorst, J. | Wilkerson Corporation* |

* Company affiliation has changed since work with the Project Group.

On 30 January 1976 the NFPA Standard was submitted to ANSI Standards Committee B93 for promulgation as an ANSI Standard and to the Board of Standards Review for ANSI's Public Review. This review closed on 27 April 1976. Balloting concluded on 8 June 1976 and resulted in one negative vote. The negative vote was resolved on 19 September 1977. Approval by the ANSI Board of Standards Review was granted on 30 March 1978.

ANSI/B93.39

On 30 January 1976, ANSI Standards Committee B93 was composed of the following: O. J. Maha, Chairman; John R. Luecke, Co-Secretary; J. Tishkowski, Co-Secretary.

AMERICAN SOCIETY OF AGRICULTURAL ENGINEERS
   E. H. Fletcher

AMERICAN SOCIETY OF LUBRICATION ENGINEERS
   M. M. Gurgo

AMERICAN SOCIETY OF MECHANICAL ENGINEERS
   R. Hildebrandt
   T. R. Curran (Alternate)

AMERICAN SOCIETY FOR TESTING AND MATERIALS
   J. D. Lykins
   J. J. Rothrock (Alternate)

CONSTRUCTION INDUSTRY MANUFACTURERS ASSOCIATION
   G. Stewart
   H. T. Larmore (Alternate)

FLUID CONTROLS INSTITUTE
   H. H. Kaemmer
   E. Bianchi (Alternate)

FLUID POWER SOCIETY
   M. Allen
   R. D. Burgess, Sr.
   W. H. Dreher
   R. W. Hanpeter
   A. Hehn
   R. Read
   A. Tucker

FLUID SEALING ASSOCIATION
   R. Prachel
   J. Scannell (Alternate)

INDUSTRIAL TRUCK ASSOCIATION
   C. D. Gibson

INSTRUMENT SOCIETY OF AMERICA
   A. I. Kutz

JOINT INDUSTRY COUNCIL
   R. Muhl

MATERIAL HANDLING INSTITUTE
   J. C. McPherson
   W. Chichester (Alternate)

MOTOR VEHICLE MANUFACTURERS ASSOCIATION
   J. Phillipson

NATIONAL FLUID POWER ASSOCIATION
   O. J. Maha
   J. L. Fisher, Jr.
   W. Forster
   W. Kudlaty
   Z. J. Lansky
   J. J. Pippenger

NATIONAL MACHINE TOOL BUILDERS ASSOCIATION
   E. Loeffler

RUBBER MANUFACTURERS ASSOCIATION
   W. J. Atwell
   N. J. Cyphers
   E. J. McCarthy

SOCIETY OF AUTOMOTIVE ENGINEERS
   W. A. Hertel
   E. L. Falendysz
   H. Schultz
   D. B. Shore
   W. L. Snyder
   D. Prevallet
   R. W. White

SOCIETY OF MANUFACTURING ENGINEERS
   J. Wood

U. S. COAST GUARD
   G. A. Casimir

U. S. DEPARTMENT OF DEFENSE
   C. A. Nazian
   H. Y. Smith
   P. Hopler (Alternate)

INDIVIDUAL MEMBER
   Professor E. C. Fitch, Jr.
   J. Johnson

# REFERENCES

1. American National Standard Glossary of Terms for Fluid Power, ANSI/B93.2-1971, and Supplements thereto. (ISO/DP 5598)

2. SI units and recommendations for the use of their multiples and of certain other units, ISO 1000-1973.

3. International Standard Fluid Power Systems and Components - Graphic Symbols, ISO 1219-1976.

4. NFPA Recommended Standard Method for Verifying the Fatigue and Static Pressure Ratings of the Pressure Containing Envelope of a Metal Fluid Power Component, NFPA/T2.6.1-1974, and appropriate Supplements (NFPA/T3.12.10-1976 and NFPA/T3.12.11-19xx).

5. American Society of Mechanical Engineers Supplement to ASME Power Test Codes; Part 5, Measurement of Quantity of Materials; Chapter 4, Flow Measurement; PTC 19.5:4-1959.

# REQUIREMENTS FOR PRESENTATION OF CATALOG DATA, FLUID COMPATIBILITY, CLEANING MEDIA, MARKINGS AND DIMENSIONAL IDENTIFICATION CODES, AND PRESSURE DROP CHARACTERISTICS FOR FLUID POWER AIR LINE FILTERS

## INTRODUCTION

In pneumatic fluid power systems, power is transmitted and controlled thru a gas under pressure within an enclosed circuit. Air line filters are used to remove solid and liquid contaminants from the gas, thus providing clean air for air-operated devices connected downstream of the filter.

1. SCOPE

   1.1 To include the minimum catalog rating data of flow, temperature, pressure and pressure drop required for manually drained, industrial type fluid power air line filters.

   1.2 To include specifications for listing compatible fluids and cleaning media for manually drained, industrial type fluid power air line filters.

   1.3 To include standard connecting port markings and standard dimensional codes for manually drained, industrial type fluid power air line filters.

   1.4 To include a pressure drop test procedure and methods of determining and presenting data for manually drained industrial type fluid power air line filters.

2. PURPOSE

   2.1 To provide comparative information.

   2.2 To aid in accomplishing proper component application.

   2.3 To establish a standard pressure drop test procedure.

3. TERMS AND DEFINITIONS

   For definition of terms used, see Reference No. 1.

4. UNITS OF MEASUREMENT

   4.1 Customary US units are used.

4.2 Approximate conversions to the International System of Units (SI) are given per Reference No. 2 and appear in parentheses after their Customary US counterparts.

NOTE: The flow rate of compressed gas is specified in standard cubic decimetres per second ($dm_n^3/s$) to differentiate between volume flow at compressed and standard conditions. When an official convention for showing this distinction is approved by ISO, any revisions required in this document will be made.

## 5. GRAPHIC SYMBOLS

Graphic symbols are used in accordance with Reference No. 3.

## 6. MINIMUM CATALOG DATA

### 6.1 Manufacturer's Rated Flow Rates

Specify in scfm ($dm_n^3/s$) for primary pressures of 35, 90 and 150 psi (2.5, 6.3 and 10 bar).

NOTE: At the option of the manufacturer, values of the Manufacturer's Minimum and Manufacturer's Maximum Rated Flow Rates may be specified for values of primary pressure in addition to those listed.

### 6.2 Temperature Rating

Specify the allowable range of temperature in °F (°C) to which filter materials may be exposed for the maximum pressure rating as specified in clause 6.3.

EXAMPLE: 35°F (2°C) to 180°F (82°C). This temperature rating relates to all functional characteristics of the filter.

### 6.3 Pressure Rating

Establish and verify pressure ratings per Reference No. 4, where applicable.

### 6.4 Pressure Drop

Measure and report pressure drop data in accordance with Section 10 of this recommended standard.

## 7 FLUID COMPATIBILITY AND CLEANING MEDIA

### 7.1
Unless otherwise noted on the filter label, use materials of construction that are compatible with "normal" compressed air and mineral-based lubricants within the range of temperatures and pressures specified in clauses 6.2 and 6.3.

NOTE: "Normal" compressed air is the air typically used in industrial shop air lines and includes significant concentrations of dirt, water, non-synthetic oil and pipe scale.

7.2 List materials of construction on instruction sheet.

7.3 Specify grossly non-compatible fluids which might be present in a compressed air system and note that this is a partial listing. Consult manufacturer if in doubt.

7.4 Recommended cleaning procedure on instruction sheet and on bowl.

7.5 Place warning on bowl referencing non-compatible fluids listed on instruction sheet.

## 8. CONNECTING PORT MARKINGS

8.1 As a minimum, provide a directional arrow indicating in-to-out flow direction.

NOTE: The words "inlet", "in", "outlet", "out", etc., are also acceptable. This represents current industry practice; however, in view of the trend toward uniform port identification, it may be subject to change in the future.

8.2 Unless otherwise noted, make all connecting port markings clear and legible.

## 9. EXTERNAL DIMENSIONS (MAXIMUM)

Use the dimensional identification code shown in Figure 1 (Dimensions are in inches (cm).)

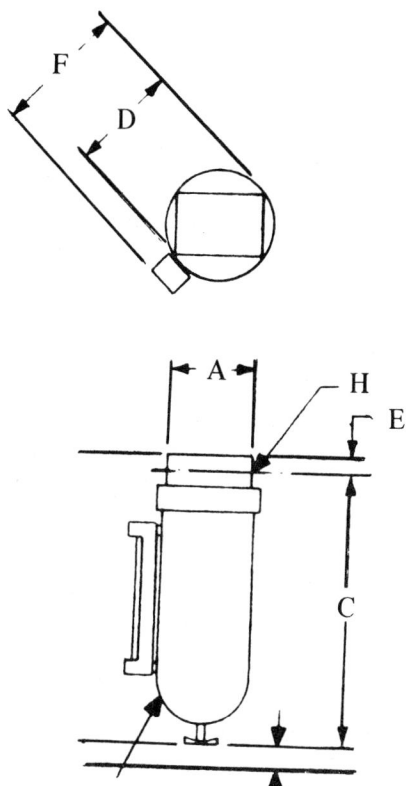

A - Distance between faces of inlet and outlet ports

B - Clearance required to remove reservoir, if applicable

C - Centerline of port to maximum extremity of filter with drain device in its maximum extended position

D - Overall width of filter at maximum point excluding sight glass, if applicable

E - Centerline of port to top extremity of filter

F - Maximum width of filter including sight gage and/or other protrusions

H - Port size

NOTE: If the inlet and outlet ports are not on a common centerline, dimensions "C" and "E" relate to the outlet port. The relative vertical displacement of inlet and outlet ports will be indicated by dimension "G".

FIGURE 1 - External dimensions of an air line filter.

## 10. PRESSURE DROP TEST PROCEDURE AND DATA PRESENTATION

### 10.1 Test Equipment

10.1.1 Use a pneumatic circuit similar to that in Figure 2.

10.1.2 Select, arrange and operate the control devices and measurement instrumentation for this test following industry accepted standards for good practice such as Reference No. 5.

### 10.2 Test Conditions Accuracy

Set up and maintain test conditions accuracy within the limits in Table 1:

TABLE 1 - Test conditions accuracy

| Test Condition | Customary US Unit | SI Unit | Maintain within (±) of Actual |
|---|---|---|---|
| Flow | scfm | $dm_n^3/s$ | 5% |
| Pressure | psi | bar | 2% |
| Temperature | °F | °C | 15°F (9°C) |

### 10.3 Test Procedures

10.3.1 Maintain the temperature of the air supply at 68°F (20°C) within the limits in Table 1.

10.3.2 Control the inlet air pressure within the limits set in Table 1 of the nominal gage pressure setting at all flow conditions.

10.3.3 Test at nominal inlet pressure settings of 35, 90 and 150 psi (2.5, 6.3 and 10 bar). Do not exceed the rated pressure of the filter.

NOTE: Tests at additional inlet pressure settings may be added at the option of the manufacturer.

10.3.4 If the pressure drop across the filter can be significantly affected by the use of different types of elements, show the pressure drop characteristics for each filter element.

10.3.5 Make pressure drop measurements with filters in the condition as received by the user.

10.3.6 At each nominal setting of inlet pressure, control the flow rate over the range of the Manufacturer's Rated Flow Rates per clause 6.1.

10.3.7 If operating life affects the pressure drop characteristics of the filter, additional pressure drop data can be provided at the option of the manufacturer.

NOTE: The range of flow rates shown for the pressure drop characteristics at each value of primary pressure corresponds to the Manufacturer's Rated Flow Rates per clause 6.1 for the filter, unless otherwise noted.

10.3.8 Select values of inlet pressures that do not exceed the pressure rating defined in clause 6.3.

10.3.9 Measure pressure drop by either point-to-point or continuous data recording.

10.3.9.1 For continous data recording, increase the flow rate at a sufficiently slow rate to insure that essentially steady state conditions exist.

10.3.9.2 If discrete changes are made in air flow, verify that steady state conditions exist prior to recording.

## 10.4 Test Data Accuracy

Select and maintain instrumentation so that test data accuracy is within the limits in Table 2:

TABLE 2 - Test Data accuracy

| Quantity | Customary US Unit | SI Unit | Maintain within (±) of Actual |
|---|---|---|---|
| Flow | scfm | $dm_n^3/s$ | 5% |
| Pressure | psi | bar | 2% |

## 10.5 Criteria for Acceptance of Test Data

10.5.1 Verify that the pressure drop as a function of flow rate for each value of inlet pressure is equal to or less than the reported value in Figure 3.

10.5.2 Verify that essentially steady state conditions exist as flow rate increases for continuous data recording.

10.5.3 Verify that steady state conditions exist prior to recording if discrete changes are made in air flow.

## 10.6 Data Presentation

10.6.1 Present pressure drop data in graphic form similar to that in Figure 3 as a function of flow rate for each value of inlet pressure.

10.6.2 Make the flow limits of the graph for each inlet pressure consistent with the Rated Flow Rates per clause 6.1.

10.7 Summary of Designated Information

The following designated information is needed when applying this recommended test procedure to a particular application or use:

10.7.1 Rated pressure.

10.7.2 Flow rate.

10.7.3 Inlet pressure.

10.8 Justification Statement

This document formalizes practices and equipment requirements that have gained general acceptance and are currently being used by a majority of the manufacturers who participated in the development of this recommended standard.

10.9 Test/Production Similarity

Utilize managerial controls necessary to maintain substantial similarity between test and production components or elements.

## 11. IDENTIFICATION STATEMENT

Use the following statement in catalogs and sales literature when electing to comply with this voluntary standard:

"Method of presenting catalog data, listing fluid compatibility and cleaning media, and port marking and dimensional identification codes and pressure drop characteristics conform to American National Standard ANSI/B93.39-1978."

ANSI/B93.39

## 12. KEY WORDS

The following Key Words, useful in indexes and in information retrieval systems, are suggested for this recommended standard:

air line filter, catalog data

air line filter, cleaning media

air line filter, dimensional identification codes

air line filter, fluid power

air line filter, markings

air line filter, pressure drop characteristics

data presentation, air line filter

filter, air line

fluid power

testing, air line filter

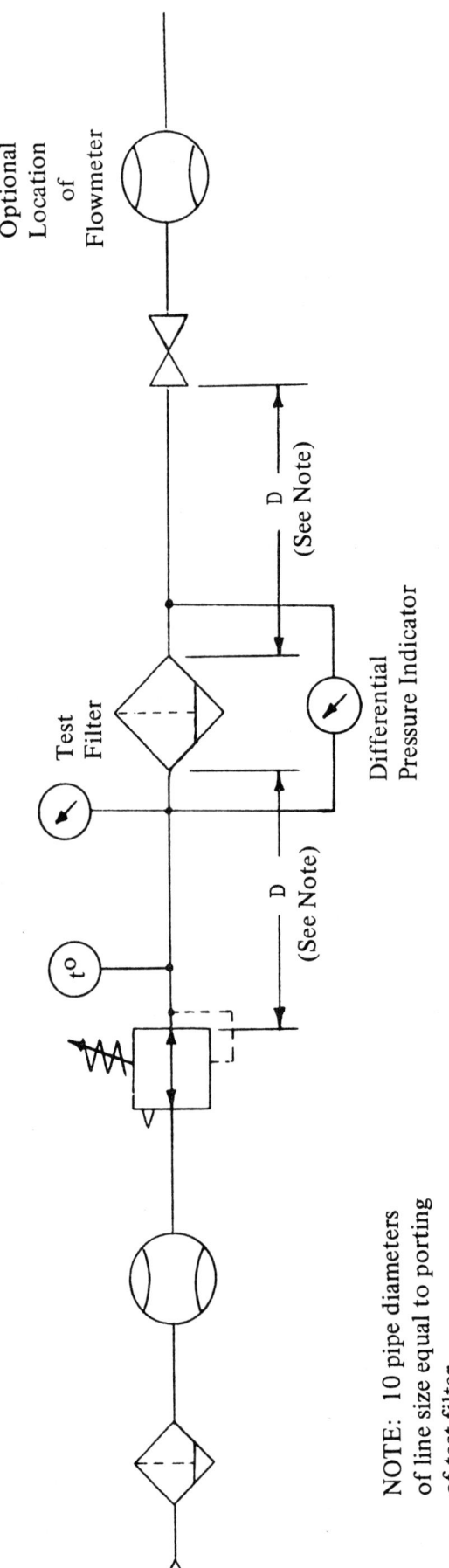

FIGURE 2 - Pressure drop test circuit

NOTE: 10 pipe diameters of line size equal to porting of test filter

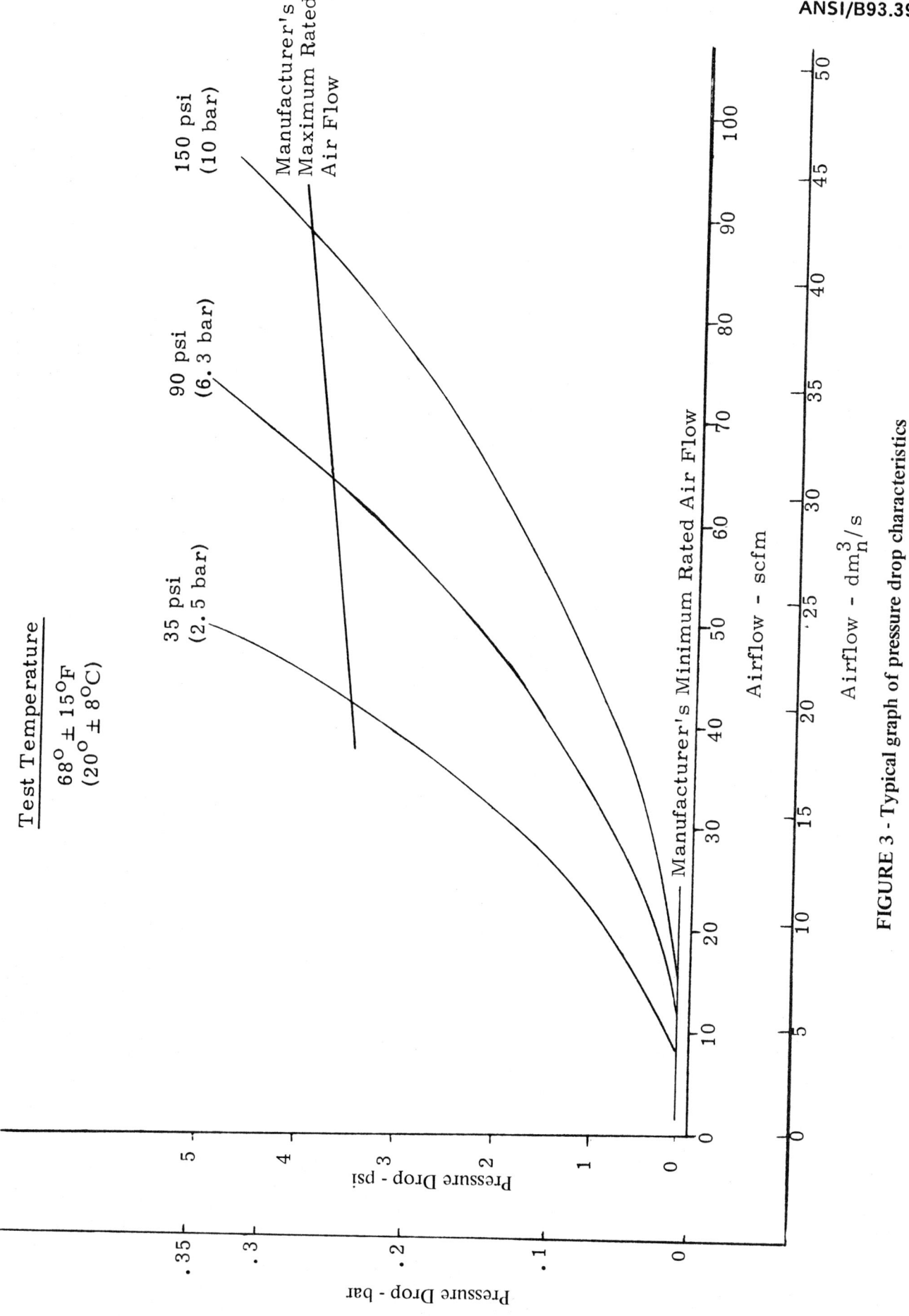

FIGURE 3 - Typical graph of pressure drop characteristics

# NOTES

ANSI/B93.13M-1981

## AN INDUSTRY STANDARD FOR FLUID POWER

American National Standard

Pneumatic fluid power -

Pressure regulators -

Industrial type
(NFPA/T3.12.3M R1-1978)

Approved as an ANSI Standard
17 August 1981

**Descriptors:** dimensions, air line pressure regulator; fluid power; port identification, air line pressure regulator; pressure regulator, air line; pressure regulator, fluid power industrial type; rating, air line pressure regulator; testing, air line pressure regulator; testing, flow characteristics; testing, regulator characteristic; testing, relief flow characteristics.

published by
# NATIONAL FLUID POWER ASSOCIATION, INC.

3333 N. Mayfair Road / Milwaukee, WI 53222 / 414-778-3344 / TLX 26898

# ANSI/B93.13

**FOREWORD**

This Foreword is not part of American National Standard Pneumatic fluid power - Pressure regulators - Industrial type, ANSI/B93.13M-1981 (a revision of ANSI/B93.13-1977 (NFPA/T3.12.3M R1-1978)).

NFPA Recommended Standard T3.12.3-1970 was approved by the Board of Directors on 15 January 1970, and was accepted as an ANSI standard by the American National Standards Institute on 14 July 1971.

In accordance with the NFPA procedure which requires the review of existing NFPA standards at five-year intervals following the date of publication, the Pneumatic FRL Section reviewed NFPA/T3.12.3-1970 on 17 March 1976 and agreed that the document required revision to reflect current practice and to update the pneumatic data parameters given in it. The Title, Scope and Purpose for the new project NFPA/T3.12.3M R1, was approved by the Technical Board on 7 April 1976.

The first meeting of the Project Group was held on 29 June 1976. Draft No. 1, which resulted from that meeting, was circulated to the Project Group on 3 August 1976. It was reviewed on 19 October 1976, and Draft No. 2 was circulated on 15 November 1976. Draft No. 2 was reviewed at the Project Group's 2 December 1976 meeting, and Draft No. 3 was circulated on 22 December 1976. Consensus was reached on 17 January 1977 following review of Draft No. 3. Project Group Chairman John Berninger (Parker Hannifin Corporation) forwarded the Final Working Draft to Headquarters on 31 January 1977, and the NFPA Technical Staff prepared the General Review Draft on 16 February 1977.

The comments on the General Review Draft were discussed at the Project Group meeting held 19 April 1977. Each comment was evaluated and drafts of answers were completed. Chairman Berninger submitted written replies of the committee's findings to each who commented. The NFPA Technical Board granted approval to the ballot on 11 May 1977, and the NFPA Technical Staff prepared the Ballot Draft on 27 May 1977.

Five negative votes were received when the ballot closed on 14 July 1977. Chairman Berninger held several Project Group meetings through which four of the negative ballots were resolved through editorial modifications. These modifications brought the document into agreement with existing standards on metric usage, graphic symbols conventions and, most importantly, flow measurement techniques. The remaining negative ballot objected to the document on the basis that it is a conglomerate document and should be broken into several separate standards. Both the FRL Section and the Technical Board discussed this objection and, on the basis of the FRL Section recommendation, agreed to maintain the document as one standard.

The NFPA Technical Board recommended final approval of the document at its 4 May 1978 meeting and the NFPA Board of Directors granted final approval at its 7 June 1978 meeting.

Project Group members who developed this standard:

**John Berninger**
Project Chairman
Parker Hannifin Corp.

**William Dewberry**
Project Vice Chairman
Watts Fluid Power

**John Pousma**
Project Secretary
C. A. Norgren Co.

**John Colter\***
Section Chairman
Watts Fluid Power

**John Lytle\***
Technical Auditor
Ross Operating Valve Co.

**John R. Luecke\***
Director of National Technical Services
National Fluid Power Association

**Richard Bailey**
C. A. Norgren Co.

**Carroll Grigsby**
Schrader Bellows Division

**Ken Russell\***
Wilkerson Corp.

It was intended that NFPA Recommended Standard T3.12.3M R1, be submitted to ANSI for promulgation as an American National Standard.

On 25 July 1979, ANSI/B93.13M (a revision of ANSI/B93.13-1977) was submitted to ANSI Committee B93 for ballot. Balloting closed 22 August 1979 with one negative comment. The committee responded to this negative comment.

Subsequently, ANSI B93.13M was forwarded to the ANSI Board of Standards Review for approval on 10 July 1981. Approval was granted 17 August 1981.

The membership roster for the Standards Committee B93 at the time of ballot was:

**Jack McPherson**
Chairman

**Daniel Shore**
Vice Chairman

---

\*Company affiliation has changed

ANSI/B93.13

**Dixie L. Prevost***
Co-Secretary

**William Toth**
Co-Secretary

**American Society of Agricultural Engineers**
Ed Fletcher

**American Society for Engineering Education**
William Smith

**American Society of Lubrication Engineers**
M. M. Gurgo

**American Society of Mechanical Engineers**
Robert Hildebrandt
Thomas R. Curran (alternate)
Frank Yeaple (alternate)

**Compressed Air and Gas Institute**
D. E. Bonn
John Addington (alternate)

**Construction Industry Manufacturers Association**
Glenn Stewart

**Fluid Controls Institute**
Herbert H. Kaemmer
Eric Bianchi

**Fluid Power Distributors Association**
Thomas H. Neff

**Fluid Power Society**
Marsh Allen
Edward C. Briggs
Ray Fiedler
Robert W. Hanpeter
Anton Hehn
Richard Read
Allen Tucker

**Fluid Sealing Association**
Ronald Prachel
John Scannell (alternate)

**Industrial Truck Association**
C. D. Gibson

**Instrument Society of America**
Aaron I. Kutz

**Joint Industry Council**
Robert Muhl

**Material Handling Institute**
Jack McPherson
Williard Chichester (alternate)

**Motor Vehicle Manufacturers Association**
Jim Phillipson

**National Fluid Power Association**
James L. Fisher, Jr.
Walter Forster
Z. J. Lansky
Otto Maha
John Pippenger
A. O. Roberts

**National Machine Tool Builders Association**
John B. Deam

**Power Crane and Shovel Association**
(to be named)

**Rubber Manufacturers Association**
John R. Loder
William Hertel
E. J. McCarthy (alternate)

**Society of Automotive Engineers**
William Hertel
Eugene Falendysz
Henry Schultz
D. B. Shore
W. L. Snyder
David Prevallet
Robert White

**Society of Manufacturing Engineers**
Raymond Grisdale

**U. S. Department**
Henry Schaefer
William Coyne (alternate)

**Individual Members**
Dr. E. C. Fitch, Jr.
Jack Johnson

---

*Company affiliation has changed

# ANSI/B93.13

**Pneumatic fluid power -
Pressure regulators -
Industrial type**

## 0 INTRODUCTION

In pneumatic fluid power systems, power is transmitted and controlled thru a gas under pressure within an enclosed circuit. Air line pressure regulators transform a fluctuating air pressure supply to provide relatively constant reduced output pressure.

Relieving type regulators automatically vent excess regulated (secondary) pressure.

## 1 SCOPE AND FIELD OF APPLICATION

1.1 This National Standard is intended to establish:

— A method of test for steady state conditions.

— A method of rating for steady state conditions, port marking, and dimensional identification for fluid power industrial type air line pressure regulators producing secondary pressures of 0-250 psig (0-17.2 bar).

1.2 This National Standard is intended to:

— Provide comparative information in standard form.

— Aid in accomplishing good product application.

— Assist in the establishment of meaningful ratings.

## 2 REFERENCES

ANSI/B93.2-1971, and Supplement ANSI/B93.2A-1978, **American National Standard Glossary of Terms for Fluid Power.**

ANSI Y14.17-1966, **Fluid power diagrams.**

ANSI Y32.10-1967, **Graphic symbols for fluid power diagrams.**

ISO 1000-1981, **SI units and recommendations for the use of their multiples and of certain other units.**

ISO 1219-1976, **Fluid power systems and components - Graphic symbols.**

ANSI/B36.10-1975, **American National Standard Welded and Seamless Wrought Steel Pipe.**

ISA-S39.4-1974, **Instrument Society of America Standard Control Valve Capacity Test Procedure for Compressible Fluids.**

## 3 TERMS AND DEFINITIONS

For definition of terms not defined below, see ANSI/B93.2 and ANSI/B93.2A.

**3.1 flow characteristics curve:** A graphical representation of the change in secondary (regulated) pressure with changes in the rate of air flow while the primary pressure is held constant.

**3.2 pressure regulation characteristics curve:** A graphical representation of secondary (regulated) pressure variation caused by changes in primary (supply) pressure, at a constant rate of air flow.

**3.3 relief flow characteristics curve.** A graphical representation of the relationship between relief flow rate and secondary (regulated) pressure, above the regulator setting while the primary pressure is held constant.

# 4 UNITS OF MEASUREMENT

**4.1** Customary US units are used.

**4.2** Approximate conversions to SI units are given in accordance with ISO 1000 and appear in parentheses after their Customary US counterparts.

**4.3** The pressure term "bar" as used in this recommended standard is relative to the atmosphere (gage).

# 5 LETTER SYMBOLS

The following letter symbols are used in this document:

- psig    pounds per square inch, gage (bar)
- scfm    flow in standard cubic feet per minute ($dm^3_n/s$ - flow in cubic decimetres per second, normalized to standrd atmospheric conditions).

# 6 GRAPHIC SYMBOLS

Graphic symbols are used in accordance with ISO 1219.

# 7 RATING

## 7.1 Temperature

Specify the degrees Fahrenheit, °F (degrees Celsius, °C) the normal operating temperature limits within the pressure range in accordance with 7.2, e.g., -40° to 180°F (-40° to 82°C).

**NOTE:** This rating does not imply that the performance characteristics of the regulator are the same throughout the specified temperature range.

## 7.2 Pressure

**7.2.1** Specify a maximum inlet pressure that is in conformity to applicable NFPA pressure rating standards.

**7.2.2** Specify pilot, gage or outlet port pressure limitations individually if less than pressure stated in 7.2.1.

**7.2.3** State the reduced pressure range within the maximum pressure and temperature ranges in accordance with 7.2.1 and 7.2.2.

**7.2.4** Designate zero at the lower end of the range where the regulator is capable of being adjusted to zero output pressure.

**NOTE:** This rating does not imply that the regulator may be successfully employed on applications requiring finite adjustment at the low end of the range.

**7.2.5** State the recommended adjustment range in place of or in addition to the reduced pressure range.

**EXAMPLE:** Reduced pressure range, 0-125 psig (0-8.6 bar). Recommended adjustment range, 10-125 psig (0.7-8.6 bar).

**7.2.6** Construct manually adjustable regulators so that secondary pressure increases with clockwise rotation of the adjusting mechanism.

**NOTE:** The component may have an indication of pressure adjustment direction.

## 7.3 Materials of Construction

**7.3.1** Use materials of construction compatible for normal air service within the temperature and pressure limitations in accordance with 7.1 and 7.2.

**NOTE:** Normal air, as used here, is filtered compressed air that is in ordinary factory pipe lines. It may include small amounts of water and non-synthetic lubricating oil.

**7.3.2** List major materials of construction and any protective coatings where applicable in product literature.

**7.3.3** Caution the user if the regulator is not suitable for use with fluids other than normal compressed air.

# 8 PORT IDENTIFICATION

**8.1** Provide clear, legible, and permanent port identifications on the component.

**8.2** Use the words "In", "Inlet", "Out", "Outlet", "Pilot", and/or a direction arrow, indicating the in-to-out flow direction.

**8.3** Side connection port identification is not required for compliance with this recommended standard.

# 9 DIMENSIONAL DESIGNATIONS

**9.1** Identify the maximum envelope dimensions in the manner shown in figure 1.

**9.1.1** Dimension "A" indicates port to port dimension, not maximum regulator diameter.

**9.1.2** Dimension "B" indicates the maximum height at zero pressure adjustment position. (Maximum varies depending on adjustment and spring range.)

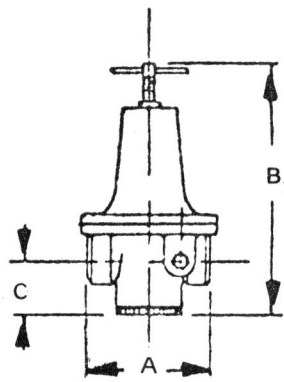

Figure 1 - Side view

**9.2** Provide a top view as shown in figure 2 where regulator configurations are not symmetrical or where other dimensions exceed port dimensions.

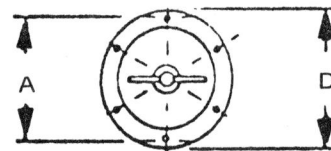

FIGURE 2 — Top view

## 10 TEST PROCEDURE

**10.1** General

**10.1.1** Conduct a series of three tests as described below for establishing a performance characteristics of a regulator.

**10.1.2** Prepare a test circuit as shown in figure 3, into which test regulators can be installed for conducting the three tests.

    a) Note the different location requirements and options for the flowmeter (only one location required per test).

    b) Check out all components to be certain they do not leak and that their pressure ratings are equal to or greater than the rating of the test regulator.

**10.1.3** Test Conduits

    a) Attach the test conduit directly to the test regulator (without additional fittings).

**NOTE:** If the test regulator does not have international pipe threads, attach test conduits with fittings or connectors normally required for proper attachment

    b) Use test conduits as shown in figure 4 having an internal diameter within the limits of ANSI/B36.10 schedule 40 pipe of a size that corresponds to the port size of the test regulator.

**NOTE:** If this is not practical (as when British Standard Pipe threads or other alternative threads are used), state the type and size of conduit used.

**10.1.4** Use pressure, temperature, and flow recording instruments which have been calibrated to the accuracies given below:

**Pressure:** $\pm 2\%$ of data point

**Flow:** $\pm 5\%$ of maximum tested flow

**Temperature:** $\pm 2°F$ ($\pm 1.1°C$) of data point

**10.1.5** Filter the air used for testing before admitting it into the test circuit (25 micron (25 micrometre) recommended).

**NOTE:** Removal of moisture and atomized lubricants is an additional preferred treatment.

**10.1.6** Conduct the tests with air temperature and the ambient temperature within a range of 65° to 85°F (18° to 29°C).

**10.1.7** Provide identification of the regulator being tested, specifying port size and pressure range.

**10.1.8** Consider test results to be representative of similarly manufactured regulators.

**10.1.9** Perform retesting when design changes affect performance characteristics.

**10.2** Flow Characteristics Test

**10.2.1** Install the test regulator in the circuit shown in figure 3 and close supply, bypass, and flow control valves.

**10.2.2** Adjust primary pressure at inlet regulator $R_1$ to a test level selected from table 1 and maintain throughout the test.

**10.2.3** Adjust control regulator $R_C$ for zero secondary pressure.

**10.2.4** Adjust test regulator $R_T$ for zero secondary pressure and flow and open the supply valve, ensuring that pressure $P_1$ is the same as pressure $P_0$.

**10.2.5** Adjust test regulator $R_T$ to generate a secondary set point pressure at $P_2$ for the lowest value in the selected category of table 1.

**10.2.6** Open flow control valve in a number of incremental steps, continually readjusting inlet regulator $R_1$ to maintain test level at $P_1$.

**10.2.7** Record pressures $P_1$ and $P_2$, flow rate, and temperatures $T_1$ and $T_2$ for each incremental step of increasing flow.

**10.2.8** Obtain data from sufficient incremental steps to ensure that a factual representation of performance can be determined. Closely spaced data points are required in the low flow region.

**10.2.9** Close the flow control valve in a number of incremental steps, continually readjusting inlet regulator $R_1$ to maintain test level at $P_1$.

**10.2.10** Record pressures $P_1$ and $P_2$, flow rate, and temperatures $T_1$ and $T_2$ for each incremental step of decreasing flow.

**10.2.11** Record data in accordance with 10.2.9 and 10.2.10 until the flow thru the test regulator stops.

NOTE: Obtain data from sufficient incremental steps to ensure that a factual representation of performance can be determined. Closely spaced data points are required in the low flow region.

**10.2.12** Repeat 10.2.5 up to and including 10.2.11 for each secondary setpoint pressure in the selected category of table 1.

**10.2.13** Record ambient temperature at the conclusion of testing.

**10.2.14** Compute the flow rate in scfm ($dm_n^3/s$) from the recorded data of flow, pressure, and temperature.

**10.2.15** Plot a rectilinear graph of secondary pressure (ordinate) vs. secondary flow (abscissa) for either (or both) the increasing or decreasing flow conditions for the levels of secondary setpoints (see sample in figure 6).

**10.2.16** Identify primary pressure $P_1$, inlet temperature $T_1$, and flow condition (increasing or decreasing) as constants on the graph.

**10.2.17** As an option, repeat 10.2.2 up to and including 10.2.16 at other primary pressures in table 1.

**10.3 Relief Flow Characteristics Test**

NOTE: It may be desirable to perform this test immediately after completing each test setting of secondary pressures in the Flow Characteristics Test, and starting at 10.3.7 (provided that flowmeter location $F_1$ was selected for the Flow Characteristics Test).

**10.3.1** Install the test regulator in the circuit shown in figure 3 and close supply, bypass, and flow control valves.

**10.3.2** Be certain that the line from the relief outlet is directed into the water, and seal all openings (such as screw threads or joints) with appropriate material such as clay or mastic.

**10.3.3** Adjust primary pressure at inlet regulator $R_1$ to a test level selected from table 1 and maintain throughout the test.

**10.3.4** Adjust control regulator $R_C$ for zero secondary pressure.

**10.3.5** Adjust test regulator $R_T$ for zero secondary pressure and flow and open the supply valve, ensuring that pressure $P_1$ is the same as $P_0$.

**10.3.6** Adjust test regulator $R_T$ to generate a secondary set point pressure at $P_2$ for the lowest value in the selected category of table 1.

**10.3.7** Readjust control regulator $R_C$ to generate a secondary pressure to match that of the test regulator secondary setpoint (i.e., $P_3$ should be the same as $P_2$).

**10.3.8** Open the bypass valve.

Note: Flow control valve remains closed during data recordings.

**10.3.9** Increase control regulator setting gradually such that pressure $P_2$ increases.

**10.3.10** Record the pressure at which bubbles appear in the water from the test regulator relief outlet.

**10.3.11** Remove hose from water and remove any restriction that might have been added to relief flow path for bubble point check.

**10.3.12** Continue to increase secondary pressure $P_2$ in a number of incremental steps using control regulator $R_C$.

**10.3.13** If necessary, readjust the inlet regulator $R_1$ to maintain the test level at $P_1$.

**10.3.14** Record pressures $P_2$ and $P_1$, flow rate, and temperatures $T_1$ and $T_2$ at each step.

**10.3.15** Record data in accordance with 10.3.12 up to and including 10.3.14 until the secondary pressure $P_2$ approaches the primary pressure $P_1$ as closely as possible.

NOTE: Obtain data from sufficient incremental steps to ensure that a representative performance can be determined.

**10.3.16** Return the control regulator $R_C$ to its zero flow and pressure position and simultaneously open the flow control valve slowly to exhaust the secondary air.

**10.3.17** Close the flow control and bypass valves when $P_2$ and $P_3$ return to their approximate initial values.

**10.3.18** Repeat 10.3.6 up to and including 10.3.17 for each secondary set point pressure in the selected category of table 1.

**10.3.19** Record ambient temperature at the conclusion of testing.

**10.3.20** Compute flow rate in scfm ($dm_n^3/s$) from the recorded data of flow, pressure, and temperatures.

**10.3.21** Plot a rectilinear graph of secondary pressure (ordinate) vs. normalized relief flow (abscissa) for the levels of secondary setpoints.

**NOTE:** As a option, this may be placed on the same performance graph as used for the Flow Characteristics Test results, using the negative abscissa for Relief Flow (see figure 6).

**10.3.22** As an option, repeat 10.3.3 up to and including 10.3.21 at other primary pressures shown in table 1.

**10.4** Regulation Characteristic Test

**10.4.1** Install the test regulator in the circuit shown in figure 3 and close supply, bypass, and flow control valves.

**10.4.2** Adjust primary pressure at inlet regulator $R_1$ to the maximum possible primary test level in table 1 without exceeding the inlet pressure rating of the test regulator $R_T$.

**10.4.3** Adjust control regulator $R_C$ for zero secondary setting.

**10.4.4** Adjust test regulator $R_T$ for zero output and open the supply valve, ensuring that pressure $P_1$ is the same as $P_O$.

**10.4.5** Simultaneously adjust test regulator $R_T$ and the flow control valve until:

   a) Secondary pressure $P_2$ reaches the maximum value in the selected category from table 1, and

   b) Secondary flow reaches 10% of the value corresponding to the test regulator port size and secondary pressure $P_2$ from table 2.

**10.4.6** Readjust inlet regulator $R_1$ if necessary to maintain the test level at $P_1$ and readjust test regulator and flow control valve as required.

**10.4.7** Record the conditions of pressure $P_1$ and $P_2$, flow rate, and temperatures $T_1$ and $T_2$ when the system has stabilized.

**10.4.8** Reduce the inlet pressure with regulator $R_1$ in a number of incremental steps, readjusting the flow control valve to maintain a constant, normalized flow rate.

**10.4.9** Record pressures $P_1$ and $P_2$, flow rate, and temperatures $T_1$ and $T_2$ for each incremental step.

**10.4.10** Record data in accordance with 10.4.8 and 10.4.9 until the primary pressure equals the secondary setpoint.

**NOTE:** Obtain data from sufficient incremental steps to ensure that a representative performance can be determined.

**10.4.11** As an option, raise primary pressure back to its maximum initial test level and repeat 10.4.5 up to and including 10.4.10, this time maintaining a secondary flow rate equal to the value shown in table 2.

**10.4.12** Repeat 10.4.5 up to and including 10.4.11 for the other secondary setpoints in the selected category of table 1.

**10.4.13** Compute the flow rate in scfm ($dm_n^3/s$) from the recorded data of flow, pressure, and temperature.

**10.4.14** Plot a rectilinear graph of secondary pressure (ordinate) vs. primary pressure (abscissa) for the levels of flow rate tested (see sample in figure 7).

**10.4.15** Identify inlet temperature $T_1$ and flow rate scfm ($dm_n^3/s$) on the graph.

## 11. TEST/PRODUCTION SIMILARITY

Utilize managerial controls necessary to maintain substantial similarity between test and production components or elements.

## 12. IDENTIFICATON STATEMENT

Use the following statement in catalogs and sales literature when electing to comply with this voluntary standard:

"Port identification, dimensional designations, and methods for determining performance characteristics conform to American National Standard, ANSI/B93.13M-1981, **Pneumatic fluid power - Pressure regulators - Industrial type.**"

ANSI/B93.13

Table 1 - Secondary pressure setpoint values

| RANGE OF MAXIMUM ADJUSTABLE LIMITS | | PRIMARY PRESSURE LEVELS | | |
|---|---|---|---|---|
| | Starting with and up to, but not including | 100 psig (6.3 bar) | 150 psig (10 bar) | 230 psig (16 bar) |
| A | 0 to 15 psig (0 to 1 bar) | 3,6,9,12 psig (0.2,0.4,0.63,0.8 bar) | Same as 100 psig (6.3 bar) | Same as 100 psig (6.3 bar) |
| | 15 to 19 psig (1.0 to 1.25 bar) | 6,9,12,15 psig (0.4,0.63,0.8,1.0 bar) | Same as 100 psig (6.3 bar) | Same as 100 psig (6.3 bar) |
| | 19 to 24 psig (1.25 to 1.6 bar) | 6,9,12,15, (18) psig (0.4,0.63,0.8,1.0,1.25 bar) | Same as 100 psig (6.3 bar) | Same as 100 psig (6.3 bar) |
| | 24 to 30 psig (1.6 to 2.0 bar) | 6,9,15,22 psig (0.4,0.63,1.0,1.6 bar) | Same as 100 psig (6.3 bar) | Same as 100 psig (6.3 bar) |
| B | 30 to 38 psig (2.0 to 2.5 bar) | 6,15,22,30 psig (0.4,1.0,1.6,2.0 bar) | Same as 100 psig (6.3 bar) | Same as 100 psig (6.3 bar) |
| | 38 to 48 psig (2.5 to 3.15 bar) | 15,22,30,37 psig (1.0,1.6,2.0,2.5 bar) | Same as 100 psig (6.3 bar) | Same as 100 psig (6.3 bar) |
| | 48 to 60 psig (3.15 to 4.0 bar) | 15,22,30,37, (45) psig (1.0,1.6,2.0,2.5, (3.15) bar) | Same as 100 psig (6.3 bar) | Same as 100 psig (6.3 bar) |
| C | 60 to 80 psig (4.0 to 5.0 bar) | 15,30,45,60 psig (1.0,2.0,3.15,4.0 bar) | Same as 100 psig (6.3 bar) | Same as 100 psig (6.3 bar) |
| | 80 to 100 psig (5.0 to 6.3 bar) | 30,45,60,75 psig (2.0,3.15,4.0,5.0 bar) | Same as 100 psig (6.3 bar) | Same as 100 psig (6.3 bar) |
| D | 100 to 125 psig (6.3 to 8.0 bar) | 30,45,60,75 psig (2.0,3.15,4.0,5.0 bar) | 30,45,60,75, (90) psig [2.0,3.15,4.0,5.0, (6.3) bar] | 30,45,60,75, (90) psig [2.0,3.15,4.0,5.0, (6.3) bar] |
| | 125 to 160 psig (8.0 to 10.0 bar) | 30,45,60,75, (90) psig (2.0,3.15,4.0,5.0, bar) | 30,60,90,120 psig (2.0,4.0,6.3,8.0 bar) | 30,60,90,120,150 psig [2.0,4.0,6.3,8.0, (10.0) bar] |
| | 160 to 200 psig (10 to 12.5 bar) | 30,45,60,75, (90) psig (2.0,3.15,4.0,5.0, bar) | 30,60,90,120 psig (2.0,4.0,6.3,8.0 bar) | 60,90,120,150, (180) psig [4.0,6.3,8.0,10.0, (12.5) bar] |

continued .../

- 325 -

ANSI/B93.13

Table 1 - Secondary pressure setpoint values (concluded)

| RANGE OF MAXIMUM ADJUSTABLE LIMITS | PRIMARY PRESSURE LEVELS | | |
|---|---|---|---|
| | 100 psig (6.3 bar) | 150 psig (10 bar) | 230 psig (16 bar) |
| Starting with and up to, but not including | | | |
| E  200 to 250 psig (12.5 to 16.0 bar) | 30,45,60,75, (90) psig (2.0,3.15,4.0,5.0, bar) | 30,60,90,120 psig (2.0,4.0,6.3,8.0 bar) | 60,90,120,150,180, (210) psig (4.0,6.3,8.0,10.0,12.5, bar) |

NOTES:

1. All pressure values are gage pressure.
2. Categories A, B, C, D, E, are preferred design groups.
3. Values in parenthesis are optional test levels.
4. Pressure values in Customary US and SI units are listed as separate test systems, with only close approximations shown and not equivalent values.

ANSI/B93.13

Table 2 - Flow rates to be used for regulation characteristics tests

| Applied Pressure psig [bar] | 1/8 | 1/4 | 3/8 | 1/2 | 3/4 | 1 | 1-1/4 | 1-1/2 | 2 |
|---|---|---|---|---|---|---|---|---|---|
| | \multicolumn{9}{c}{Nominal Pipe Size — Inches} | | | | | | | | |
| | \multicolumn{9}{c}{100% Maximum Recommended Flow scfm [dm$_n^3$/s]} | | | | | | | | |
| 1.5 | 0.27 | 0.60 | 1.34 | 2.48 | 3.70 | 6.99 | 14.2 | 21.6 | 41.7 |
| [0.2] | [0.18] | [0.41] | [0.91] | [1.69] | [2.53] | [4.79] | [9.73] | [14.8] | [28.5] |
| 3.0 | 0.39 | 0.89 | 1.97 | 3.66 | 5.47 | 10.32 | 21.1 | 32.0 | 61.6 |
| [0.4] | [0.28] | [0.62] | [1.39] | [2.59] | [3.86] | [7.30] | [14.8] | [22.5] | [43.5] |
| 6.0 | 0.60 | 1.36 | 3.02 | 5.60 | 8.36 | 15.8 | 32.2 | 48.8 | 94.2 |
| 9.0 | 0.79 | 1.78 | 3.96 | 7.34 | 11.0 | 20.7 | 42.2 | 64.0 | 123 |
| [0.63] | [0.38] | [0.85] | [1.89] | [3.50] | [5.22] | [9.88] | [20.1] | [30.5] | [58.9] |
| [0.8] | [0.44] | [1.00] | [2.23] | [4.14] | [6.18] | [11.7] | [23.8] | [36.1] | [69.7] |
| 12.0 | 0.97 | 2.18 | 4.85 | 8.99 | 13.4 | 25.4 | 51.7 | 78.4 | 151 |
| [1.0] | [0.52] | [1.18] | [2.63] | [4.88] | [7.28] | [13.8] | [28.0] | [42.5] | [82.1] |
| 15.0 | 1.14 | 2.57 | 5.72 | 10.6 | 15.8 | 29.9 | 61.0 | 92.4 | 178 |
| 18.0 | 1.31 | 2.96 | 6.57 | 12.2 | 18.2 | 34.4 | 70.1 | 106 | 205 |
| [1.25] | [0.62] | [1.40] | [3.12] | [5.78] | [8.63] | [16.3] | [33.2] | [50.4] | [97.2] |
| 22.0 | 1.53 | 3.46 | 7.69 | 14.3 | 21.3 | 40.3 | 82.1 | 124 | 240 |
| [1.6] | [0.75] | [1.70] | [3.79] | [7.03] | [10.5] | [19.9] | [40.4] | [61.3] | [118] |
| [2.0] | [0.91] | [2.04] | [4.55] | [8.44] | [12.6] | [23.8] | [48.5] | [73.6] | [142] |
| 30.0 | 1.97 | 4.46 | 9.92 | 18.4 | 27.5 | 51.9 | 106 | 160 | 310 |
| [2.5] | [1.09] | [2.46] | [5.49] | [10.2] | [15.2] | [28.8] | [58.5] | [88.8] | [171] |
| 37.0 | 2.35 | 5.33 | 11.9 | 22.0 | 32.8 | 62.0 | 126 | 192 | 369 |
| 45.0 | 2.79 | 6.32 | 14.0 | 26.1 | 38.9 | 73.4 | 149 | 227 | 438 |
| [3.15] | [1.34] | [3.01] | [6.71] | [12.5] | [18.6] | [35.2] | [71.5] | [109] | [209] |
| [4.0] | [1.65] | [3.72] | [8.30] | [15.4] | [23.0] | [43.5] | [88.4] | [134] | [259] |
| 60.0 | 3.61 | 8.15 | 18.1 | 33.7 | 50.2 | 94.9 | 193 | 293 | 566 |
| [5.0] | [2.02] | [4.56] | [10.2] | [18.9] | [28.1] | [53.3] | [108] | [164] | [317] |
| 75.0 | 4.42 | 9.99 | 22.2 | 41.3 | 61.5 | 116 | 237 | 359 | 694 |
| 90.0 | 5.23 | 11.8 | 26.3 | 48.8 | 72.8 | 138 | 280 | 425 | 821 |
| [6.3] | [2.50] | [5.64] | [12.6] | [23.3] | [34.8] | [65.9] | [134] | [203] | [393] |
| 100.0 | 5.77 | 13.0 | 29.0 | 53.8 | 80.3 | 152 | 309 | 469 | 905 |
| [8.0] | [3.13] | [7.05] | [15.7] | [29.2] | [43.6] | [82.5] | [168] | [255] | [491] |
| 120.0 | 6.85 | 15.5 | 34.4 | 63.9 | 95.4 | 180 | 367 | 557 | 1075 |
| [10.0] | [3.87] | [8.72] | [19.4] | [36.1] | [53.9] | [102] | [207] | [315] | [607] |
| 150.0 | 8.47 | 19.1 | 42.6 | 79.0 | 118 | 223 | 454 | 688 | 1329 |
| [12.5] | [4.79] | [10.8] | [24.1] | [44.7] | [66.7] | [126] | [257] | [390] | [752] |
| 180.0 | 10.1 | 22.8 | 50.7 | 94.1 | 140 | 265 | 541 | 820 | 1583 |
| 210.0 | 11.7 | 26.5 | 58.8 | 109 | 163 | 308 | 627 | 951 | 1836 |
| 230.0 | 12.8 | 28.9 | 64.2 | 119 | 178 | 336 | 685 | 1039 | 2005 |
| [16.0] | [6.08] | [13.7] | [30.6] | [56.7] | [84.7] | [160] | [326] | [494] | [954] |
| 250.0 | 13.9 | 31.3 | 69.7 | 129 | 193 | 365 | 743 | 1127 | 2175 |

These flows are based upon the following pressure drops in 100 feet of Schedule 40 Wrought Steel and Iron Pipe @ 70°F (21°C).

| Drop in Gauge Pressures | Pipe Sizes |
|---|---|
| 10% | 1/8", 1/4", 3/8", 1/2" |
| 5% | 3/4", 1", 1-1/4", 1-1/2", 2" |

ANSI/B93.13

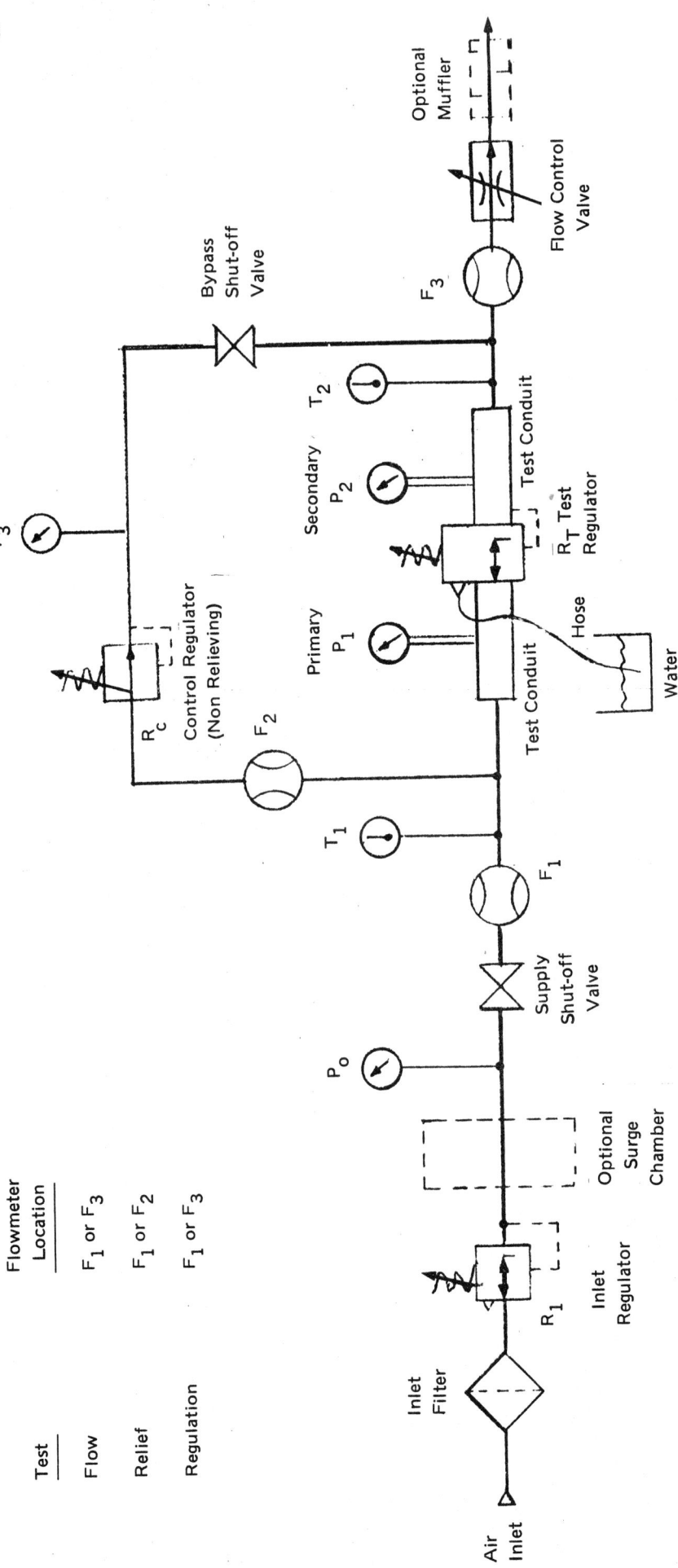

Figure 3 - Test circuit

ANSI/B93.13

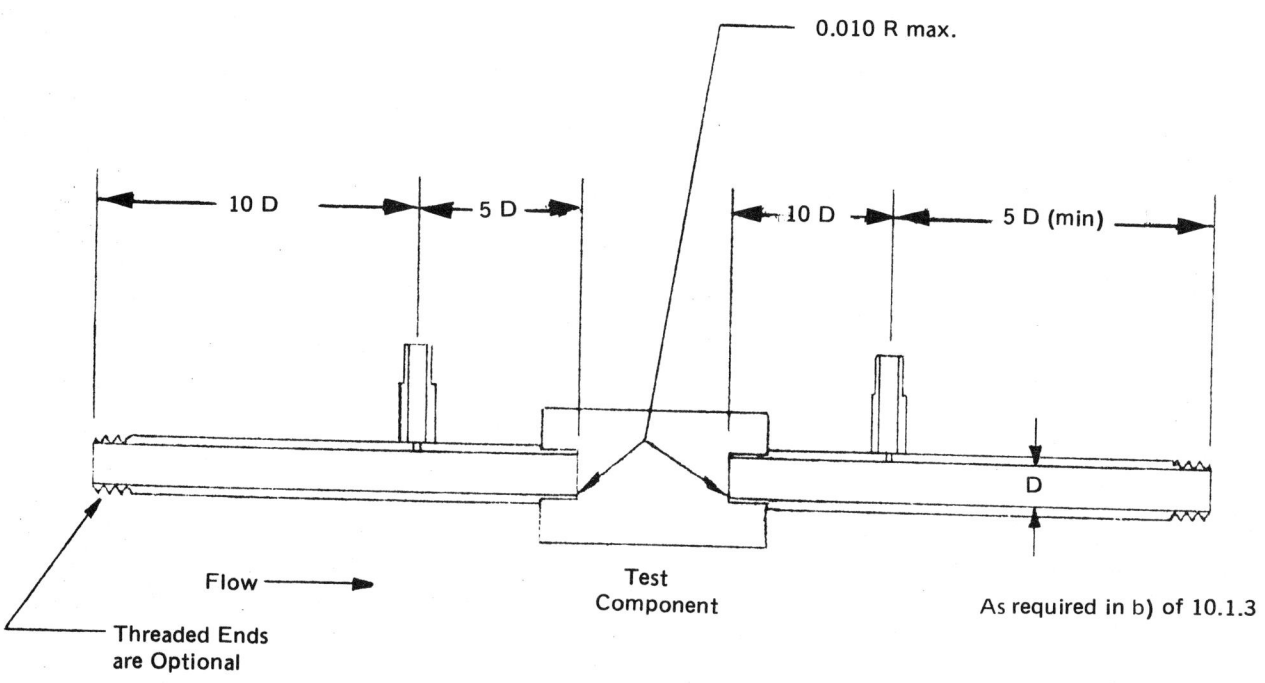

Figure 4 - **Test conduits**

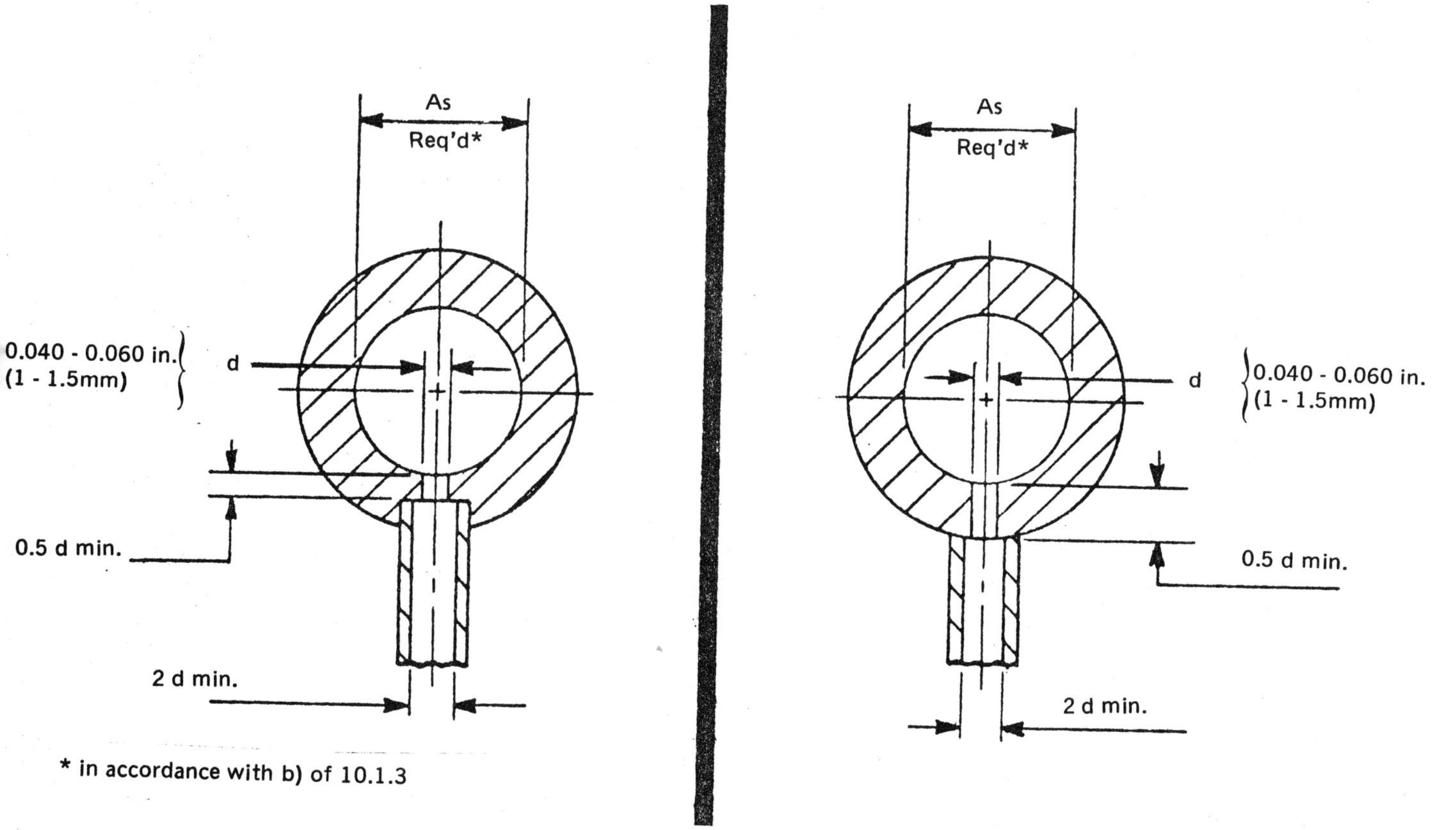

Figure 5 - **Pressure tap detail**

ANSI/B93.13

Primary pressure = 150 psig (10.3 bar)
Inlet air temperature = 72°F (22.3°C)
Decreasing flow condition

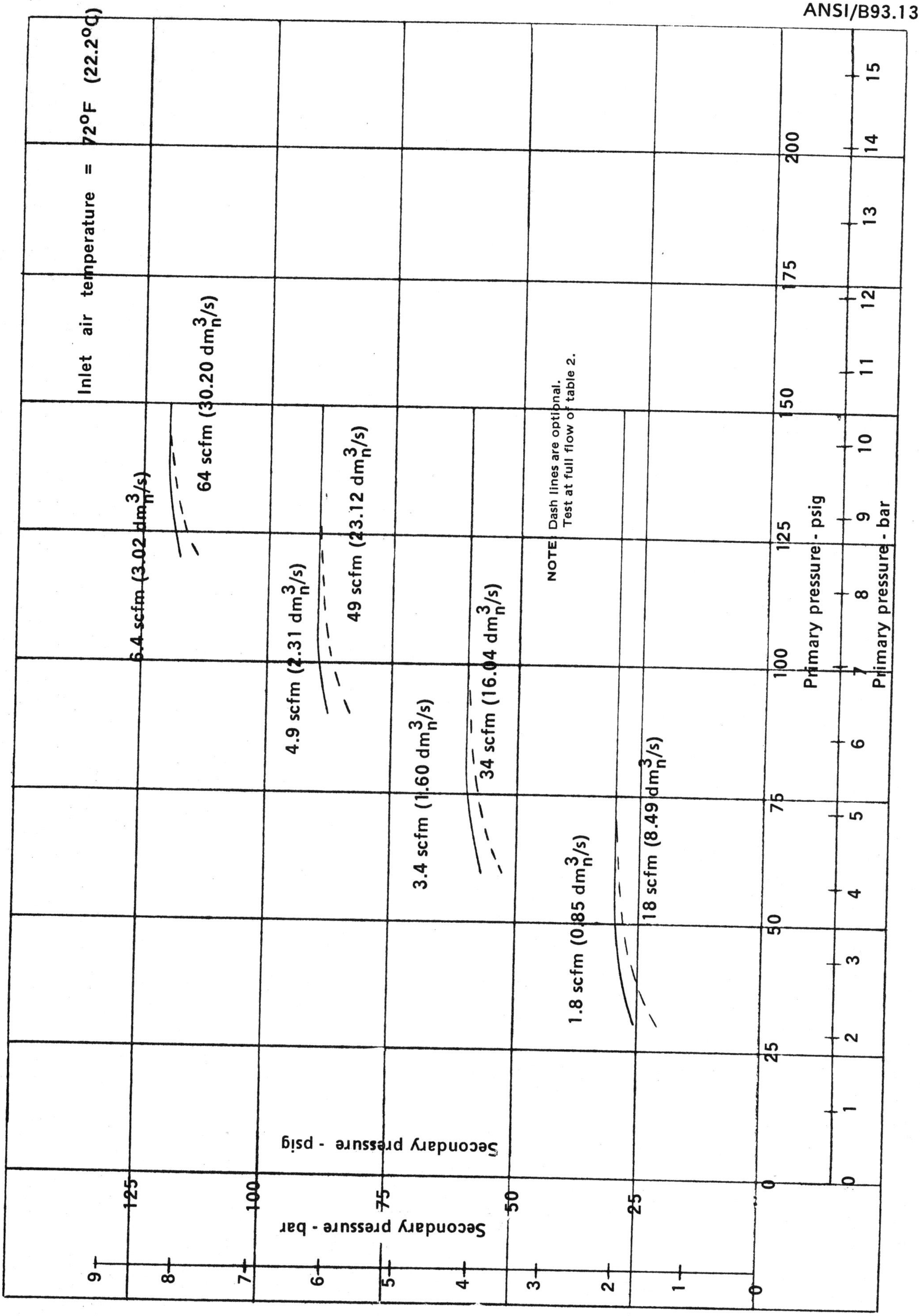

ANSI/B93.13

Figure 7 - Regulation Characteristics

ANSI/B93.13

# APPENDIX

## to ANSI/B93.13M-1981

**2 REFERENCES**

NFPA/T2.10.1M-1978, **National Fluid Power Association Standard Metric Units for Fluid Power Applications.**

ANSI/Y32.10-1967 (R1979), **American National Standard Graphic Symbols for Fluid Power Diagrams.**

ISO 5598*, **Fluid power systems and components - Vocabulary.**

ANSI/B93.33M-1974 (R1981)

## AN INDUSTRY STANDARD FOR FLUID POWER

American National Standard

Interfaces for

4-Way General Purpose

Industrial Pneumatic Directional Control Valves

(NFPA/T3.21.1M-1973)

Approved as an ANSI Standard
17 January 1974

published by

## NATIONAL FLUID POWER ASSOCIATION, INC.

3333 N. Mayfair Road  /  Milwaukee, WI 53222  /  414-778-3344  /  TLX 26898

## FOREWORD

This Foreword is not part of American National Standard Interfaces for 4-Way General Purpose Industrial Pneumatic Directional Control Valves, ANSI/B93.33-1974.

In response to general industry request, NFPA undertook this project in February 1966. It was originally assigned to a Project Group of the NFPA Valve Section. In January 1968 the Project Group was transferred under the newly formed Pneumatic Valve Section.

At the Project Group's first meeting, and during all the Project Group's subsequent deliberations, the work by the Welding Development Department of the Ford Motor Company was constantly being taken into account. At least six of the Project Group's drafts were of Ford origin. Ford Motor Company representatives personally presented Ford's viewpoints to the Project Group on several occasions.

The Project Group work proceeded thru a series of three surveys and seventeen drafts, and thru several changes in the Project Scope. Consensus was reached on 16 November 1971 on the interface dimensions and configurations for ¼, ½ and 1 inch nominal sizes.

Notwithstanding the original intent to achieve standardization of the entire sub-base, including electrical connecting means, the Project Group terminated their immediate efforts with the completion of the "interface" portion of the sub-base. Standardization of the complete sub-base and the electrical connecting means and the pilot will be handled at a later date after the results of the interface standardization are evalutated.

The NFPA Pneumatic Valve Section (at their 16 November 1971 meeting - which immediately followed the Project Group meeting) accepted the Project Group's consensus and authorized its being submitted to the United States Technical Advisory Group to ISO in preparation for the Paris January 1972 meeting of ISO/TC 131/SC 5.

Following General Review, the Project Group reviewed the comments received on 7 March 1972. In response to unresolved comments, the Project Group prepared Drafts No. 18 and 19. These drafts were reviewed on 18 April and 9 June 1972 respectively. The Project Group reached consensus on the new drafts on 7 September 1972 and the Second General Review Draft was prepared on 19 September 1972.

The Project Group resolved the comments received in the review on 28 November 1972. At this time, in response to comments, the group agreed to specify that the electrical connector be optional. The Ballot Draft was prepared on 20 December 1972.

Balloting closed on 5 February 1973 with these results, "13 Aye", "7 Nay", and "12 Abstentions".

At the 23 May 1973 NFPA Technical Board meeting, it was decided that T3.21.1 would be approved with the condition that the 6 companies who indicated agreement at the two meetings after ballot will confirm in writing by 11 June 1973 their withdrawal of their negative ballot. Confirmation was accomplished.

The NFPA Board of Directors approved this document as NFPA Recommended Standard T3.21.1-1973 on 19 July 1973.

Members of the NFPA Project Group responsible for the development of this standard included:

**Jerry Komendera**
Project Chairman
(May 1967 to present)
Automatic Valve Corp.

**Z. J. Lansky**
Project Chairman
(February 1966 to May 1967)
Parker Hannifin Corp.

**Ernest Cole**
Project Secretary
(April 1972 to present)
Bellows Valvair

**Les Malinowski**
Project Secretary
(February 1966 to April 1972)
Parker Hannifin Corp.

**Robert Peterson**
Hydraulic Valve Section Chairman
(July 1968 to July 1971)
Racine Hydraulics Div.*

**Allen Bower**
Hydraulic Valve Section Chairman
(July 1961 to July 1968)
Fluid Controls, Inc.*

**Clifford Brake**
Pneumatic Valve Section Chairman
(July 1970 to present)
Scovill Fluid Power Div.

**Wil Aslan**
Pneumatic Valve Section Chairman
(July 1968 to July 1970)
Alkon Products Corp.

**James I. Morgan**
Secretariat
National Fluid Power Assn.

**A. Ackerman**
MAC Valves, Inc.

**J. Bowbin**
Miller Fluid Power

**E. Brinkel**
Rexnord, Inc.

**C. Carlson**
Versa Products, Inc.

**D. Claydon**
Ross Valve Co.

**J. Curnow**
Sperry Vickers

**C. Davis**
Rexnord, Inc.

**J. Fisher**
Bellows-Valvair

**D. Florence**
The Aro Corp.

**G. Goepfrich**
Skinner Electric Valve

**R. Grassl**
Galland Henning

**M. Greenwood**
Alkon Products

**C. Grigsby**
Scovill Fluid Power

**A. Hinz**
Ross Operating Valve

**T. Johnson**
Automatic Switch Co.

**G. Kriel**
The Aro Corp.

**J. Lytle**
Ross Operating Valve

**D. McGeachy**
Numatics, Inc.

**C. McLean**
Miller Fluid Power

**A. Moffat**
Componetrol, Inc.

**J. Moritz**
Logansport Machine

**P. Olson**
WABCO

**D. Seabrook**
Miller Fluid Power

**J. Sisbarro**
Automatic Switch Co.

**L. Swickley**
Bellows-Valvair

**J. Trevethan**
Ross Operating Valve

**C. Tuerk**
C. A. Norgren Co.

**C. Ward**
Bellows-Valvair

**R. Zbell**
Ross Operating Valve

*Company affiliation has changed since work with the Project Group.

On 20 August 1973 the NFPA Recommended Standard was submitted concurrently to ANSI Standards Committee B93 for ballot, and to the ANSI Board of Standards Review for ANSI General Review whereby more than 10,000 potential commentors are informed of this proposal.

The ANSI B93 ballot was entirely favorable. The General Review resulted in one comment, which was acted upon by the ANSI Board of Standards Review, and ANSI approval was granted on 17 January 1974.

On 20 August 1973 Standards Committee B93 was comprised of the following:

**O. J. Maha**
Chairman

**James I. Morgan**
Co-Secretary

**R. Thomas Northrup**
Co-Secretary

**AMERICAN SOCIETY OF AGRICULTURAL ENGINEERS**
E. H. Fletcher

**AMERICAN SOCIETY OF LUBRICATION ENGINEERS**
H. Kaufman

**AMERICAN SOCIETY OF MECHANICAL ENGINEERS**
H. Parsons
T. R. Curran (Alternate)

**AMERICAN SOCIETY FOR TESTING AND MATERIALS**
J. D. Lykins

**AUTOMOBILE MANUFACTURERS ASSOCIATION**
R. Mustonen

**CONSTRUCTION INDUSTRY MANUFACTURERS ASSOCIATION**
G. Stewart
H. T. Larmore (Alternate)

**FLUID CONTROLS INSTITUTE**
A. W. Churchill
E. A. Blanchi (Alternate)

**FLUID POWER SOCIETY**
T. Goldoftas
R. Henke
M. E. Long
F. L. Mackin
W. R. Smith
L. G. Soderholm
F. Yeaple

**FLUID SEALING ASSOCIATION**
W. Krucke

**INDUSTRIAL TRUCK ASSOCIATION**
C. D. Gibson
R. T. McNeely (Alternate)

**INSTRUMENT SOCIETY OF AMERICA**
A. I. Kutz

**JOINT INDUSTRY COUNCIL**
R. Muhl

**MATERIAL HANDLING INSTITUTE**
J. McPherson
W. Chichester (Alternate)

**NATIONAL FLUID POWER ASSOCIATION**
O. J. Maha
J. L. Fisher, Jr.
W. R. Forster
W. J. Kudlaty
Z. J. Lansky
J. J. Pippenger

**NATIONAL INDUSTRIAL LEATHER ASSOCIATION**
E. R. Rath

**NATIONAL MACHINE TOOL BUILDERS ASSOCIATION**
J. I. Ehrhardt
E. Loeffler (Alternate)

**POWER CRANE AND SHOVEL ASSOCIATION**
W. M. Shook

**RUBBER MANUFACTURERS ASSOCIATION**
W. J. Atwell
N. J. Cyphers (Alternate)
A. J. Jeffcott (Alternate)

**SOCIETY OF AUTOMOTIVE ENGINEERS**
W. A. Hertel
R. E. Lyons
E. L. Falendysz
J. E. Wieschel (Alternate)

**SOCIETY OF MANUFACTURING ENGINEERS**
R. E. Willette

**U. S. COAST GUARD**
LCDR G. A. Casimir

**U. S. DEPARTMENT OF DEFENSE**
C. A. Nazian
H. Y. Smith
P. Hopler (Alternate)

**INDIVIDUAL MEMBERS**
Prof. E. C. Fitch
Prof. J. Johnson

**REFERENCES**

1. American National Standard Glossary of Terms for Fluid Power, ANSI/B93.2-1971, and Supplements thereto. (ISO/TC 131/SC 1 [USA-2] 3).

2. SI units and recommendations for the use of their multiples and of certain other units, ISO 1000-1973.

3. American National Standard Symbols for Marking Electrical Leads and Ports on Fluid Power Valves, ANSI/B93.9-1969. (ISO/TC 131/SC 5 [USA-3] 10).

ANSI/B93.33

# INTERFACES FOR 4-WAY GENERAL PURPOSE
# INDUSTRIAL PNEUMATIC DIRECTIONAL CONTROL VALVES

## INTRODUCTION

In pneumatic fluid power systems, power is transmitted and controlled thru a gas under pressure within an enclosed circuit. A pneumatic directional control valve's primary function is to direct or prevent flow thru selected passages.

Sub-base (or sub-plate) type valves consist of two principle parts -- the sub-base or mounting portion and the operating valve portion. All piping is permanently connected to the sub-base. The operating valve portion is bolted to the sub-base; it can easily be removed for service or replacement with simple tools. Air passages between the sub-base and operating valve portion are sealed by resilient sealing means. The mating surfaces of the sub-base and operating valve are called the "interface".

Users of pneumatic valves benefit when valves of the same nominal size from various manufacturers have a common interface -- and therefore, can be interchanged when servicing or replacement is required.

This document is one of many worldwide documents relating to pneumatic sub-bases (sub-plates).

1. SCOPE

    1.1 To include standard interfaces for a series of six sizes of sub-base type 4-way pneumatic industrial general purpose directional control valves. The interface configuration will apply to modular and/or individual sub-bases.

    1.2 To include the interface fluid power passages, hold down bolt locations and sizes, and provision for an optional electrical connector. (Details of the electrical connector will be set forth in a separate document.)

    1.3 This Standard only applies to the dimensional criteria of products manufactured in conformance with this standard. It does not apply to their functional characteristics.

2. PURPOSE

    2.1 To achieve dimensional interchangeability.

    2.2 To achieve simplification of variety.

3. TERMS AND DEFINITIONS

    For definitions of terms used, see Reference No. 1).

ANSI/B93.33

4. UNITS

    4.1    The International System of Units (SI) is used in accordance with Reference No. 2.

    4.2    Rounded conversions to "Customary US" units are given in parentheses following their SI counterpart. The following dimension adjustment tabulation gives an example of how the conversions were achieved for interface type 1:

| Established dimension in inches | Dimension millimetre equivalent | Dimension millimetres (rounded off) | Dimension inch equivalent |
|---|---|---|---|
| 2 | 50.8 | 50 | 1.968 |
| 1-1/4 | 31.75 | 32 | 1.260 |
| 1/2 | 12.7 | 13 | .512 |
| 7/8 | 22.225 | 22 | .866 |
| 1/4 | 6.35 | 6 | .236 |

5. PORT IDENTIFICATION

Use the numbers 1 thru 5 which identify the fluid passages on the interfaces to also identify the associated sub-base port in accordance with Reference No. 3.

6. USE OF OPTIONAL HOLES

The use of external pilot supply hole, remote pilot signal holes and electrical service location is optional.

7. IDENTIFICATION STATEMENT

Use the following statement in catalogs and sales literature when electing to comply with this voluntary standard:

"Interface conforms to American National Standard ANSI/B93.33-1974, Type ____."

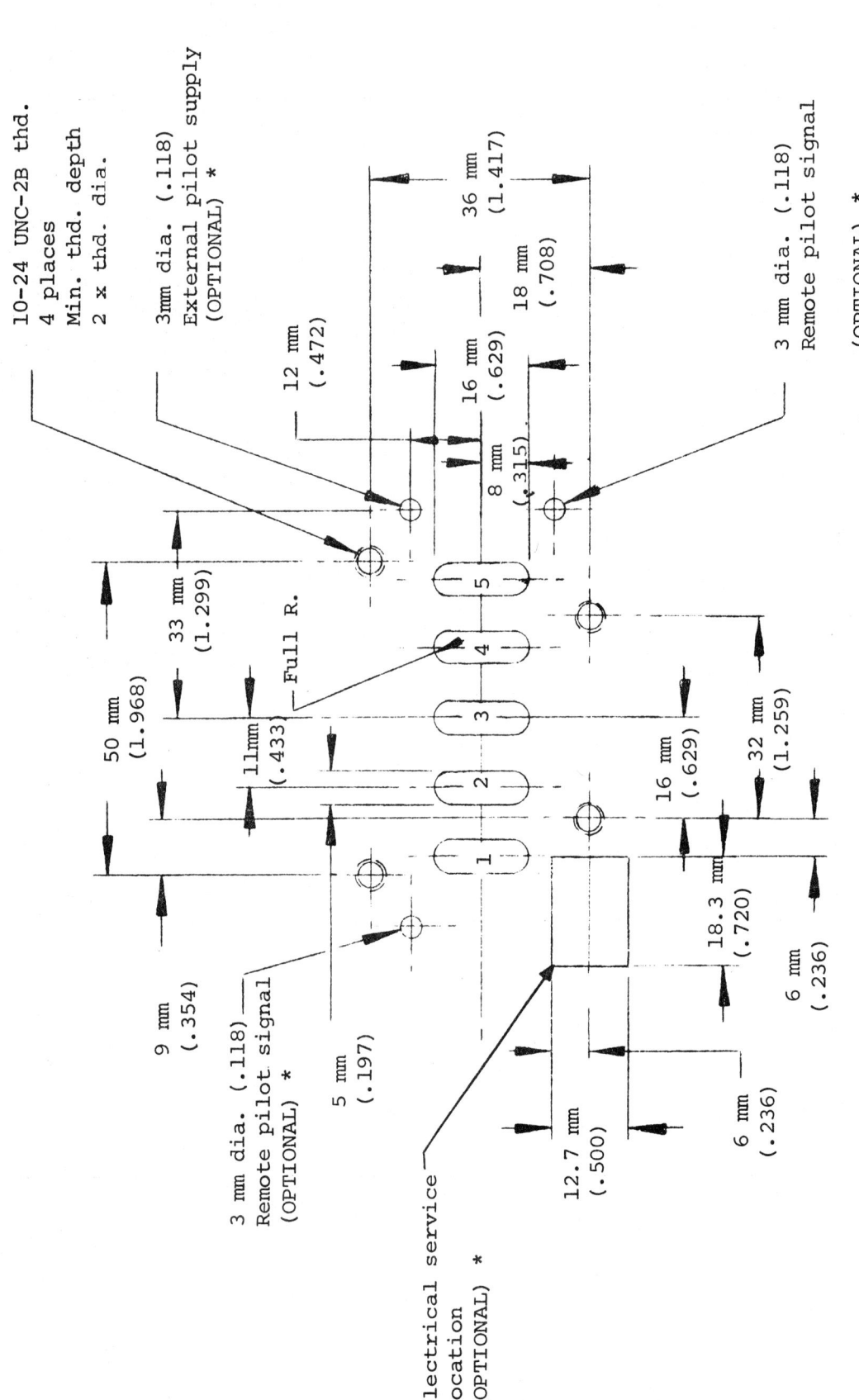

FIGURE 1 — Type 1 interface — top view.

* See Paragraph 6

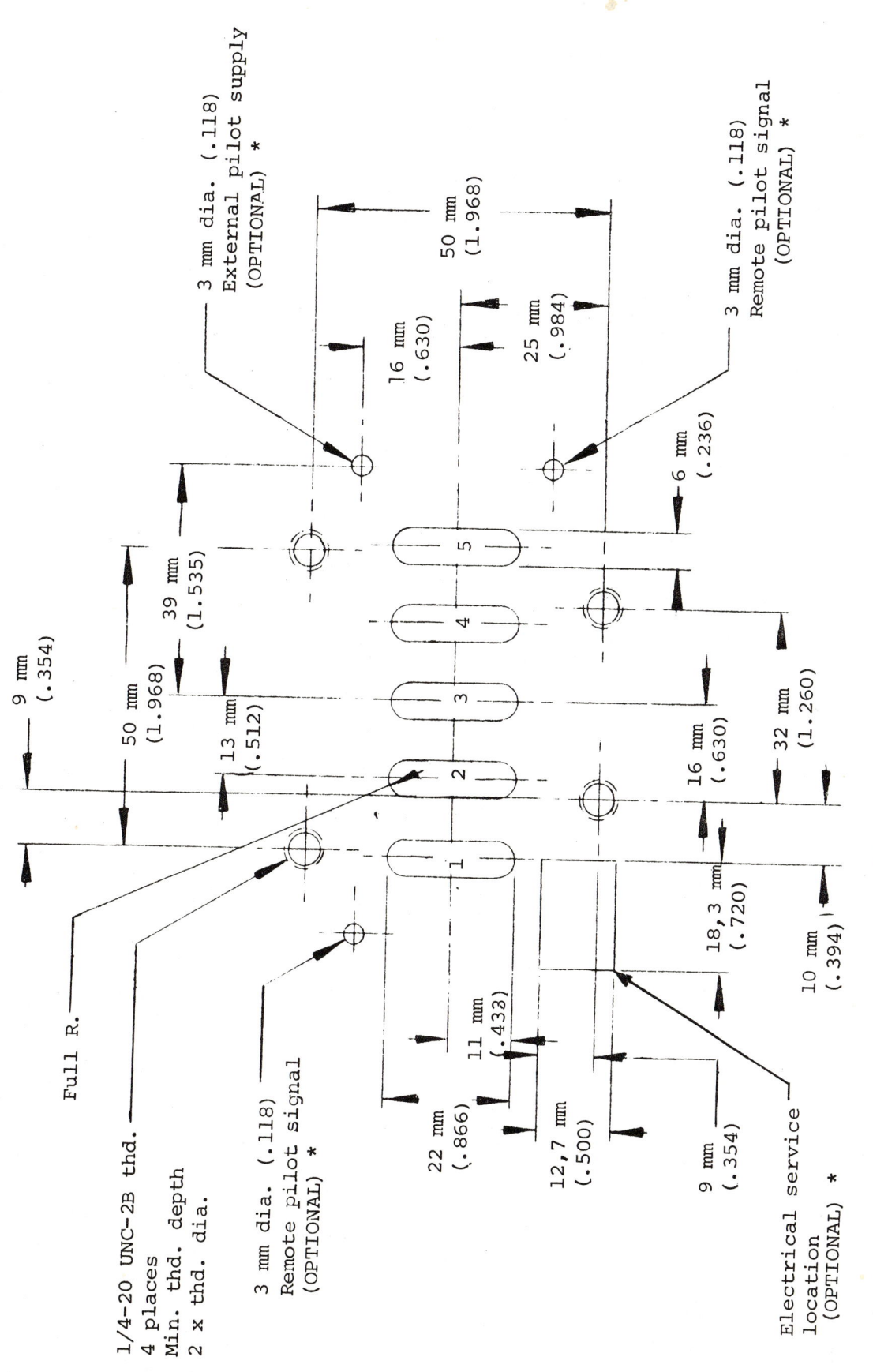

FIGURE 2 - Type 2 interface - top view.

* See Paragraph 6

ANSI/B93.33

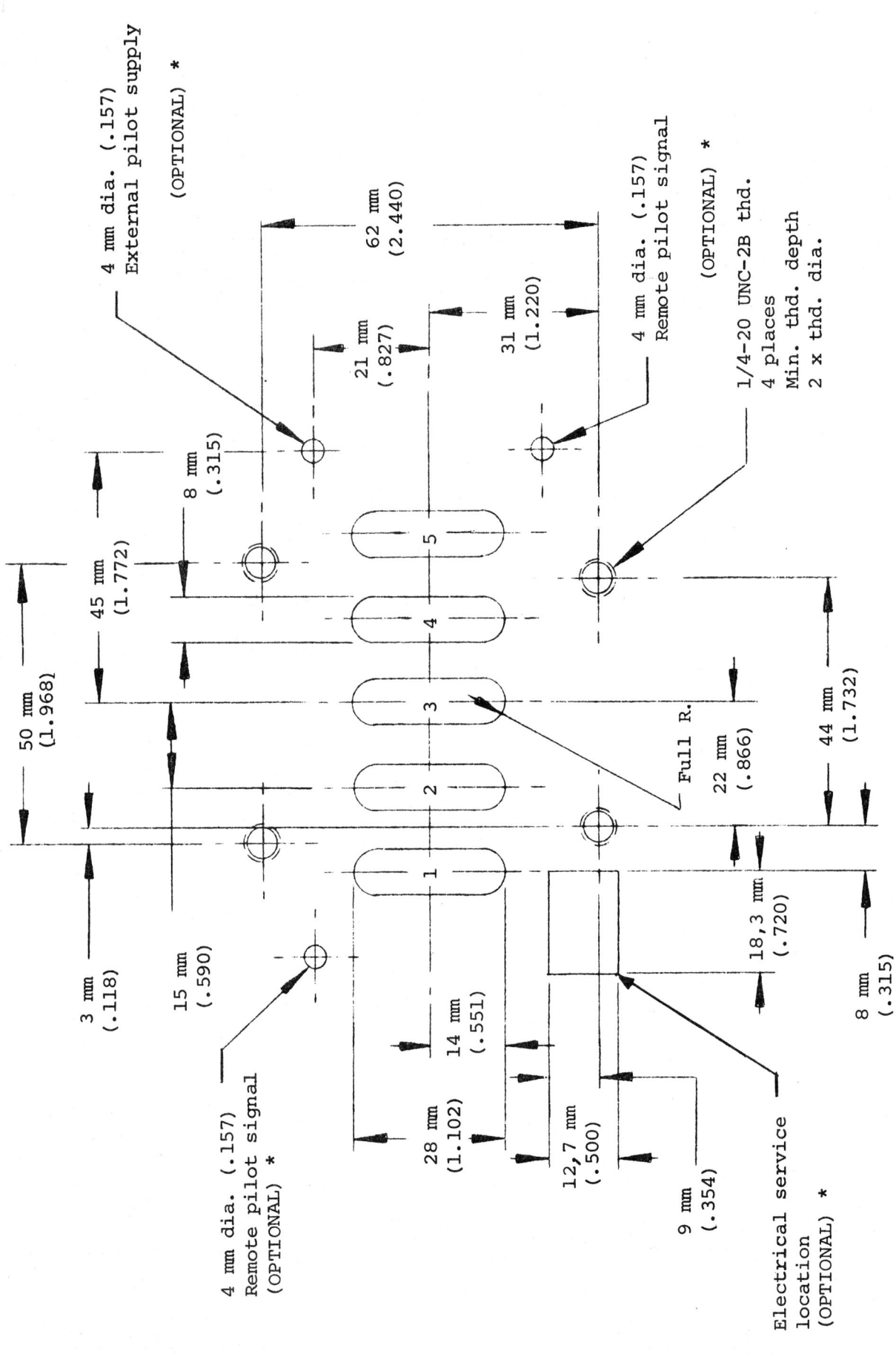

FIGURE 3 - Type 4 interface - top view.

* See Paragraph 6

ANSI/B93.33

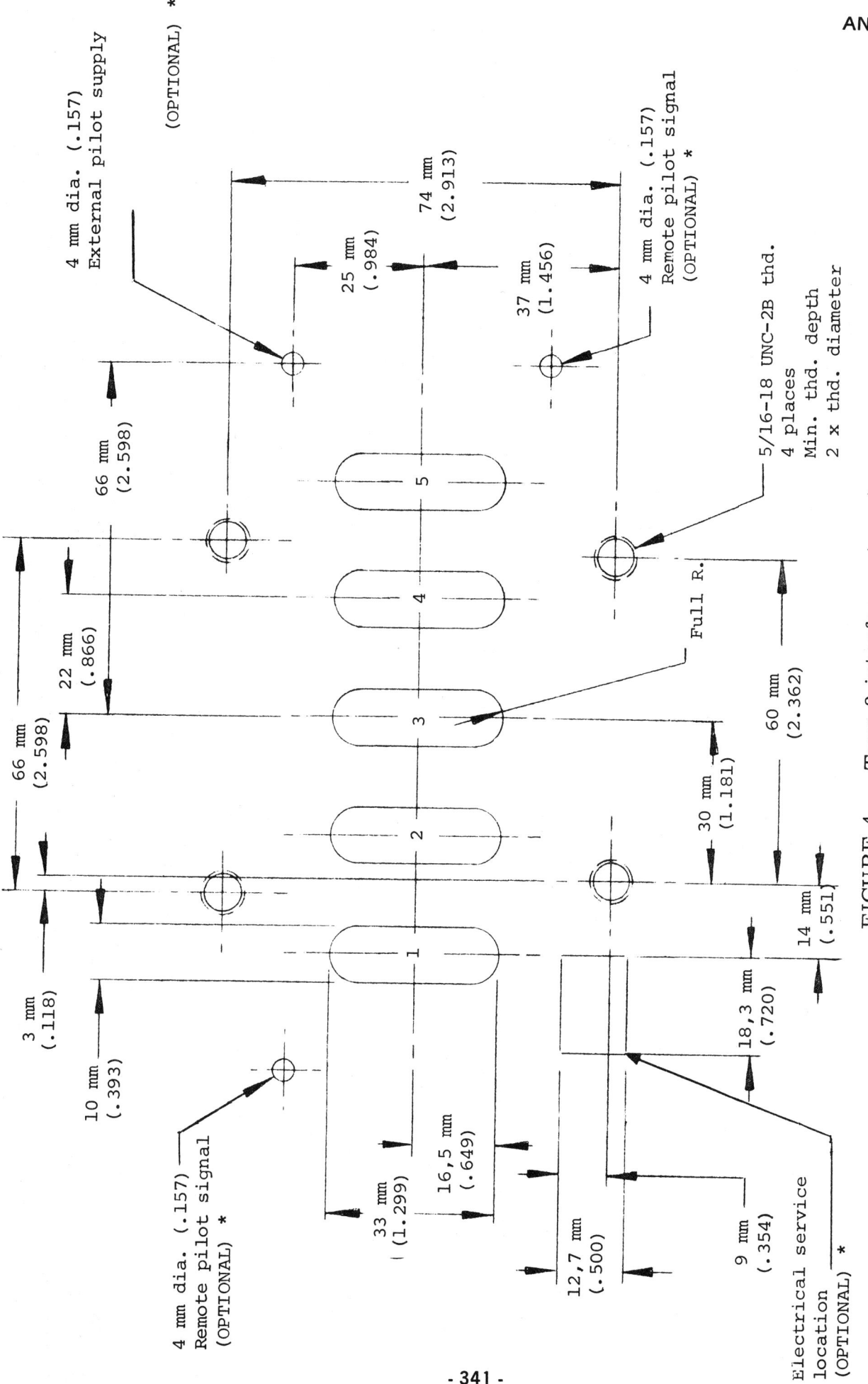

FIGURE 4 - Type 8 interface - top view.

* See Paragraph 6

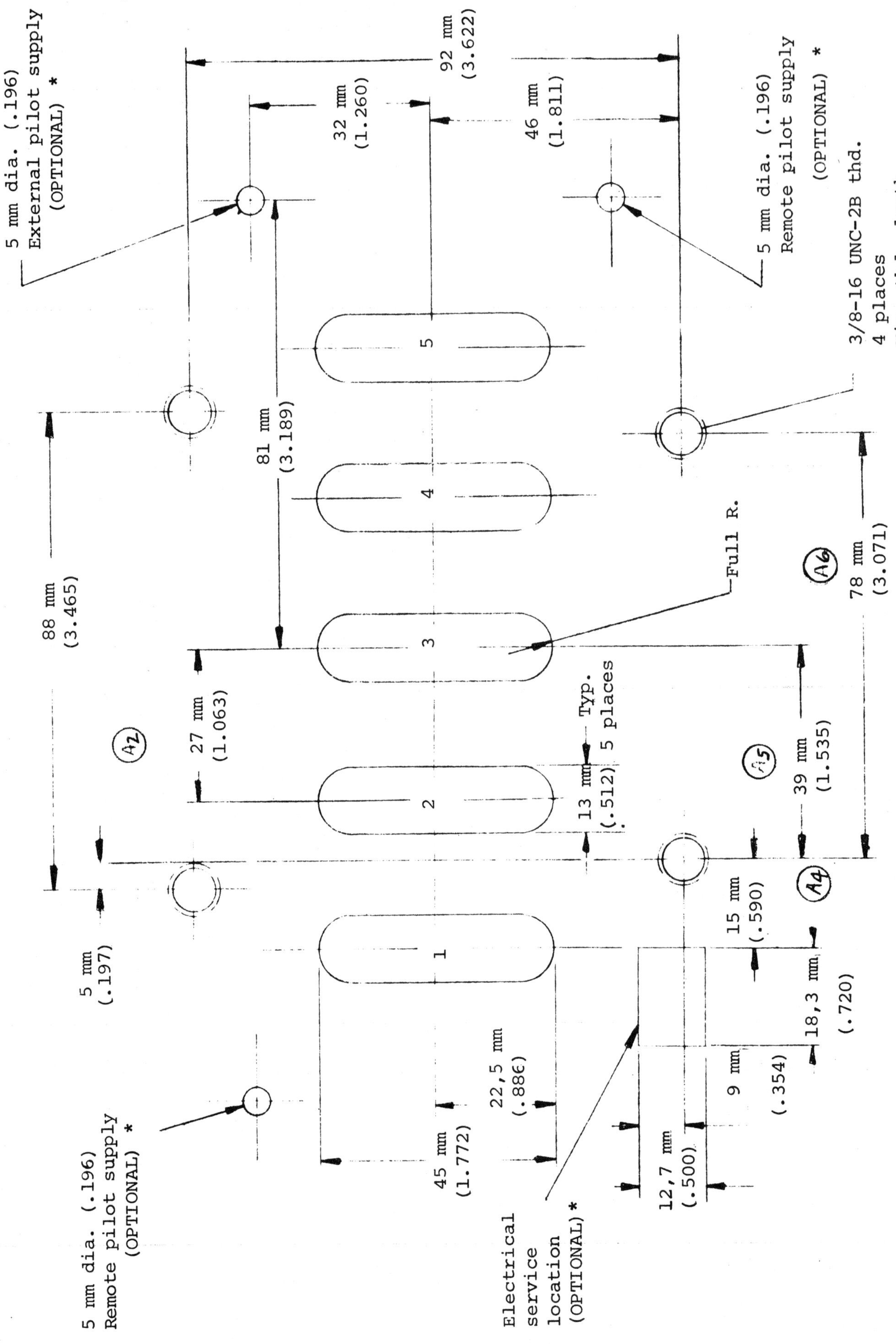

ANSI/B93.33

FIGURE 5 – Type 12 interface – top view.

* See Paragraph 6

ANSI/B93.33

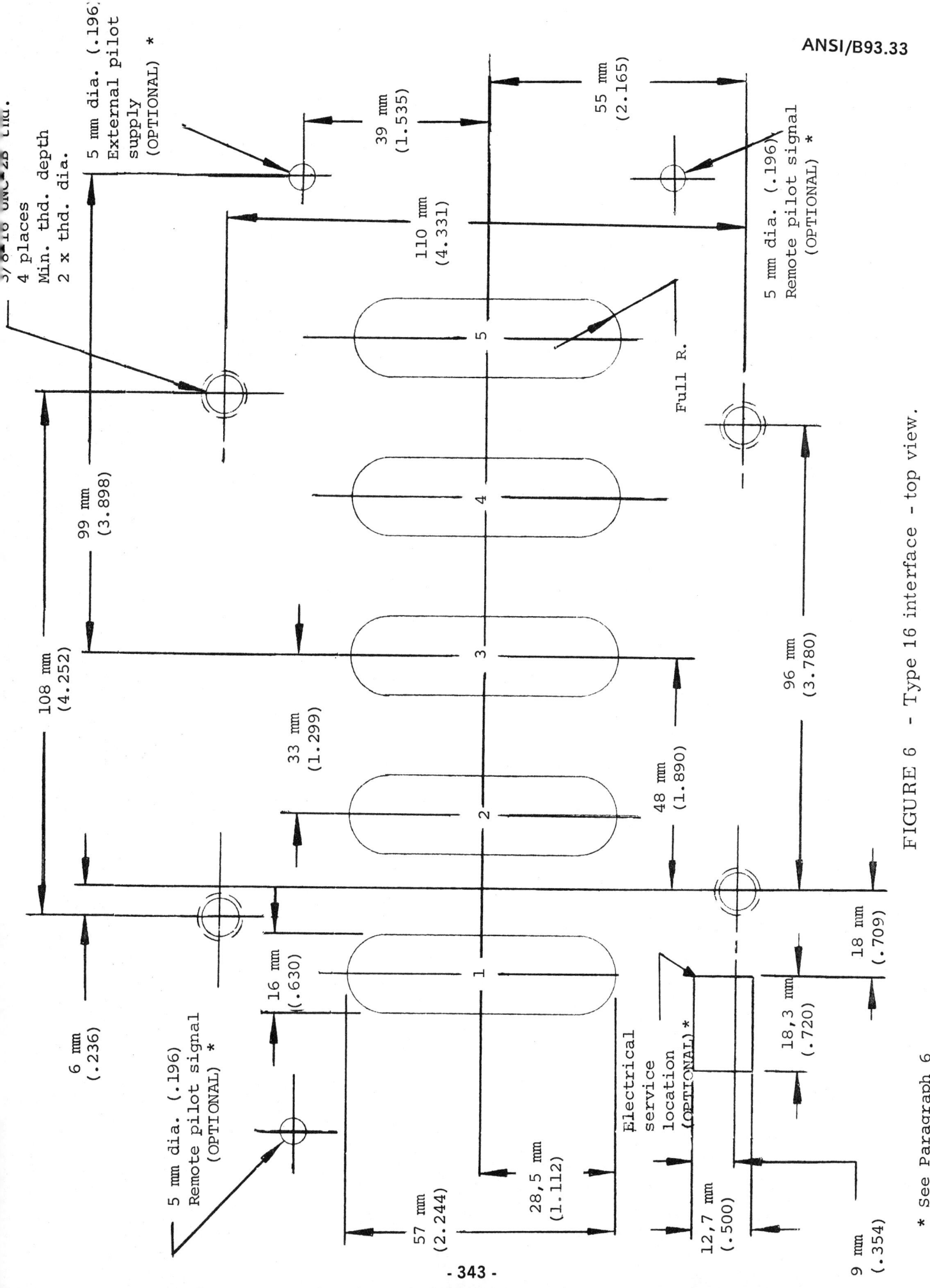

FIGURE 6 - Type 16 interface - top view.

* See Paragraph 6

# APPENDIX

## To ANSI/B93.33-1974

REFERENCES

ISO 5599/1-1978: Pneumatic fluid power five port directional control valves mounting surfaces.

DIN 24 341: Pneumatic valves; mounting surface and connecting plates for directional control valves.

C.N.O.M.O. 06 05: Mounting surface drawings for pneumatic poppet type directional control valves.

C.N.O.M.O. 06 05: Mounting surface drawings for pneumatic spool type directional control valves.

CETOP RP 32P (PB 22.10): Subplates for directional control valves.

**NFPA Recommended Standard
T3.21.7M-1976 (R1981)**

---

## AN INDUSTRY STANDARD FOR FLUID POWER

Defining Interface Surfaces

For Each Pneumatic Valve Interface

In NFPA Recommended Standard T3.21.1-1973

Approved as an NFPA Recommended Standard
3 June 1976

---

published by
# NATIONAL FLUID POWER ASSOCIATION, INC.
3333 N. Mayfair Road / Milwaukee, WI 53222 / 414-778-3344 / TLX 26898

NFPA/T3.21.7

## FOREWORD

This Foreword is not part of NFPA Recommended Standard for Defining Interface Surfaces for Each Pneumatic Valve Interface in NFPA Recommended Standard T3.21.1-1973, NFPA/T3.21.7-1976.

This project originated at the T3.21 Pneumatic Valve Section meeting on 5 February 1974. At that time, Section Chairman C. Brake (Scovill) appointed J. Komendera (Automatic Valve) as Chairman of Project Group T3.21.7. Chairman Brake volunteered to write a TSP for the project, which he submitted to NFPA Headquarters on 5 March 1974.

The NFPA Technical Board gave approval to the initiation of Project T3.21.7-19xx on 12 March 1974, even though the formal TSP was not available for the Board's review. This was done in view of the urgent need to continue action on this project.

At the T3.21 meeting on 21 May 1974, it was agreed to submit a revision to the Scope of the TSP, calling for definition of minimum width and length of the finished gasket interface, minimum centerline spacing of manifold mounting, pads, and surface finish of the valve interface. The revised TSP was approved by the Technical Board on 5 June 1974.

Draft No. 1 of T3.21.7 was written on 9 July 1974, and was reviewed on 9 September 1974. The Final Working Draft was submitted to Headquarters on 10 September 1974. The NFPA Technical Staff prepared the General Review Draft on 26 December 1974.

General Review comments were resolved at the T3.21 Section meeting on 1 October 1975. Technical Board approval to ballot was granted on 5 November 1975. The NFPA Technical Staff prepared the Ballot Draft on 12 November 1975.

Three negative ballots were received during the balloting period. Following discussion by the Project Group and Section Chairmen, two of the negative ballots were resolved on 17 March 1976 by making minor editorial modifications in the document. The third negative ballot objected to this standard on the basis of pending ISO work. Both Chairmen judged the third ballot to be irresolvable due to its nature, and so advised the submittor.

One negative ballot was received after the balloting period had closed. It was reviewed by the Project Group and Section Chairmen but was

continued . . .

NFPA/T3.21.7

## FOREWORD (continued)

not answered, an option allowed by NFPA procedures for late ballots. In addition, the written substantiating statement for one negative ballot was received after the balloting period had closed; in accordance with NFPA procedures, the ballot was recorded as an abstention, and did not receive a reply from the Project Group or Section Chairman.

On 7 April 1976, the NFPA Technical Board reviewed the irresolvable negative ballot, the negative ballots which had been received late, and the Technical Auditor's recommendation that the document be approved. The Technical Board concurred with the Technical Auditor, and voted to recommend to the Board of Directors that this document be approved as an NFPA Recommended Standard. The NFPA Technical Staff prepared the Edited Ballot Draft on 15 April 1976.

The NFPA Board of Directors approved this document as NFPA Recommended Standard T3.21.7-1976 on 3 June 1976.

NFPA/T3.21.7

## PROJECT GROUP MEMBERS WHO DEVELOPED THIS STANDARD

| | | |
|---|---|---|
| Komendera, J. B. | Project Chairman | Automatic Valve Corp. |
| Grigsby, Carroll | Project Secretary | Scovill Fluid Power Div. |
| Lytle, John | Section Chairman (1975 to present) | Ross Operating Valve Co. |
| Brake, Clifford | Section Chairman (1970 to 1975) | Scovill Fluid Power Div. |
| Culbertson, Ralph | Technical Auditor | Dynamco Inc. |
| Luecke, John R. | Director of National Technical Services | National Fluid Power Association |

| | |
|---|---|
| Arvin, P. | The Aro Corporation |
| Cole, E. | Bellows International |
| Hoffman, R. | Parker Hannifin Corp. |
| Sisbarro, J. | Automatic Switch Co. |

## REFERENCES

1. American National Standard Interfaces for 4-Way General Purpose Industrial Pneumatic Directional Control Valves, ANSI/B93.33-1974. (NFPA/T3.21.1-1973)

2. American National Standard Glossary of Terms for Fluid Power, ANSI/B93.2-1971, and Supplements thereto. (ISO/TC 131/SC 1 (USA-2) 3)

3. SI units and recommendations for the use of their multiples and of certain other units, ISO 1000-1973.

4. Technical drawings - Method of indicating surface finish on drawings, ISO/R 1302-1974.

NFPA/T3.21.7

# DEFINING INTERFACE SURFACES FOR EACH
# PNEUMATIC VALVE INTERFACE
# IN NFPA RECOMMENDED STANDARD T3.21.1-1973

## INTRODUCTION

In pneumatic fluid power systems, power is transmitted and controlled thru a gas under pressure within an enclosed circuit. A pneumatic directional control valve's primary function is to direct or prevent flow thru selected passages.

Sub-base (or sub-plate) type valves consist of two principal parts -- the sub-base or mounting portion and the operating valve portion. All piping is permanently connected to the sub-base. The operating valve portion is bolted to the sub-base; it can easily be removed for service or replacement with simple tools. Air passages between the sub-base and operating valve portion are sealed by resilient sealing means. The mating surfaces of the sub-base and operating valve are called the "interface".

A manifold is a type of construction in which the valve is mounted to a plate which provides multiple connection ports for two or more valves. Manifolds offer economy of both space and interconnection simplicity. In order to insure complete interchangeability, it is necessary to establish certain dimensions relating to the interface surface and the manifold.

1. SCOPE

    To include:

    1.1   Minimum width and length of the finished gasket surface for each interface size established in Reference No. 1.

    1.2   Minimum centerline spacing of manifold mounting pads for each interface size established in Reference No. 1.

    1.3   Finish requirements for the gasket surface for each interface size established in Reference No. 1.

NFPA/T3.21.7

2. PURPOSE

2.1 To achieve interchangeability.

2.2 To achieve simplification of variety of valve configurations.

3. TERMS AND DEFINITIONS

For definitions of terms used, see Reference No. 2.

4. UNITS OF MEASUREMENT

4.1 The International System of Units (SI) is used in accordance with Reference No. 3.

4.2 Rounded conversions to Customary US units are given in parentheses following their SI counterparts.

4.3 The following dimension adjustment tabulation gives an example of how the conversions (clause 4.2) were achieved for interface type 1:

| Established Dimension In Inches | Dimension Millimetre Equivalent | Dimension Millimetres (Rounded Off) | Dimension Inch Equivalent |
|---|---|---|---|
| 2 3/8 | 60.325 | 60 | 2.362 |
| 2 23/32 | 69.056 | 69 | 2.717 |
| 3 3/8 | 85.725 | 86 | 3.386 |
| 1 23/32 | 43.656 | 44 | 1.732 |
| 1 21/32 | 42.069 | 42 | 1.654 |

5. LETTER SYMBOLS

A     mimimum distance from port 3 to the edge of the gasketed surface nearest the optional electrical service location for each interface size

B     minimum distance from port 3 to the edge of the gasketed surface nearest the optional external pilot supply port for each interface size

| | |
|---|---|
| L | minimum length of the gasketed surface for each interface size |
| $W_{CL}$ | minimum spacing between centerlines of the mounting pads of adjacent manifolds for each interface |
| $W_I$ | minimum width of the gasketed surface for each interface size |

## 6. SURFACE FINISH AND FLATNESS

6.1 Provide a surface for a gasket to a roughness grade of N7 (1.6 μm, 63 μin in RMS) per Reference No. 4.

6.2 Provide a surface for a gasket to a flatness within 0.10 mm per 100 mm (.001 in per in), with exception of the functional openings specified in Reference No. 1.

TABLE 1 - Gasketed surface dimensions (in mm (in))

| Interface Size | $W_I$ | L | A | B |
|---|---|---|---|---|
| 1 | 60 (2.362) | 86 (3.386) | 44 (1.732) | 42 (1.654) |
| 2 | 66 (2.598) | 96 (3.780) | 48 (1.890) | 48 (1.890) |
| 4 | 78 (3.071) | 106 (4.173) | 52 (2.047) | 54 (2.126) |
| 8 | 90 (3.543) | 150 (5.906) | 75 (2.953) | 75 (2.953) |
| 12 | 108 (4.252) | 180 (7.087) | 90 (3.543) | 90 (3.543) |
| 16 | 126 (4.961) | 216 (8.504) | 108 (4.252) | 108 (4.252) |

6.3 Ensure that the minimum spacing between centerlines of the mounting pads $W_{CL}$ of adjacent manifolds is in accordance with Table 2.

TABLE 2 - Manifold spacing (in mm (in))

| Interface Size | $W_{CL}$ |
|---|---|
| 1 | 69 (2.717) |
| 2 | 78 (3.071) |
| 4 | 90 (3.543) |
| 8 | 104 (4.094) |
| 12 | 126 (4.961) |
| 16 | 144 (5.669) |

NFPA/T3.21.7

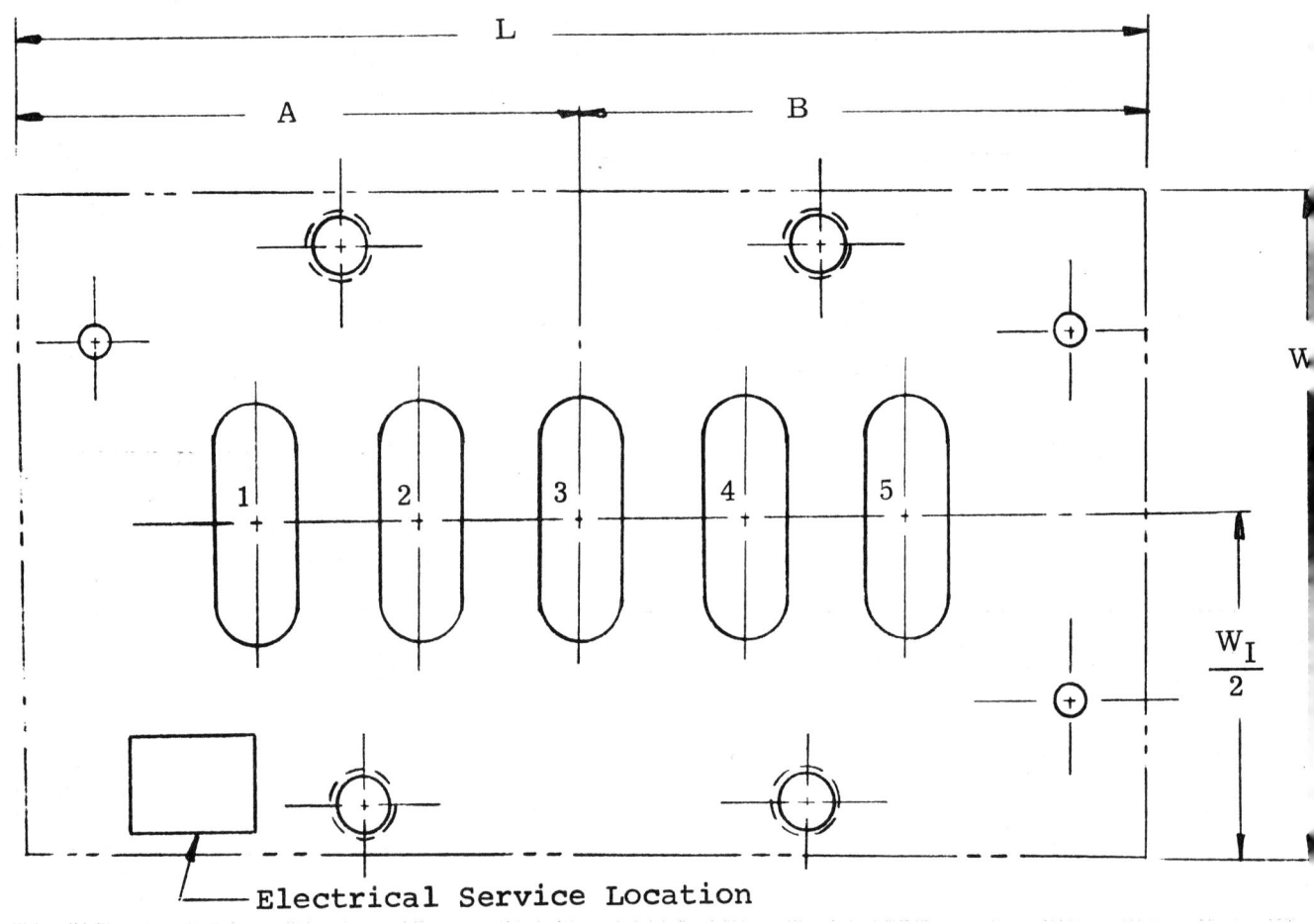

FIGURE 1 - Minimum logitudinal length and minimum width of gasketed surface.

7. IDENTIFICATION STATEMENT

   Use the following statement in catalogs and sales literature when electing to comply with this voluntary standard:

   "Interface surface conforms to NFPA Recommended Standard, NFPA/T3.21.7-1976."

8. KEY WORDS

   The following Key Words, useful in indexes and in information retrieval systems, are suggested for this recommended standard:

   dimensions, gasketed surface        interface, pneumatic valve
   dimensions, manifold spacing        interface surfaces, definition of
   fluid power                         valve, pneumatic fluid power

NFPA Recommended Standard
T3.21.9M-1976 (R1981)

## AN INDUSTRY STANDARD FOR FLUID POWER

Definition of Port Communication for The

Fluid Power Pneumatic Valve Interface to

NFPA Recommended Standard T3.21.1 with the Valve

In Position in Response to A Remote Pilot Signal

Or Electrical Energization

Approved as an NFPA Recommended Standard
22 September 1976

published by
# NATIONAL FLUID POWER ASSOCIATION, INC.

3333 N. Mayfair Road / Milwaukee, WI 53222 / 414-778-3344 / TLX 26898

## FOREWORD

This Foreword is not part of NFPA Recommended Standard Definition of Port Communication for the Fluid Power Pneumatic Valve Interface to NFPA Recommended Standard T3.21.1 With the Valve in Position in Response to a Remote Pilot Signal or Electrical Energization, NFPA/T3.21.9-1976.

At the 28 May 1974 meeting of the Pneumatic Valve Section, T3.21, a Project Group was formed to determine valve orientation to air or solenoid actuation.

The project's Title, Scope and Purpose (TSP) was approved at the 5 June 1974 meeting of the NFPA Technical Board.

The initial meeting of the Project Group was held on 8 July 1974, at which time a consensus was reached. The NFPA Technical Staff prepared the General Review Draft on 13 August 1974.

At its 16 April 1975 meeting, the Section voted to use the proposed ISO port identification system in this document and to add a drawing (Figure 1) to clarify identification.

Comments resulting from General Review were resolved on 8 May 1975, and the proposed recommended standard was approved for ballot by the NFPA Technical Board on 21 August 1975. The Ballot Draft was prepared on 28 August 1975.

Negative ballots were discussed at a Project Group meeting on 5 April 1976, and were resolved at the 28 April 1976 Section meeting. The NFPA Technical Staff prepared the Edited Ballot Draft on 3 May 1976.

On 11 August 1976, the NFPA Technical Board voted unanimously to recommend that this document be approved as an NFPA Recommended Standard. The NFPA Board of Directors granted final approval to NFPA/T3.21.9-1976 on 22 September 1976.

NFPA/T3.21.9

## PROJECT GROUP MEMBERS WHO DEVELOPED THIS STANDARD

| | | |
|---|---|---|
| Olson, Paul E. | Project Chairman (1975 to present) | WABCO Fluid Power Div. |
| Lytle, John | Project Chairman (1974 to 1975) Section Chairman (1975 to present) Section Vice Chairman (1974 to 1975) | Ross Operating Valve Co. |
| Brake, Clifford | Section Chairman (1970 to 1975) | Schrader Fluid Power Div. |
| Chenoweth, Robert | Technical Auditor | Abex Corporation |
| Luecke, John | Director of National Technical Services | National Fluid Power Association |

| | |
|---|---|
| Arvin, P. | The Aro Corporation |
| Grigsby, C. | Schrader Fluid Power Division |
| McGeachy, D. | Numatics, Inc. |

## REFERENCES

1. American National Standard Glossary of Terms for Fluid Power, ANSI/B93.2-1971, and Supplements thereto. (ISO/TC 131/SC 1 (USA-2) 3)

2. American National Standard Interfaces for 4-Way General Purpose Industrial Pneumatic Directional Control Valves, ANSI/B93.33-1974. (NFPA/T3.21.1-1973)

3. American National Standard Symbols for Marking Electrical Leads and Ports on Fluid Power Valves, ANSI/B93.9-1969. (NFPA/T3.5.2-1968)

4. International Organization for Standardization Port Identification Code for Pneumatic and Hydraulic Control Valves and Other Components, ISO/TC 131/SC 5/WG 1 (Secr.-2), 14.

NFPA/T3.21.9

DEFINITION OF PORT COMMUNICATION FOR THE

FLUID POWER PNEUMATIC VALVE INTERFACE TO

NFPA RECOMMENDED STANDARD T3.21.1 WITH THE VALVE

IN POSITION IN RESPONSE TO A REMOTE PILOT SIGNAL

OR ELECTRICAL ENERGIZATION

INTRODUCTION

In pneumatic fluid power systems, power is transmitted and controlled thru a gas under pressure within an enclosed circuit. A pneumatic directional control valve's primary function is to direct or prevent flow thru selected passages.

Sub-base (or sub-plate) mounted valves consist of two principal parts -- the sub-base or mounting portion and the operating valve portion. All piping is permanently connected to the sub-base. The operating valve portion is bolted to the sub-base; it can easily be removed for service or replacement with simple tools.

Users of pneumatic valves benefit when valves of the same nominal size from various manufacturers have a common interface -- and therefore can be interchanged when servicing or replacement is required.

It is important that actuation of the valve by a specific signal results in the same flow paths thru the valve and the sub-base in any standard valve/sub-base combination.

It must be emphasized that there are several accepted standard systems for port marking. (See Reference Nos. 3 and 4.) This recommended standard uses one system only to identify portions of the interface so as to improve understanding. This system need not be used for port marking and identification. Reference No. 3 is to be consulted for standard port marking.

It is the intent of the Project Group which developed this recommended standard to incorporate the contents of this document in Reference No. 2 at its next revision.

NFPA/T3.21.9

1. SCOPE

   1.1 To define which ports are placed in communication when a remote pilot signal port is pressurized or de-pressurized.

   1.2 In a single solenoid assembly, to define which ports are placed in communication with electrical energization and which ports are placed in communication with electrical de-energization.

   1.3 In a double solenoid valve assembly:

   1.3.1 To define which ports are placed in communication when the valve is controlled by a specific solenoid.

   1.3.2 To define which ports are placed in communication when the valve is controlled by the other specific solenoid.

2. PURPOSE

   2.1 To achieve functional interchangeability of valve assemblies.

   2.2 To provide safety for personnel.

   2.3 To promote long life of the equipment.

   2.4 To assure proper function of other system components.

   2.5 To promote proper service and maintenance.

   2.6 To provide universal understanding of fluid power valve function.

3. TERMS AND DEFINITIONS

   For definitions of terms used, see Reference No. 1.

# NFPA/T3.21.9

4. ## CONTROL PORT IDENTIFICATION

   In accordance with Reference No. 4, identify remote pilot signal ports as 12 and 14, where 12 is the remote signal port nearer interface port 2, and 14 is the remote signal port nearer interface port 4. (See Figure 1.)

5. ## SOLENOID IDENTIFICATION

   5.1 In accordance with Reference No. 4, solenoid 14 is the solenoid which when energized produces the same port communication as would be obtained by pressurizing remote pilot signal 14.

   5.2 In accordance with Reference No. 4, solenoid 12 is the solenoid which when energized produces the same port communication as would be obtained by pressurizing remote pilot signal 12.

   5.3 In accordance with Reference No. 4, identify the external pilot supply as "X".

6. ## APPLICATION

   6.1 Air Actuated Valves

   6.1.1 Single remote pilot signal operated.

   6.1.1.1 Use remote pilot signal port 14.

   6.1.1.2 When a valve is positioned by a remote pilot signal pressure at 14, ports 1 and 4 are in communication and ports 2 and 3 are in communication.

   6.1.1.3 For a two-position valve in the normal depressurized position, thru action of a return spring or return pressure, ports 1 and 2 are in communication and ports 4 and 5 are in communication.

   6.1.2 Double remote pilot signal operated.

   6.1.2.1 Use remote pilot signal ports 12 and 14.

6.1.2.2  When a valve is positioned by a remote pilot signal pressure at 12, ports 1 and 2 are in communication and ports 4 and 5 are in communication.

6.1.2.3  When the valve is positioned by a remote pilot signal pressure at 14, ports 1 and 4 are in communication and ports 2 and 3 are in communication.

6.2  Solenoid Actuated Valves

6.2.1  Single solenoid operated.

6.2.1.1  Use solenoid 14.

6.2.1.2  When the valve is positioned by solenoid 14 energization, ports 1 and 4 are in communication and ports 2 and 3 are in communication.

6.2.1.3  For a two-position valve in the normal de-energized position thru the action of a return spring or return pressure, ports 1 and 2 are in communication and ports 4 and 5 are in communication.

6.2.2  Double solenoid operated.

6.2.2.1  Use solenoids 12 and 14.

6.2.2.2  When the valve is positioned by solenoid 12 energization, ports 1 and 2 are in communication and ports 4 and 5 are in communication.

6.2.2.3  When the valve is positioned by solenoid 14 energization, ports 1 and 4 are in communication and ports 2 and 3 are in communication.

6.3  Solenoid and Remote Pilot Signal Actuated Valves

6.3.1  Use solenoid 14 and remote pilot signal port 12.

6.3.2  When the valve is positioned by solenoid 14 energization, ports 1 and 4 are in communication and ports 2 and 3 are in communication.

NFPA/T3.21.9

6.3.3 When the valve is positioned by remote signal pilot pressure at 12, ports 1 and 2 are in communication and ports 4 and 5 are in communication.

7. IDENTIFICATION STATEMENT

Use the following statement in catalogs and sales literature when electing to comply with this voluntary standard:

"Valve/Interface Orientation conforms to NFPA Recommended Standard, NFPA/T3.21.9-1976."

8. KEY WORDS

The following Key Words, useful in indexes and in information retrieval systems, are suggested for this recommended standard:

fluid power

interface, pneumatic valve

port communication, defining

valve, air actuated

valve, pneumatic fluid power

valve, remote pilot signal actuated

valve, solenoid actuated

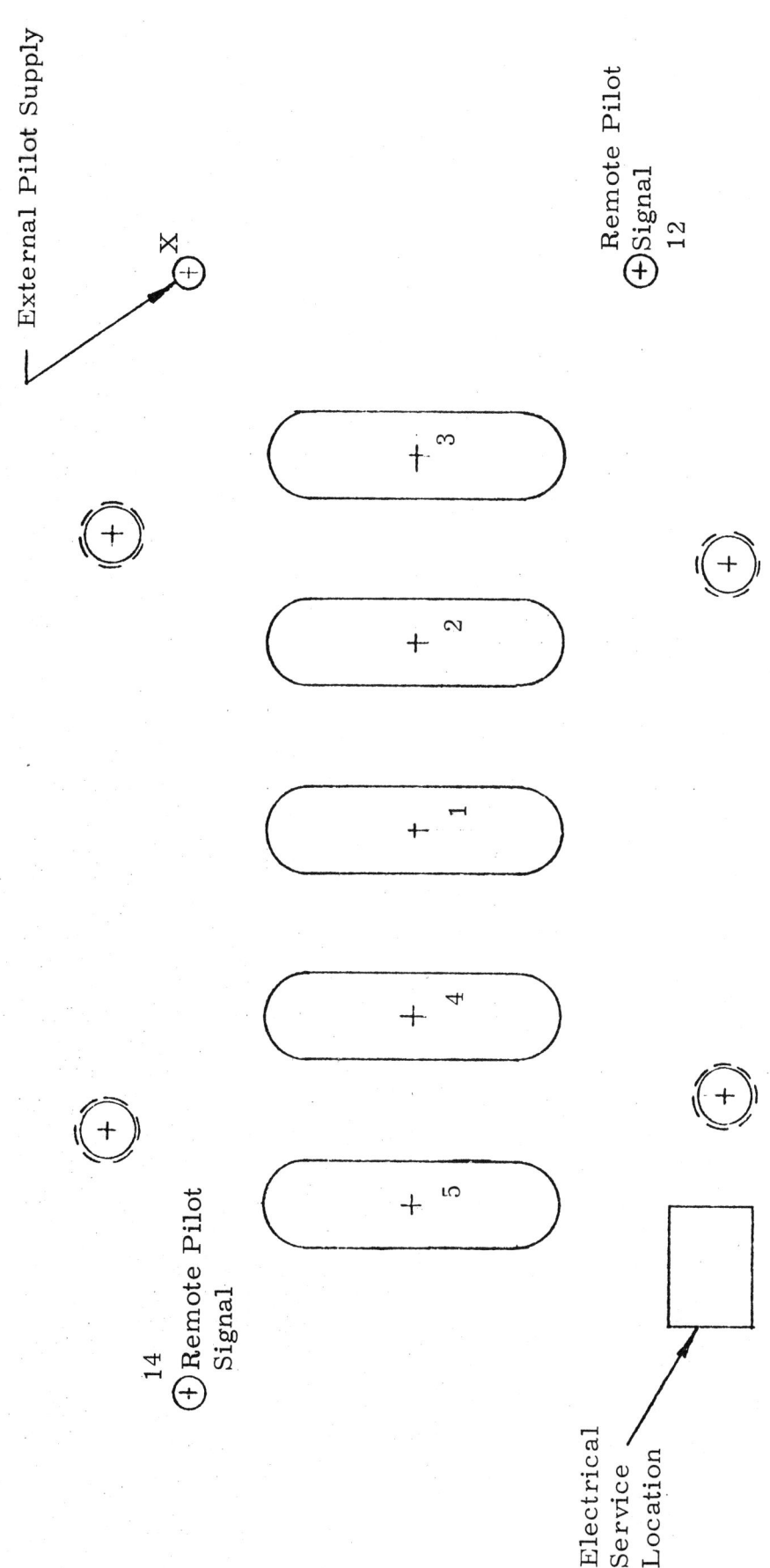

FIGURE 1 - Interface port identifications

# NOTES

ANSI/B93.67M-1983

## AN INDUSTRY STANDARD FOR FLUID POWER

American National Standard

Pneumatic fluid power - Five-port directional control valves -

Mounting surfaces - Optional electrical connector -

Dimensions and requirements
(Metric Only)
(NFPA/T3.21.6M-1982)

Approved as an ANSI Standard
16 May 1983

**Descriptors:** dimensions, electrical connector; electrical connector, valves; fluid power; requirements, electrical connector; valve, 4-way pneumatic directional control; valve, pneumatic; pneumatic fluid power.

published by
## NATIONAL FLUID POWER ASSOCIATION, INC.

3333 N. Mayfair Road / Milwaukee, WI 53222 / 414-778-3344 / TLX 26898

# ANSI/B93.67

## FOREWORD

This Foreword is not part of NFPA Recommended Standard - Pneumatic fluid power - Five-port directional control valves - Mounting surfaces - Optional electric connections - Dimensions and requirements, ANSI/B93.67M-1983 (technically identical to ISO/DIS 5599/3).

At its 7 March 1972 meeting, NFPA Project Group T3.21.1M recommended that a pneumatic valve subbase include an electrical wiring opening. Both the location and size of the electrical opening were unanimously agreed upon at the 19 April 1972 meeting of T3.21.1M.

A proposed electrical plug design was submitted by AMP, Inc. at the 9 June 1972 meeting, but it was not possible at subsequent meetings to get agreement on a plug-in. On 28 November 1972, Project Group T3.21.1M eliminated the electrical connector from consideration in NFPA/T3.21.1M, although both the size and location of the opening are included in NFPA/T3.21.1M-1973 (ANSI/B93.33M-1974 (R1981)).

This Project, T3.21.6M, originated at the 28 November 1972 meeting of T3.21.1M, and was formally approved by the NFPA Technical Board on 12 December 1972. A Project Group Chairman was named in June 1973, and Project Group Members were named at the October 1973 meeting of the Pneumatic Valve Section.

The first meeting of T3.21.6M was held on 27 November 1973, at which time the group reached agreement on general specifications for intermateability and electrical characteristics.

Additional Project Group meetings were held on 21 January 1974, 4 February 1974, 21 March 1974 and 16 May 1974. During these meetings, the design and specifications were further refined, and consensus was reached at the 16 May 1974 Project Group meeting.

The Final Working Draft was submitted to Headquarters on 5 June 1974, and the NFPA Technical Staff prepared the General Review Draft on 29 January 1975. Comments resulting from General Review were resolved on 16 May 1975.

The Technical Board granted approval to ballot on 21 August 1975, pending concurrence by the Technical Auditor that all General Review comments had been resolved. Concurrence was received on 12 September 1975.

Nine negative ballots were subsequently received, primarily because of possible patent infringement problems. In the fall of 1976, eight of these negative ballots were resolved when AMP, Inc. granted a waiver.

During this same period of time, the ISO standard for air valves was nearing completion and a proposal for an electrical connector had been prepared for that interface standard. This electrical connector proposal was nearly identical to the NFPA type and a delay in Project T3.21.6M occurred to study the ISO proposal.

At its meeting of 12 October 1977, the T3.21 Pneumatic Valve Section voted to modify Project T3.21.6M to be in accordance with the ISO connector.

Between 1978 and 1980, a project was conducted by the USA TAG to have Underwriters Laboratories examine the proposed electrical connector and the related proposed ISO standard. This was completed with favorable results and a final report issued in January, 1981. Project Chairman Berninger reported these results to the Pneumatic Valve Section on 25 March 1981 and it was decided Project T3.21.6M was to proceed.

Project T3.21.6M was then reviewed and Working Draft No. 2 was completed and forwarded to Headquarters. Headquarters Staff circulated Draft No. 2 (technically identical to ISO/DIS 5599/3) for comment on 14 October 1981.

Chairman Berninger requested comments from the T3.21 Section at their 19 October 1981 meeting. After discussion, it was moved and carried unanimously to resolve any comments received and then send the document out for Second General Review.

The comment period for Draft No. 2 ended 14 November 1981. The comments received were resolved and incorporated into the proposed document and forwarded to Headquarters.

Headquarters Staff prepared the document for Second General Review on 16 April 1982. Second General Review closed 17 May 1982 with one comment. After the resolution of the comment, the Technical Board granted approval to ballot on 26 May 1982.

The Second Ballot Draft was prepared by Headquarters Technical Staff on 23 July 1982. Successful balloting concluded 23 August 1982 with 2 comments. Comments were reviewed and resolved.

T3.21.6M was forwarded to the Technical Board where it received approval on on 16 September 1982. Final approval was granted by the Board of Directors on 20 October 1982.

Project Group Members who developed this standard:

**John Berninger**
Project Chairman (1980-present)
Parker Hannifin Corp.

**John Lytle***
Project Chairman
(1972-1974) (1977-1978)
Section Chairman (1975-1978)
Ross Operating Valve Company

**Carroll Grigsby**
Section Chairman (1978-present)
Schrader Bellows Division

**Robert Hoffman**
Project Chairman (1975-1976)
Parker Hannifin Corp.

---

* Company affiliation has changed

**Clifford Brake**
Section Chairman (1970-1975)
Schrader Bellows Div./Scovill Inc.

**Robert Entwisle**
Section Vice Chairman
Automatic Switch Co.

**Henry Barthe**
Technical Auditor
Schroeder Brothers Corp.

**Allen E. Tucker**
Director of Technical Services
National Fluid Power Association

**E. Cole**
Schrader Bellows Division

**W. Forster\***
WABCO Fluid Power Div.

**R. Huss**
AMP, Inc.

**M. Kavanaugh\***
Rexnord, Inc.

**J. Komendera**
Automatic Valve Corp.

**M. Magera**
Underwriters Laboratories

On 21 January 1983, ANSI/B93.67M was submitted to ANSI Committee B93 for ballot. Balloting closed 1 April 1983 with unanimous approval.

ANSI/B93.67M was forwarded to ANSI Board of Standards Review on 19 April 1983 and granted approval 16 May 1983.

The membership roster for the Standards Committee B93 at the time of ballot:

**Jack C. McPherson**
Chairman

**Daniel B. Shore**
Vice Chairman

**Allen E. Tucker**
Co-Secretary

**William G. Wagner**
Co-Secretary

**American Society of Agricultural Engineers**
Ed Fletcher

**American Society for Engineering Education**
Wm. R. Smith

**American Society of Lubrication Engineers**
(To be named)

**Compressed Air & Gas Institute**
David E. Bonn
John Addington (alternate)

**Construction Industry Manufacturers Association**
Glenn Stewart

**Fluid Controls Institute**
Jude Pauli
Eric Bianchi (alternate)

**Fluid Power Distributor Association**
Thomas Neff

**Fluid Power Society**
Robert L. Firth
Carroll Grigsby
Robert W. Hanpeter
Marty King
Richard Read
Ronald Smith
Alan Tiedman

**Fluid Sealing Association**
John Scannell
Alex Pilecki (alternate)

**Instrument Society of America**
(To be named)

**Joint Industry Council**
Robert Muhl

**Material Handling Institute**
Jack C. McPherson
Willard Chichester (alternate)

**National Fluid Power Association**
Richard N. Bailey
John Bowbin
Walter Forster
Z. J. Lansky
Paul Schacht

**National Machine Tool Builders Association**
John B. Deam

**Rubber Manufacturers Association**
William A. Hertel
John R. Loder
E. J. McCarthy (alternate)

**Society of Automotive Engineers**
William A. Hertel
John T. Parrett
Henry Schultz
Daniel B. Shore
W. L. Snyder
Robert W. White

**U. S. Dept. of Defense**
William P. Coyne

**Company Members**
L. L. Schmaltz
Don McGeachy
John Welker (alternate)

**Individual Members**
Dr. E. C. Fitch, Jr.
Otto Maha
John J. Pippenger
A. O. Roberts
Jack Walrad
Tom Wanke
Frank Yeaple

ksg

---

\*Company affiliation has changed.

ANSI/B93.67

**Pneumatic fluid power - Five-port directional control valves Mounting surfaces - Optional electrical connector - Dimensions and requirements**

## 0 INTRODUCTION

In pneumatic fluid power systems, power is transmitted and controlled thru a gas under pressure within an enclosed circuit. A pneumatic directional control valves primary function is to direct or prevent flow thru selected passages.

The equipment for pneumatic distribution and control can be either directly mounted onto the pipelines or mounted on subplates allowing quicker dismantling which promotes equipment interchangeability.

Subbase (or subplate) mounted valves consist of two principal parts — the subbase or mounting portion and the operating valve portion. Much of the piping, and often all of the piping, is permanently connected to the subbase. When the operating valve portion is bolted to the subbase, it can easily be removed for service or replacement with simple tools. The mating surfaces of the subbase and operating valve are call the "interface."

When pneumatic fluid power subbase type valves are electrically operated, it is sometimes desireable to locate an electrical connector at the interface of the valve body and subbase. Users of pneumatic valves benefit when the electrical connector is standardized, allowing electrical interchangeability between valves of various manufacturers.

## 1 SCOPE AND FIELD OF APPLICATION

1.1 This National Standard specifies:

— The dimensional requirements for mateability of the optional electrical connector used with mounting surfaces for pneumatic directional control valves with 5 service ports as described in B93.33M-1974 (R1981) and ISO 5599/1.

— The necessary requirements for safety and interchangeability.

1.3 This Standard only applies to the dimensional criteria of products manufactured in conformance with this standard. It does not apply to their functional characteristics.

1.2 This National Standard proposes to:

— Achieve dimensional interchangeability.

## 2 REFERENCES

ANSI/B93.33M-1974 (R1981) (NFPA/T3.21.1M-1973), American National Standard Interfaces for 4-way General Purpose Industrial Pneumatic Directional Control Valves.

ANSI/B93.2-1971 and Supplement ANSI/B93.2A-1978, American National Standard Glossary of Terms for Fluid Power.

NFPA/T2.10.1M-1978, National Fluid Power Association Recommended Standard Metric Units for Fluid Power Aplications.

ANSI/B93.9M-1969 (R1981)(NFPA/T3.5.2M-1968), American National Standard Symbols for Marking Electrical Leads and Ports on Fluid Power Valves.

UL - 429 Third Edition-1982, Underwriter's Laboratories Standard for Safety - Electrically Operated Valves.

C22.2 No. 139; C22.2 No. 153; and C22.2 No. 42 (old) Canadian Standards Association.

FB10-1973, National Electrical Manufacturers Association - General Standards for Plugs, Receptacles and Connectors for the Pin and Sleeve Type.

FB11-1973, National Electrical Manufacturers Association - Plugs, Receptacles and Connectors of the Pin and Sleeve Type for Hazardous Locations.

ISO 5599/1-1978, International Standard Pneumatic fluid power - 5 Port directional control valves - Mounting surfaces - Part 1: General.

## 3 DEFINITIONS

For definitions of other terms used see ANSI/B93.2 and Supplement ANSI/B93.2A.

**3.1 electrical connector:** A two-piece instrument (contact and housing) which, when joined, provides electrical and mechanical continuity.

**3.2 contact:** A current carrying component used at a detachable junction of an electrical circuit.

**3.3 socket:** A contact with an opening or hollow designed to be the mechanical holder of a pin contact.

**3.4 pin:** A pointed contact designed to mate with a socket contact.

**3.5 housing:** A device designed to orientate, secure, and insulate contacts.

## 4 ELECTRICAL CONNECTOR

See figure 1 for illustration of the electrical connector showing pin, socket and housing.

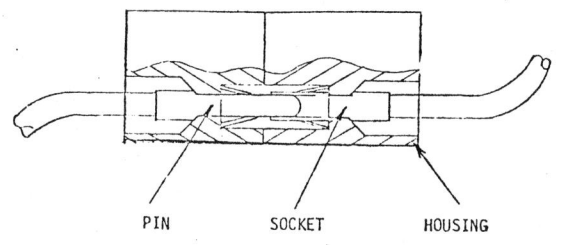

FIGURE 1 - Electrical connector

## 5 ELECTRICAL CONNECTOR SPECIFICATIONS

Design the connector to be readily removed if not required and to meet the following additional requirements.

**5.1** Use electrical connectors rated at 300 V (volts) AC or DC.

**5.2** Use electrical connectors rated at 2 A (amps) maximum inrush. Use leads with insulation suitable for 105°C and 600 volts.

**5.3** Contacts

**5.3.1** Provide four live contacts and one earth contact. Earth contact to make first, break last by 1.5 mm.

**5.3.2** Use pin contacts with diameters of 2.03 mm to 2.18 mm.

**5.3.3** Use socket contacts with an opening diameter that friction fit on the pin contacts.

**5.4** Wire Leads

**5.4.1** Use 0.75 $mm^2$ to 2.5 $mm^2$ wire.

**5.4.2** Use same color insulation on diagonally opposite leads.

**5.4.3** Use a green color insulation for earth connection.

**5.5** Orientation

**5.5.1** Refer to figure 2 for the dimensions of the electrical connector.

**5.5.2** Use contact pattern of four rectangular and center earth. Locate contacts No. 1 and 4 parallel to valve port's slots and toward the valve interior.

**5.5.3** Use contacts No. 1 and 3 for single solenoid valves.

**5.5.4** Use contacts No. 2 and 4 for the second solenoid of a double solenoid valve.

**5.5.5** Use contact No. 5 for earth.

**5.5.6** Where polarity of contacts is a requirement, use No. 1 and No. 2 as positive; use No. 3 and No. 4 as negative.

**5.5.7** Retain each connector half in its respective location when the valve body/base are disconnected.

## 6 IDENTIFICATION STATEMENT

Use the following statement in test reports, catalogs and sales literature when electing to comply with this National Standard:

"Electrical connector conforms to ANSI/B93.67M-1983, **Pneumatic fluid power - Five-port directional control valves - Mounting surfaces - Optional electrical connectors - Dimensions and requirements.**"

CORNER RADIUS

| Base Size | R4 | R3 | R2 | R1 |
|---|---|---|---|---|
| 1-2-3 | 2,6/3,0 | 2 Max. | 2,6/3,0 | ,2 Max. |
| 4-5-6 | 2,6/3,0 | 2,6/3,0 | 2,6/3,0 | 2,6/3,0 |

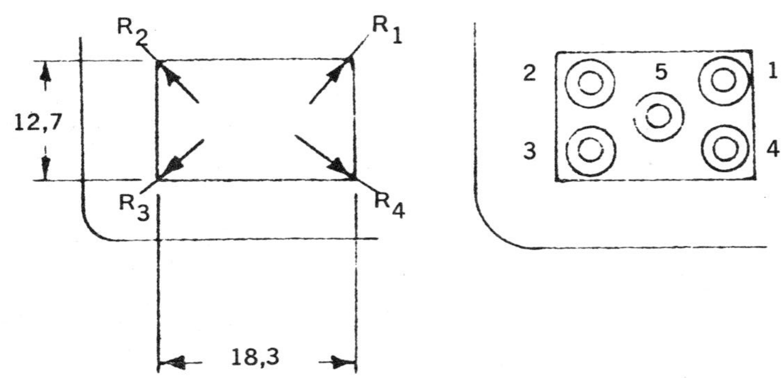


Maximum Envelope Limits of Connector — Interior Contours Indicated with * are Permissible

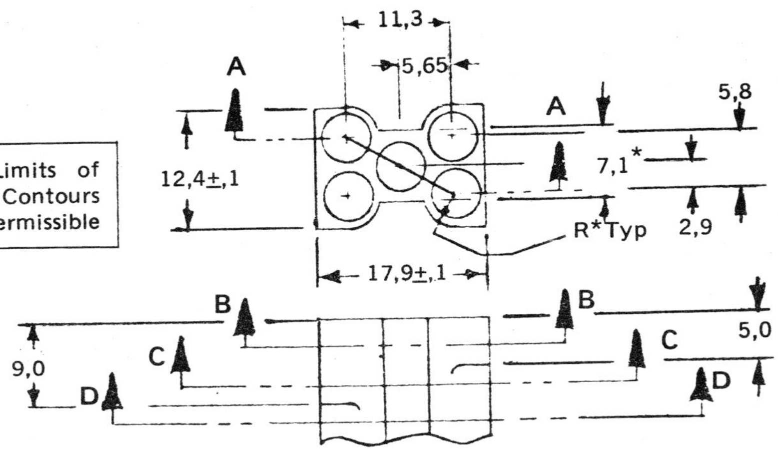


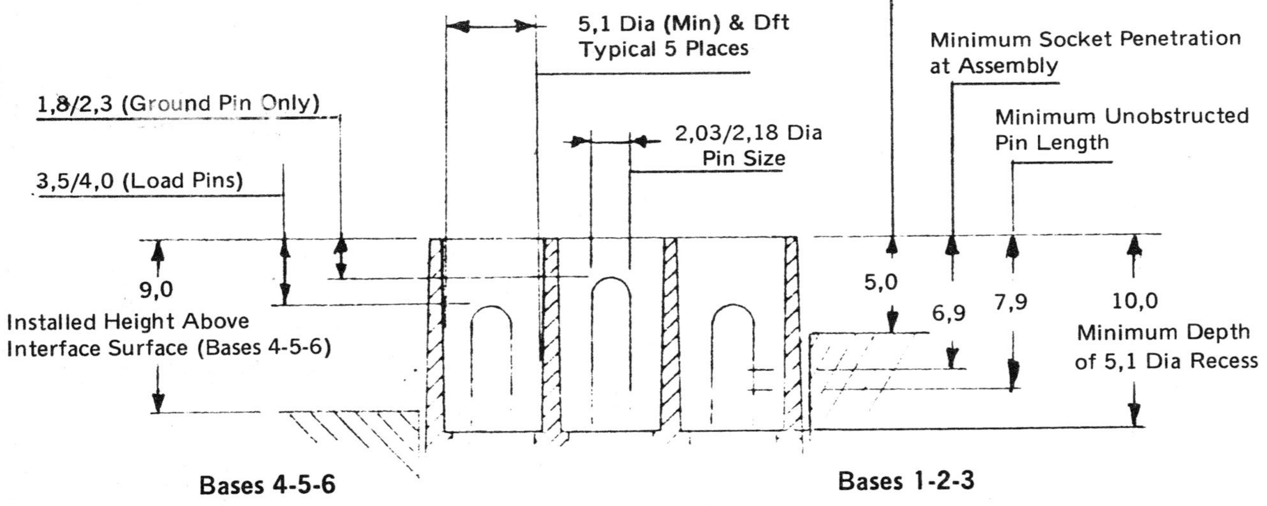


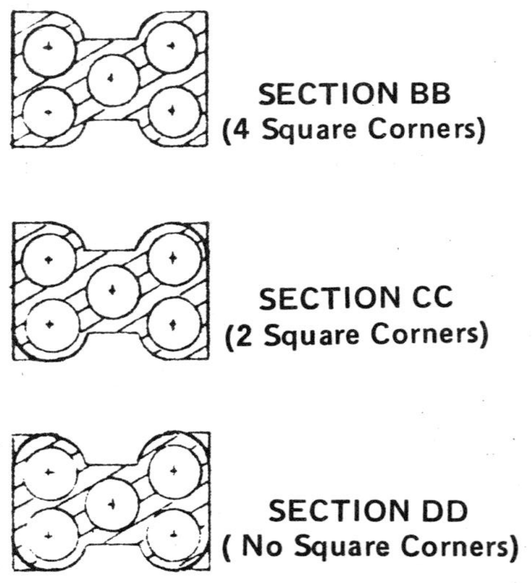


Figure 2 - Dimensions of electrical connector required for mateability

# APPENDIX

## to ANSI/B93.67M-1983

## 2  REFERENCES

ISO 5598, International Standard Fluid power systems and products - Vocabulary.

ISO 1000-1981, International Standard SI units and recommendations for the use of their multiples and of certain other units.

- NOTES -

ANSI/B93.45M-1982

---

## AN INDUSTRY STANDARD FOR FLUID POWER

American National Standard

Pneumatic fluid power -

Compressed air dryers -

Methods for rating and testing
(NFPA/T3.27.3M R1-1981)

SPONSORS
National Fluid Power Association, Inc.
Compressed Air & Gas Institute

Approved as an ANSI Standard
19 August 1982

---

**Descriptors:** pneumatic fluid power; compressed air dryer; water vapor; ambient temperature; inlet temperature; cooling water temperature; inlet pressure dew point; outlet pressure dew point; pressure drop.

---

published by
# NATIONAL FLUID POWER ASSOCIATION, INC.
3333 N. Mayfair Road / Milwaukee, WI 53222 / 414-778-3344 / TLX 26898

ANSI/B93.45

**FOREWORD**

This Foreword is not part of American National Standard - Pneumatic fluid power - Compressed air dryers - Methods for rating and testing, ANSI/B93.45M-1982.

This project was initiated at the 17 June 1970 Section meeting of T3.27. Draft No. 1 was written on 6 December 1972. It was reviewed at the Section meeting of 25 April 1973. Draft No. 2 resulted from that meeting. The Section reviewed Draft No. 2 at meetings held on 27 November 1973, 21 May 1974 and 11 March 1975. Consensus was reached at the 30 September 1975 T3.27 meeting, and the Final Working Draft was forwarded to NFPA Headquarters on 14 November 1975. The NFPA Technical Staff prepared the General Review Draft on 10 December 1975.

General Review comments were answered on 19 March 1976. On 7 April 1976, under the direction of the Compressed Air Dryer Section, NFPA Headquarters circulated copies of this document to compressed air dryer manufacturers who are not NFPA members. The manufacturers were encouraged to offer suggestions for the document's further development, and several did so.

The NFPA Technical Staff prepared the Ballot Draft on 9 September 1976. Negative comments resulting from ballot were resolved at the Compressed Air Dryer Section meeting on 19 October 1976.

On 10 November 1976, the NFPA Technical Board voted to recommend to the Board of Directors that this document be approved as an NFPA Recommended Standard. The NFPA Board of Directors granted final approval to NFPA/T3.27.3-1974 on 14 December 1976.

The NFPA Recommended Standard was submitted to ANSI Standards Committee B93 for promulgation as an ANSI Standard. Due to negative comments, the document was temporarily withdrawn from ANSI, and a Project Group made up of members of the NFPA Compressed Air Dryer Section and CAGI (Compressed Air and Gas Institute) was formed to resolve the negative comments. On 8 February 1980, both organizations met and subsequently agreed to jointly sponsor a revised document. Through this joint effort to resolve the negative comments, the committee combined T3.27.2 and T3.27.3 with additional information from the CAGI Committee.

*Company affiliation has changed.
†Deceased

A TSP was written to initiate the revision and was approved by the Technical Board on 7 May 1980. Upon review of additional comments received by CAGI, T3.27.3M R1 was distributed for General Review on 6 February 1981.

After completion of the General Review and resolution of comments received, the document was presented to the Technical Board for approval to ballot. Approval was granted by the Technical Board on 13 May 1981.

T3.27.3M R1 was forwarded to Headquarters and Headquarters Technical Staff prepared the document for ballot on 18 June 1981.

The ballot closed 27 July 1981 with some comments. The comments received were resolved and the document was forwarded to the Technical Board for approval. Approval was granted 15 September 1981.

Subsequently, T3.27.3M R1-1981 was granted final approval by the Board of Directors on 23 September 1981.

It was intended that the NFPA Recommended Standard, T3.27.3M R1 be submitted to ANSI B93 for promulgation as an American National Standard.

Project Group members who developed this standard:

**John Bowbin**
Chairman
Miller Fluid Power Corp.

**John Pousma**
Vice Chairman
C. A. Norgren Co.

**James Rheinheimer**
Secretary
Wilkerson Corp.

**Paul Schacht**
Racine Hydraulics Div./Dana Corp.
Technical Auditor

**James C. White***
Director of Technical Services
National Fluid Power Association

**John Addington**
Compressed Air & Gas Institute

ANSI/B93.45

**F. Ribble**
Wilkerson Corp.

**G. Behrens**
Gas Drying Inc.

**J. Kuppe**
Arrow Pneumatics

**J. McCloskey**
Hankison Corp.

**M. Strosser**
Compressed Air & Gas Institute

On 8 January 1982, ANSI/B93.45M was submitted to ANSI Committee B93 for ballot. Balloting closed 15 March 1982. A negative comment was received late with no substantiating documenting evidence. Consequently, the document was forwarded to ANSI Board of Standard Review for approval on 15 June 1982.

ANSI/B93.45M was granted approval on 19 August 1982 by ANSI Board of Standards Review.

The membership roster for the Standards Committee B93 at the time of ballot:

**Jack McPherson**
Chairman

**Daniel B. Shore**
Vice Chairman

**James C. White**
Co-Secretary

**William Wagner**
Co-Secretary

**American Society of Agricultural Engineers**
Ed Fletcher

**American Society for Engineering Education**
William R. Smith

**American Society of Lubrication Engineers**
(To be named)

**Compressed Air & Gas Institute**
David E. Bonn
John Addington (alternate)

**Construction Industry Manufacturers**
Glen Stewart

**Fluid Controls Institute**
Jude Paull
Eric Blanchi (alternate)

**Fluid Power Distributors Association**
Thomas Neff

**Fluid Power Society**
Edward C. Briggs†
Ronald Brettnacher
Robert L. Firth
Carroll Grigsby
Robert W. Hanpeter
Marty King
Richard Read
Ronald Smith
Alan Tiedman

**Fluid Sealing Association**
John Scannell
Alex Pilecki (alternate)

**Instrument Society of America**
Aaron I. Kutz

**Joint Industry Council**
Robert Muhl

**Material Handling Institute**
Jack McPherson
Willard Chichester (alternate)

**National Fluid Power Association**
Richard N. Bailey
John Bowbin
James L. Fisher, Jr.
Walter Forster
Z. J. Lansky
A. O. Roberts
Paul Schacht

**National Machine Tool Builders Association**
John B. Deam

**Rubber Manufacturers Association**
William A. Hertel
John R. Loder
E. J. McCarthy (alternate)

**SAE**
William A. Hertel
Eugene Falendysz
John Parrett
Henry Schultz
Daniel B. Shore
W. L. Snyder
Robert W. White

**Society of Manufacturing Engineers**
(To be named)

**U. S. Dept. of Defense**
William P. Coyne

**Company Members**
L. L. Schmaltz
Don McGeachy
John Welker (alternate)

**Individual Member**
Dr. E. C. Fitch, Jr.
Robert Hildebrandt
Otto Maha
John J. Pippenger
Tom Wanke
Frank Yeaple

---

*Company affiliation has changed.
†Deceased

ANSI/B93.45

# Pneumatic fluid power
# Compressed air dryers
# Methods for rating and testing

## 0 INTRODUCTION

In pneumatic fluid power systems, power is transmitted and controlled thru a gas under pressure within an enclosed circuit. For reliable operation, many applications require installation of a compressed air dryer to remove water vapor and prevent condensation within the pneumatic system.

## 1 SCOPE AND FIELD OF APPLICATION

1.1 This National Standard includes:

— Standard rating conditions of inlet temperature, ambient temperature, cooling water temperatures, inlet pressure dew point, operating pressure, outlet pressure dew point and pressure drop to enable a compressed air dryer to be classified;

— Test procedures for measuring performance to determine the capabilities of compressed air dryers.

1.2 This National Standard provides:

— A uniform basis on which to rate the performance of compressed air dryers;

— A uniform basis of measuring the performance of compressed air dryers.

## 2 REFERENCES

ANSI/B93.2-1971 and Supplement ANSI/B93.2A-1978, **American National Standard, Glossary of Terms for Fluid Power.**

ANSI/Y32.10-1967 (R 1979), American National Standard **Graphic Symbols for Fluid Power Diagrams.**

## 3 CONDITIONS FOR RATING COMPRESSED AIR DRYERS

3.1 Standard rating conditions and accuracies (table 1).

3.2 Performance ratings (table 2).

3.3 Ratings.

Classify compressed air dryers tested for operation at standard rating conditions defined in 3.1 "Standard rating conditions and accuracies" according to flow capacity and compressed air outlet dew point.

NOTE: Performance rating may vary when other than standard rating conditions exist. Actual field performance can vary from the rated performance due to contamination in the compressed air, power supply, desiccant life and equipment maintenance.

## 4 METHOD FOR TESTING COMPRESSED AIR DRYERS

4.1 Test equipment (figure 1).

4.1.1 Use test equipment that is of laboratory quality and that has been calibrated prior to the start of the test procedure.

4.1.2 Use Dew Point measuring instruments certified by the instrument manufacturer to be accurate with $\pm 1°C$ ($\pm 1.8°F$) in the range the measurements are taken.

NOTE: Two or more instruments may be required.

4.2 Test procedure (figure 1).

4.2.1 Stabilize air dryers to be tested in operation prior to the taking of pressure dew point readings. Operate the dryer for a sufficient period to stabilize outlet compressed air conditions prior to observations for record.

NOTE: Stabilization shall be considered achieved when repeated observations at 15-minute intervals show duplicate results within a $\pm 2°C$ ($\pm 3.6°F$) effluent dew point.

4.2.2 Take and record observations at 15-minute intervals to permit definition of characteristics of the dryer and to permit plotting of the results.

NOTE: For heatless type desiccant dryers, take and record one measurement for each tower per cycle.

**4.2.3** Ensure the unit be in its proper operation condition prior to beginning of the test.

**4.2.4** Make no adjustments to the dryer during the test run.

**4.2.5** Measure inlet air temperature in the approximate center of the inlet air stream immediately upstream from the inlet port.

**4.2.6** Measure inlet air pressure in a static pressure tap located immediately upstream from the inlet port ahead of any elbow or pipe restrictions.

**4.2.7** Measure inlet pressure dew point from a tap immediately upstream from the inlet port.

**4.2.8** Measure pressure drop of the compressed air as the pressure difference between a static tap immediately upstream of the inlet port and a static tap immediately downstream of the outlet port.

**4.2.9** Measure cooling water inlet temperatures, if cooling water is used, at the approximate center of the inlet water stream immediately upstream from the inlet water port.

**4.2.10** Measure ambient temperature in the following manner:

    a) Refrigerated Dryers (Air-Cooled Type).

At the center of the cooling air flow entering the refrigerant condenser within 915mm (36 in) of the unit.

    b) Deliquescent Dryers, Desiccant Dryers and Refrigerated dryers (Water-Cooled Types).

Within a stablized temperature area; such an area is not affected by heat given off by the unit, within 610 mm (24 in) of the unit.

**4.2.11** Measure the flow rate of the inlet compressed air in a straight run of pipe, a minimum of ten (10) pipe diameters from any elbow or restrictions.

**NOTE:** For dual tower regenerative desiccant dryers only also measure the flow rate of the outlet compressed air in a straight run of pipe, a minimum of ten (10) pipe diameters from any elbow or restriction.

**4.2.12** Measure outlet air pressure and pressure dew point at a static pressure in a straight run of pipe a minimum of ten (10) pipe diameters from any elbow or restriction.

**4.2.13** Determine outlet compressed air dew point rating of the dryer from the arithmetic average of the pressure dew point as recorded at regular intervals during the test.

**4.2.14** Conduct tests for the following minimum period, after stabilization:

    a) Non-cycling type refrigerant air dryers two (2) hours;

    b) Cycling type refrigerant air dryers eight (8) hours;

    c) Deliquescent (absorbent) air dryers two (2) hours;

    d) Heated desiccant (adsorbent) air dryers, two complete cycles per tower, but not less than sixteen (16) hours;

    e) Heatless type desiccant air dryers for a total of twelve (12) cycles.

**4.3** Data Accuracy

Maintain in accordance with 3.1 "Standard rating conditions and accuracies" and 3.2 "Performance ratings."

**4.4** Data Presentation

Have available a record of all the following minimum test data in all test reports referencing this recommended standard:

    a) All physical values pertaining to the test (see table 1 & 2);

    b) All additional provisions and modifications pertaining to the test.

**4.5** Justification Statement

This National Standard verification procedure is based on the combined expert experiences of those who have participated in its preparation and review.

**4.6** Test/Production Similarity

Utilize managerial controls necessary to maintain substantial similarity between test and production components or elements.

## 5 IDENTIFICATION STATEMENT

Use the following statement when electing to comply with this National Standard:

"Performance data obtained and presented in accordance with ANSI/B93.45M-1982, **Pneumatic fluid power - Compressed air dryers - Methods for rating and testing.**"

Table 1 - Standard rating conditions and accuracies

| Test Condition (Compressed Air) | Standard Rating Condition | ISO 1000 Unit | Customary US Unit | Maintain Within ($\pm$) of Actual Gauge Value | Instrument Accuracy |
|---|---|---|---|---|---|
| Inlet Temperature | 37.8°C (100°F) | °C | °F | $\pm$2°C ($\pm$3.6°F) | $\pm$1°C ($\pm$1.8°F) |
| Inlet Pressure | 6.9 bar (100 psig) | bar | psig | $\pm$2% | $\pm$0.07 bar ($\pm$1 psi) |
| Inlet Pressure Dew Point | 37.8°C (100°F) | °C | °F | $\pm$2°C ($\pm$3.6°F) | $\pm$1°C ($\pm$1.8°F) |
| Ambient Temperature | 37.8°C (100°F) | °C | °F | $\pm$3°C ($\pm$5.4°F) | $\pm$1°C ($\pm$1.8°F) |
| Pressure Drop (Maximum) | 0.35 bar (5 psi) | bar | psi | Not Applicable | $\pm$0.035 bar ($\pm$0.5 psi) |
| Cooling Water Inlet Temperature | 29.4°C (85°F) | °C | °F | $\pm$3°C ($\pm$5.4°F) | $\pm$1°C ($\pm$1.8°F) |

Table 2 - Performance rating

| Test Condition (Compressed Air) | Standard Rating Condition | ISO 1000 Unit | Customary US Unit | Maintain Within ($\pm$) of Actual Gauge Value | Instrument Accuracy |
|---|---|---|---|---|---|
| Outlet Dew Point Temperature | Results at measured pressure | °C | °F | $\pm$2°C ($\pm$3.6°F) | $\pm$1°C ($\pm$1.8°F) |
| Rated Inlet Air Flow | Specify based on standard air | *dm$^3$/s | scfm | $\pm$2% | $\pm$2% full scale |
| Rated Outlet Air Flow ** | Results based on standard air | *dm$^3$/s | scfm | $\pm$2% | $\pm$2% full scale |

*Vary time to provide suitable numbers.

** Regenerative type only.

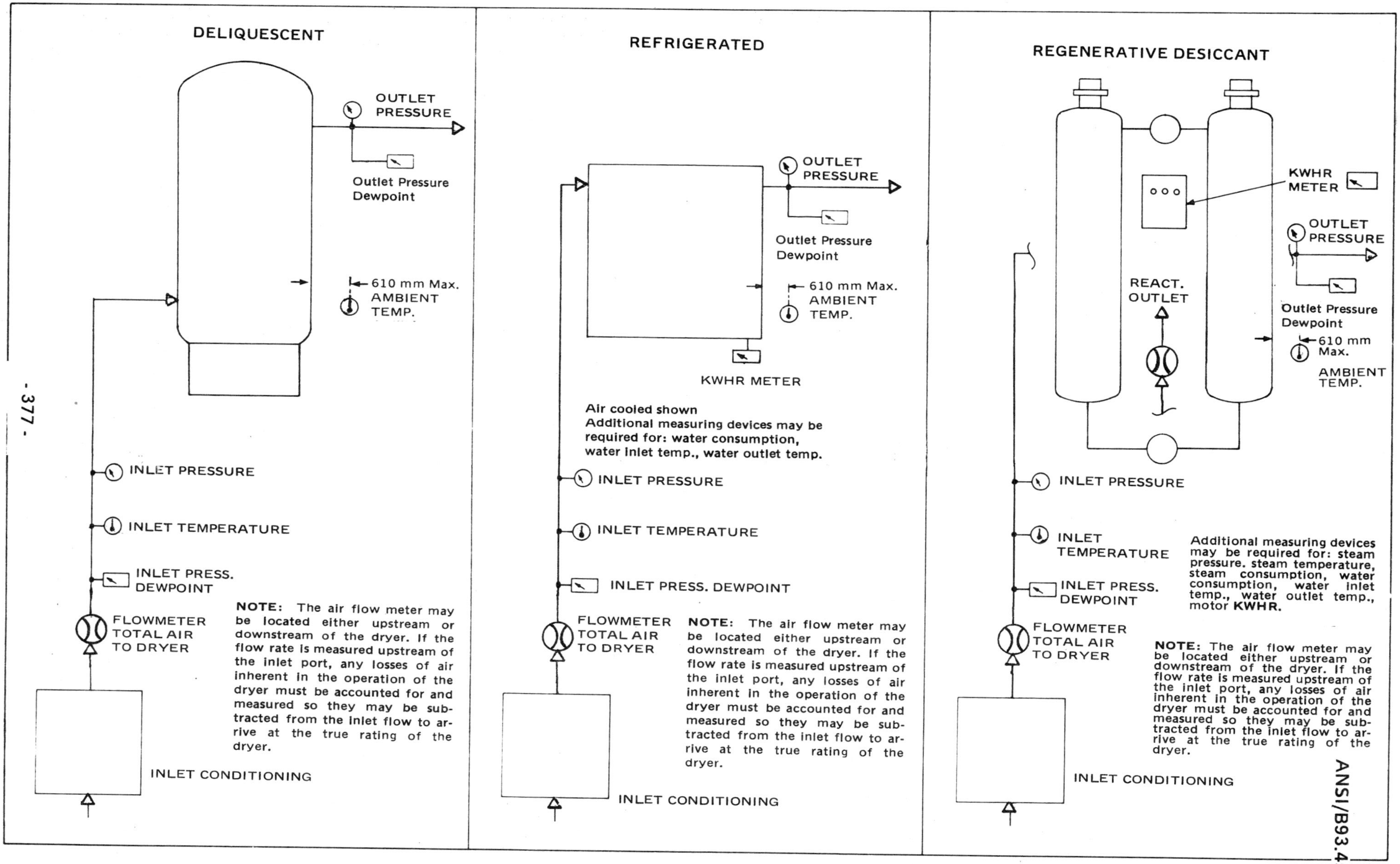

Figure 1 - **Types of dryers**

# APPENDIX to
# ANSI/B93.45M - 1982

## 1 REFERENCES

ISO 1000-1981, International Standard SI units and recommendations for the use of their multiples and of certain other units.

ANSI/ASME PTD 19.3, American National Standard Temperature Measurement, Instruments and Apparatus (Performance Test Code).

ASME PTC 19.2, American Society of Mechanical Engineers Pressure Measurement (Performance Test).

ISO/R541-1967, International Standard Measurement of Fluid Flow by Means of Orifice Plates and Nozzles.

ISA/RP 16.5, Instrument Society of America Installation, Operation, Maintenance Instructions for Glass Tube Variable Area Meters.

11-12-1959, National Electrical Manufacturers Association Standard.

11-13-1963, National Electrical Manufacturers Association Standard.

1-21-1964, National Electrical Manufacturers Association Standard.

ISO 5598*, International Standard Fluid power systems and components - Vocabulary.

NFPA/T2.10.1M-1978, National Fluid Power Association Recommended Standard Metric Units for Fluid Power Applications.

Compressed Air and Gas Institute Standard for Rating and Testing Compressed Air Dryers.

---

\* At present at the stage of draft.

## 2 DEFINITIONS

### 2.1 Air, Free

Air at ambient temperature, pressure, relative humidity and density.

### 2.2 Air, Standard

Air at temperature of 68°F (20°C) a pressure of 14.7 pounds per square inch absolute (1.01 bar) and a relative humidity of 36% (0.0750 pounds per cubic foot). In gas industries, the temperature of standard air is usually given at 60°F (15.6°C).

### 2.3 Bar

A unit of pressure based on $10^5 N/m^2$ (Newtons per square metre, approximately equal to 14.5 psig).

### 2.4 Compressed Air Dryer

A device that lowers the dew point of compressed air.

### 2.5 Types of Compressed Air Dryers

#### 2.5.1 Desiccant

A compressed air dryer which lowers the dew point of compressed air by passing the air through a bed of desiccant.

##### 2.5.1.1 Regenerative

A compressed air dryer which periodically processes the desiccant to restore its ability to adsorb water vapor.

#### 2.5.2 Deliquescent

A desiccant compressed air dryer which utilizes absorbent desiccant and required periodic addition of desiccant.

#### 2.5.3 Refrigerated

A compressed air dryer that utilizes a mechanical refrigeration device to cool compressed air, condensing and separating moisture.

## 2.6 Ambient Temperature Range

The range of temperature of the air surrounding the dryer in which the equipment will perform as recommended.

## 2.7 Capacity, Compressed Air Dryer

The amount of saturated air entering a dryer at standard rating conditions.

## 2.8 Net Outlet Capacity, Compressed Air Dryer

The amount of dry air delivered at standard rating conditions.

## 2.9 Condensation

The process of changing a vapor into liquid condensate by the extraction of heat.

## 2.10 Contact Time

The time required for a molecule in a stream of air to pass completely through a desiccant bed based on superficial bed velocity.

## 2.11 Cycle, Adsorbent Dryer

The time required for an adsorbent desiccant bed to pass through one drying period and one regenerative period.

## 2.12 Dew Point

The point (temperature) at which vapors in air condense. For practical purposes, it must be referred to a stated pressure.

## 2.13 Dew Point, Atmosphere

The dew point in the air at atmospheric pressure.

## 2.14 Dew Point, Pressure

The dew point in the air at the actual operating pressure.

## 2.15 Dew point Depression

The difference between inlet and outlet dew points of a compressed air dryer referred to the same (inlet/outlet) operating conditions.

## 2.16 Purge Flow

A flow of air to regenerate a desiccant.

## 2.17 Regeneration (Reactivation)

The process of restoring the capacity of adsorbent desiccant.

## 2.18 Regeneration, Heat

Reactivation of desiccant by increasing its temperature.

## 2.19 Regeneration, Heatless

Reactivation of desiccant without heat. It is usually done with dry air flow.

## 2.20 Compressed Air Dryer Components

### 2.20.1 Afterfilter

A filter which follows the compressed air dryer and usually for the protection of downstream equipment from desiccant dust.

### 2.20.2 Automatic Drain

A device which automatically discharges condensate from the moisture separator.

### 2.20.3 Condensing Unit

A specific refrigerant machine combination for a given refrigerant consisting of one or more power driven compressors, air or water-cooled condensers, etc.

### 2.20.4 Desiccant

Material that tends to remove moisture from compressed air.

### 2.20.5 Desiccant, Absorbent (Deliquescent)

A desiccant that dissolves into the moisture it removes from the compressed air and is slowly consumed in the process.

### 2.20.6 Desiccant, Adsorbent

A solid desiccant which is capable of removing moisture from compressed air by adherence of moisture to its surface.

### 2.20.7 Evaporator

The heat exchanger where the refrigerant absorbs heat.

### 2.20.8 Hot Gas By-Pass Valve

A modulating valve which bypasses hot refrigerant gas from the high pressure to the low pressure side of the system in order to reduce refrigeration capacity commensurate with reduction in load and to control evaporator pressure (temperature).

### 2.20.9 Moisture Separator

A device which removes liquids from an air stream.

### 2.20.10 Precooler Reheater (Air-to-Air Heat Exchanger)

A heat exchanger which lowers the temperature of the inlet air and raises the temperature of the exiting air.

# ANSI/B93.45

### 2.20.11 Prefilter

A filter which precedes the compressed air dryer and usually for the protection of the desiccant or heat transfer surfaces.

### 2.20.12 Refrigerant

A substance which produces a cooling effect by its absorption of heat while evaporating or vaporizing.

### 2.20.13 Refrigeration Compressor

The part of refrigeration system which takes refrigerant at low pressure and compresses it to a smaller volume at a higher pressure.

The following terms have been extracted from National Electric Manufacturers Association (NEMA) Standards:

### 2.21 Absorption

Absorption is the penetration of a substance into the body of another.

### 2.22 Adsorption

Adsorption is the adherence of gases, liquids or dissolved substances on the surface of solids.

### 2.23 Adiabatic Drying

Adiabatic drying is dehumidification without gain or loss of total heat.

### 2.24 Compressed Air

Air compressed is air at a pressure greater than one standard atmosphere.

### 2.25 Humidity

#### 2.25.1 
Absolute Humidity is the mass of water vapor present in unit volume, usually measured as grams per cubic metre or grains per cubic foot.

#### 2.25.2 
Relative Humidity is the ratio of the quantity of water vapor present to the quantity which would saturate at the existing temperature. It is also the ratio of the partial pressure of water vapor present to the partial pressure of saturated water vapor at the same temperatures.

### 2.26 Standard Cubic Feet Per Minute

The flow rate which, for convenience, has been referred to standard conditions. (See: Air, Standard 2.2)

## 3 UNITS OF MEASUREMENT

The following units of measurement have been extracted from International Standard ISO 1000. (See ISO 1000 for complete reference).

| Item No. in ISO/R31 | Quantity | SI Unit | Multiples of the SI unit | Units Outside the SI* |
|---|---|---|---|---|
| (1) | (2) | (3) | (4) | (5) |
| 1- 5.1 | volume | $m^3$ | $dm^3$ | L (litre) |
| 1- 6.1 | time | s (second) | | h (hour) min (minute) |
| 1-10.1 | velocity | m/s | | |
| 2- 3.1 | frequency | Hz (Hertz) | | |
| 3- 1.1 | mass | kg (kilogram) | | |
| 3- 2.1 | density (mass density) | $kg/m^3$ | $Mg/m^3$ or $kg/dm^3$ or $g/cm^3$ | $t/m^3$ or or kg/l |
| 3-11.1 | pressure | Pa (pascal) | kPa | bar = $10^5$ Pa |

* Recognized by CIPM as having to be retained.

ANSI/B93.38-1976 (R1981)

## AN INDUSTRY STANDARD FOR FLUID POWER

**This Standard is now under review for possible revision see TSP**

American National Standard

Method of Diagramming for

Moving Parts Fluid Controls

(NFPA/T3.28.9-1973)

Approved as an ANSI Standard
4 March 1976

published by
# NATIONAL FLUID POWER ASSOCIATION, INC.

3333 N. Mayfair Road / Milwaukee, WI 53222 / 414-778-3344 / TLX 26898

# ANSI/B93.38

## FOREWORD

This Foreword is not part of NFPA Recommended Standard Method of Diagramming for Moving Parts Fluid Controls, NFPA/T3.28.9-1973.

Upon the request of industry, NFPA called a General Conference on 20 July 1967 to explore the need for a standard method of diagramming for control circuits using moving parts fluid controls (valves). It was agreed that the need did exist and that NFPA should sponsor this activity.

Early in its work, the Project Group decided to offer two methods of diagramming moving parts fluid controls — a detached method and an attached method.

Wilfred Aslan was elected Chairman of the Project Group assigned to this task. Meetings were held on a regular basis with the first draft prepared through Section 5 on 19 June 1968. Additions to Draft No. 1 were presented at meetings on 21 January 1969 and 10 June 1969. At the June meeting, the Introduction, Scope and Purpose were agreed upon.

At a July meeting, Draft No. 2 dated 20 July 1969 was presented. Draft No. 3 was prepared on 21 November 1969.

Charles Bert became the new Project Chairman in 1970. On 6 January 1971 the Project Group reached consensus on the draft. The General Review Draft was prepared on 4 November 1971. The review period closed on 2 January 1972 and all comments were resolved by 8 March 1972.

In 1972 all of the NFPA Fluidics and Fluid Logic standards activities were brought together in a newly formed Fluid Logic Component Section. As a result, this project which was originally assigned Project Number T3.21.2 was given the new Project Number T3.28.9.

The Ballot Draft was prepared on 8 November 1972. This document was approved by the NFPA Technical Board on 23 May 1973. At its 19 July 1973 meeting the Board of Directors approved this document as NFPA Recommended Standard NFPA/T3.28.9-1973.

Members of the NFPA Project Group responsible for the development of this standard included:

**Charles Bert**
Project Chairman
1970 to 1973
Hill Rockford Co.

**Wilfred Aslan**
Project Chairman
1967 to 1970
I-T-E Imperial Corp.*

**Ralph Culbertson**
Section Chairman (T3.28)
1973 to present
Dynamco Inc.

**Clifford Brake**
Section Chairman (T3.21)
1970 to 1975
Scovill Fluid Power Div.

**John R. Luecke**
Director of National Technical Services
National Fluid Power Association

**R. Berg**
Rexnord, Inc.

**G. Doig**
Doig associates

**C. Grigsby**
Scovill Fluid Power Div.

**P. Huff**
Dynamco Inc.

**R. Kiehl**
Parker Hannifin Corp.

**G. McEwen**
Numatics, Inc.

**S. North**
Rexnord, Inc.

**J. Stodola**
Numatics, Inc.

**F. Yeaple**
Product Engineering

*Retired

On 16 August 1973 the NFPA Standard was submitted to ANSI Standards Committee B93 for promulgation as an ANSI Standard. Balloting concluded on 26 October 1973, and the standard was approved by the B93 Committee on 11 September 1974. Approval by the ANSI Board of Standards Review was granted on 4 March 1976.

On 11 September 1974 Standards Committee B93 was composed of the following:

**Otto J. Maha**
Chairman

**Melvin E. Long**
Vice Chairman

**James I. Morgan**
Co-Secretary

**James Tishkowski**
Co-Secreataty

**AMERICAN SOCIETY OF AGRICULTURAL ENGINEERS**
E. H. Fletcher

**AMERICAN SOCIETY OF LUBRICATION ENGINEERS**
M. M. Gurgo

**AMERICAN SOCIETY OF MECHANICAL ENGINEERS**
R. Hildebrandt
T. R. Curran (Alternate)

**AMERICAN SOCIETY FOR TESTING AND MATERIALS**
J. D. Lykins

**CONSTRUCTION INDUSTRY MANUFACTURERS ASSOCIATION**
G. Stewart
H. T. Larmore (Alternate)

**FLUID CONTROLS INSTITUTE**
H. H. Kaemmer
E. A. Bianchi (Alternate)

**FLUID POWER SOCIETY**
T. Goldoftas
R. Henke
M. E. Long
W. R. Smith
L. G. Soderholm
F. Yeaple

**FLUID SEALING ASSOCIATION**
R. Prachel
J. Scannell (Alternate)

**INDUSTRIAL TRUCK ASSOCIATION**
C. D. Gibson

ANSI/B93.38

**INSTRUMENT SOCIETY OF AMERICA**
A. I. Kutz

**JOINT INDUSTRY COUNCIL**
R. Muhl

**MATERIAL HANDLING INSTITUTE**
J. C. McPherson
W. Chichester (Alternate)

**MOTOR VEHICLE MANUFACTURERS ASSOCIATION**
J. Phillipson

**NATIONAL FLUID POWER ASSOCIATION**
O. J. Maha
J. L. Fisher, Jr.
W. R. Forster
W. J. Kudlaty
Z. J. Lansky
J. J. Pippenger

**NATIONAL MACHINE TOOL BUILDERS ASSOCIATION**
E. Loeffler

**POWER CRANE AND SHOVEL ASSOCIATION**
W. M. Shook

**RUBBER MANUFACTURERS ASSOCIATION**
W. J. Atwell
N. J. Cyphers (Alternate)
E. J. McCarthy (Alternate)

**SOCIETY OF AUTOMOTIVE ENGINEERS**
W. A. Hertel
E. L. Falendysz
H. Schultz
D. B. Shore
W. L. Snyder
D. Prevallet
R. W. White

**SOCIETY OF MANUFACTURING ENGINEERS**
J. Wood

**U.S. COAST GUARD**
LCDR G. A. Casimir

**U.S. DEPARTMENT OF DEFENSE**
C. A. Nazian
H. Y. Smith
P. Hopler (Alternate)

**INDIVIDUAL MEMBERS**
Prof. E. C. Fitch
J. Johnson

**REFERENCES**

1. American National Standard Drafting Practices for Fluid Power Diagrams, ANSI/Y14.17-1974.

2. International Standard Graphical Symbols for Hydraulic and Pneumatic Equipment and Accessories for Fluid Power Transmission, ISO/R 1219-1970. Agrees with ANSI/Y32.10-1974.

3. American National Standard Glossary of Terms for Fluid Power, ANSI/B93.2-1971, and Supplements thereto. (ISO/TC 131/SC 1 [USA-2] 3).

**BACKGROUND REFERENCES**

1. JIC Electrical Standards for Mass Production Equipment - EMP - 1-1967

2. JIC Electrical Standards for General Purpose Machine Tools - EGP - 1-1967.

3. General Motors Manufacturing Standards - Basic Electrical Standards for Industrial Equipment, June 1968.

4. General Motors Manufacturing Standards - Fluid Power Standards for Industrial Equipment, April 1967.

5. Numatics, Inc. - Bulleting 502B, The Numatrol Diagram.

6. Military Standard Graphic Symbols for Logic Diagrams, MIL-STD-806B.

7. American National Standard Graphic Symbols for Logic Diagrams, ANSI/Y32.14-1973.

ANSI/B93.38

# METHOD OF DIAGRAMMING FOR MOVING PARTS FLUID CONTROLS

## PART I - GENERAL

### INTRODUCTION

In fluid power systems power is transmitted and controlled thru a fluid (liquid or gas) under pressure within an enclosed circuit.

Moving parts fluid controls achieve control thru the use of devices having moving parts.

With the growth of controls technology, there has arisen within the technology of fluid power a major field of controls endeavor involving the use of air and other fluids as the operating media for logic controls systems.

1. SCOPE

    1.1  To include drafting practices for drawings which depict logic control circuits using moving parts fluid controls, including all power sources, all inputs, all logic functions and all outputs.

    1.2  To include control systems using positive pressure and negative pressure (vacuum).

    1.3  To provide diagrams for logic control circuits. (For power diagrams use Reference No. 1 and Reference No. 2.)

2. PURPOSE

    2.1  To develop a uniform means for diagramming moving parts fluid controls.

    2.2  To provide a communication means for industrial and educational purposes.

    2.3  To simplify design, fabrication, analysis and servicing of moving parts fluid controls.

    2.4  To promote international understanding and use of moving parts fluid control systems.

ANSI/B93.38

3. TERMS AND DEFINITIONS

For definitions of other terms used, see Reference No. 3.

3.1 Circuit, Logic Control. A circuit which gathers and processes information to signal power controls and interfaces.

3.2 Circuit, Power Control. A circuit which directs and regulates fluid power to working devices.

3.3 Control Point. The point where the logic control flow path terminates at the actuator input of a logic control, power control or interface.

3.4 Diagram, Attached Symbols. A diagram in which all functions and connections to component symbols are shown in the symbols.

3.5 Diagram, Detached Symbols. A diagram in which various functions and connections of component symbols are shown by separate symbols located in various places on the diagram.

3.6 Diagram, Detail Logic. A diagram which depicts all logic functions of a circuit including identification and description of logic components, ports and connecting flow paths.

3.7 Diagram, Ladder. A diagram in which inputs are located to the left in a vertical column and outputs in a right vertical column. Interconnecting horizontal flow paths and components give the diagram a ladder appearance.

3.8 Diagram, Logic Control. A diagram which depicts logic control of power controls and interfaces.

3.9 Diagram, Power Control. A diagram which depicts all powered devices and interfaces including identification and description of components and their effect on the system.

3.10 Flow Passage, Controlled. A flow passage whose ability to pass fluid can be changed by the influence of a signal.

3.11 Flow Path. A series of conductors and passages which convey fluid.

3.12 Fluid Memory, Off Return. Fluid memory which receives a momentary signal and produces a change of state which continues to exist after the initiating signal has disappeared, providing an input is present at the supply port of the device. Upon loss of the supply pressure, the device reverts to its initial state.

ANSI/B93.38

3.13 Fluid Memory, Retentive. Fluid memory which receives a momentary signal and produces a change of state which continues to exist after the initiating signal has disappeared regardless of the presence or absence of supply pressure to the device. The device returns to its original state only upon receipt of a second reset control signal.

3.14 Fluid Signal. Fluid pressure or flow which can be detected or sensed.

3.15 Fluid Signal, Maintained. A fluid signal which exists indefinitely until caused to disappear by a secondary control action.

3.16 Fluid Signal, Momentary. A fluid signal which exists briefly and then disappears.

3.17 Fluid Signal, Timed. A fluid signal which exists for a definite period of time and then disappears.

3.18 Logic Control Device. Any device employed in a logic control circuit.

3.19 Moving Parts Logic. The technology of achieving logic control by means of fluid devices having moving parts.

3.20 Power Control Device. Any device used in a power control circuit.

3.21 Pressure Switches. This device uses an air signal from the control system to actuate an electric switch, and thereby produces an electrical output. It may be non-adjustable (actuate at a fixed pressure level) or adjustable to actuate at an adjustable pressure level. When the air signal is applied to the device, the switch actuates. When the air signal is removed, the device resets.

3.22 Relay Valve. A logic device which receives control signals and changes flow conditions in one or more controlled flow passages.

3.23 Relay Valve, Free Floating. A relay valve wherein the internal element moves freely without restraint and normally utilizes bias pressure at one control point.

3.24 Relay Valve, One Shot. A relay valve wherein controlled flow passages immediately change conditions when a control point is pressurized by a maintained signal. After a period of time, the controlled flow passages return to their original conditions, even though the control point is pressurized. When the control signal is removed it resets for another operation.

3.25 Relay Valve, Time Delay. A relay valve which creates a time interval between the pressurizing of a control point and a change in the controlled flow passages.

3.26 Relay Valve, Time Delay after Exhausting a Control Point. A relay valve with one control point which receives a maintained signal and causes immediate actuation of controlled flow passages. When the control signal is removed, a time delay occurs before controlled flow passages are reset.

3.27 Relay Valve, Time Delay after Pressurizing a Control Point. A relay valve with one control point which receives a maintained signal and causes a time delay before the controlled flow passages are actuated. The device resets immediately upon exhausting the control point.

3.28 Relay Valve, Time Delay, Detented. A time delay relay valve having "A" and "B" control points arranged to accept and act on momentary signals. A momentary signal into the "A" control point starts actuation and a momentary signal into the "B" control point starts reset.

3.29 Relay Valve, Time Delay, Detented, Delayed Action. A detented time delay relay valve in which a momentary signal into the "A" control point creates a time delay before controlled flow passages are actuated. The device resets immediately upon receipt of a signal in the "B" control point.

3.30 Relay Valve, Time Delay, Detented, Delayed Reset. A detented time delay relay valve in which a momentary signal into the "A" control point produces immediate actuation of the controlled flow passages. A momentary signal to the "B" control point starts the reset action with a time interval before controlled flow passages reset.

3.31 Sensor. Any device which detects a condition in a system and produces a signal.

3.32 Valve, Power Control. A valve which controls fluid power operating working devices.

3.33 Valve, Rotary Selector. A valve which utilizes rotary actuation to connect the inlet to any one of a number of outlets.

3.34 Valves, Pressure Sensing. A device similar to an electrical pressure switch, in which a signal to be sensed enters a control point, and actuates a mechanism which, at the proper pressure level, causes one or more flow passages to change condition. Removal of the signal allows the pressure sensing valve to reset.

4. SYMBOL RULES

   4.1   Use symbols to show connections, flow passages and functions of the device represented.

   NOTE. - Symbols are digital, and do not indicate conditions occuring during transition from one flow condition to another.

   4.2   Do not use symbols to indicate construction, or to indicate values, such as pressure, flow rate or other component settings.

   4.3   Do not use symbols to indicate the physical location of ports, direction of shifting of spools, poppets or diaphragms, or the position of actuators on actual components.

   4.4   Rotate or reverse symbols without altering their meaning.

   4.5   Avoid using symbols with words or abbreviations.

   NOTE. - Symbols capable of crossing language barriers are presented herein.

   4.6   Draw symbols so that flow is assumed to proceed from left to right unless clearly indicated otherwise.

   NOTE. - Exhaust flow is assumed to proceed from right to left unless clearly indicated otherwise.

   4.7   Use a combination of simple symbols to make composite symbols.

   4.8   Use composite symbols to **represent complex components**.

   4.9   Use an enclosure to show functional assemblies too complex to be shown by composite symbols.

   4.10  Show each symbol with all external ports to be connected for the component to perform its function properly.

   NOTE. - Do not show exhaust ports, vent ports and unused flow passages which require no external connections.

   4.11  Draw each symbol to show the conditions existing when the machine is in the condition specified by the "Initial Conditions" statement in the Sequence of Operations.

      4.11.1   Use the Initial Conditions to show the conditions at the start of a normal automatic cycle, with the power and the control air ON.

4.11.2 Where special conditions or circuitry techniques warrant, show the Initial Conditions with the control air and power OFF the machine, and all component symbols shown accordingly.

NOTES. -

1. In cases where the particular symbols is incapable of showing the starting condition, the Initial Conditions are of no concern.

2. The basic standards symbols are designed without alphabetical or numerical designations and are thus suitable candidates for international standard symbols. Where letters and/or numbers do appear in this document, it is done for illustrative purposes only. Such letters and/or numbers should in no way be construed as being part of the basic standards symbol.

# 5. LINE TECHNIQUES

5.1 Keep line widths approximately equal.

NOTE. - Line width does not alter the meaning of the symbols.

5.2 Solid Line ——————————

Supply line to control system and all lines in the logic control system except output lines.

5.3 Dashed Lines ———  ———  ———

Indicate output lines from logic control system to power control points and interfaces.

5.4 Dotted Line — — — — — — — —

Indicates mechanical connection between devices.

5.5 Center Line ——— — ——— — ——— — ———

Indicates outline of an enclosure.

NOTE. - May also be used around a composite symbol.

5.6    Lines Crossing

   NOTE. - Intersection is not necessarily at 90° angle.

5.7    Lines Joining

5.8    Lines, Flexible (Continually flexing)

5.9    Sensing Lines

   Drawn the same as the line to which they connect.

5.10   Pivot Point

   Shown in a mechanical connection.

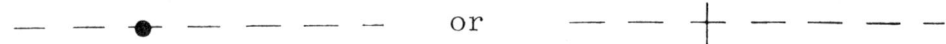

5.11   Line Identification

   Identify each line representing a conductor by a number drawn above or to the right of the line.

   5.11.1    Use consecutive numbering.

   5.11.2    Use each line identification only once in any logic control system.

   5.11.3    Use the same identification at all terminals and tie points for each line (conductor).

   5.11.4    Use the same identification for all lines (conductors) connected to the same terminal or tie point.

ANSI/B93.38

# PART II - SYMBOLS - INPUT DEVICES

## 6. SYMBOLS - MANUAL CONTROLS

### 6.1 Push Buttons

6.1.1  Terminals (ports) on a device that has an implied exhausting function which is not shown.

NOTE. - Fluid flow from left to right.

6.1.2  Terminals (ports) on a device which has no implied exhausting function. All functions are shown.

NOTE. - Filled-in circles do not denote hydraulic fluid as expressed in Section 3.5 of ANSI/Y32.10-1974. The filled-in circles are used here to indicate a device with no implied exhausting function.

6.1.3  Operator and bridge, flush extended, half guarded or guarded.

6.1.4  Operator and bridge, mushroom head or palm button.

6.1.5  Spring return action (shown by the presence of the spring symbol).

6.1.6  Detented action (shown by the absence of the spring symbol). Solid line shows start condition, dotted line shows actuated condition.

6.1.7  Detented push button, air return.

6.1.8  Latching button, spring return with manually operated mechanical latch.

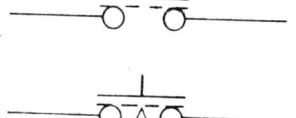

6.1.8.1   Depress and hold button down.

6.1.8.2   Engage latch.

6.1.8.3   Release button. Latch will hold button depressed.

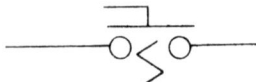

6.1.8.4   To release, pull out latch. Spring will return button to OUT.

NOTE. - Latch usually has provision for padlock, so it can be locked in latched position.

6.1.9   Key lock button with lock built into button. Additional note explains any details of key and lock action.

K

6.1.9.1   Spring return, key lock push button.

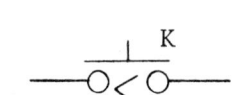

6.1.10   Two-button station, mechanically linked so that when one button is depressed, the second is OUT. Detented in both positions.

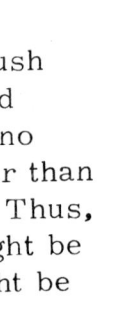

6.1.11   Identify each push button with a push button number, starting with 1 and running consecutively. There is no significance to the numbers, other than to identify the particular button. Thus, in one circuit the start button might be 1, while in another circuit it might be 5.

6.1.11.1   For Spring Return Buttons, put above the symbol the push button number and the function which the button performs in the cycle.

PB 1
Cycle Start

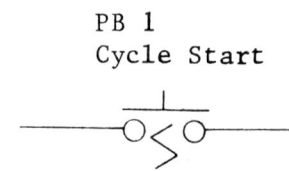

6.1.11.2   For Detented Buttons, put above the symbol the push button number and the function which the button performs in the cycle. At the side put notes telling what happens when the button is pushed in, and when it is pulled out.

PB 6
Emerg. Stop

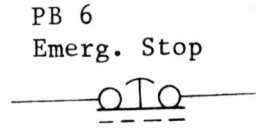

Pull-Rese
Push-Sto

6.1.11.3 For Port Numbers or letters, put under the symbol in small circles, the ports on the component which must be connected to provide the function shown.

PB 5
Clamp

6.1.12 See Table 1 for typical push button symbols.

TABLE 1. - Typical push button symbols.

| Action | Symbol | Alternate Symbol |
|---|---|---|
| 2-way, spring return, normally passing | (symbol) | |
| 2-way, spring return, normally not passing | (symbol) | |
| 3-way, spring return, normally passing | (symbol) | |
| 3-way, spring return, normally not passing | (symbol) | |
| 3-way, detented, passing at start of cycle, push to break | (symbol) | |
| 3-way, detented, not passing at start of cycle, push to make | (symbol) | |
| 3-way, 3 port used as a diverter, spring return, no exhaust in this function, alternate outlet is blocked to reverse flow. | (symbol) | (symbol) |
| 3-way, 3 port used as a selector, spring return, alternate supply is blocked | (symbol) | (symbol) |
| 4-way, 4 port 2-position, spring return, alternate output is exhausted | (symbol) | (symbol) |
| Dual path 4-way, 5 port spring return, alternate output is exhausted. | (symbol) | (symbol) |

NOTE. - Fluid flow is from left to right. Filled-in circles indicate a device which has no implied exhausting function. See clause 6.1.2.

ANSI/B93.38

6.2    Selectors and Toggles - Two-Position

    6.2.1    Operator and bridge. Show start position by solid arrow. Show actuated position by dotted arrow.

    6.2.2    Spring return action. (Shown by the presence of the spring symbol).

    6.2.3    Detented action. Show start condition by solid lines. Show actuated conditions by dotted lines.

    6.2.4    Typical two-position selector. Indicate spring return to the left position.

    6.2.5    Typical two-position selector. Indicate spring return to the right position.

    6.2.6    Typical two-position selector. Indicate detented in both positions; indicate left position as starting position.

    6.2.7    Typical two-position selector. Indicate detented in both positions; indicate right position as starting position.

    6.2.8    Identification. Identify each selector with an SV number, starting with SV1 and running consecutively. There is no significance to the numbers other than to identify the particular selector.

        6.2.8.1    Above the symbol, put the SV number and the function which the selector performs in the cycle. At each side of the arrows put the action which occurs when the operator is in that position.

SV 1
Auto-Manual Selector
Auto     Manual

        6.2.8.2    At the bottom of each port, put the port identification which provides the function shown.

ANSI/B93.38

6.3 Foot Operated - Two-Position

    6.3.1 Treadle operator and bridge, hooded or open.

    6.3.2 Single treadle spring return action (shown by the presence of the spring symbol).

    NOTE. - The spring return symbol is necessary to prevent confusion with detented operators. Such a distinction has not been necessary in electrical diagrams because there was not a push button-detented switch until quite recent times.

        6.3.2.1 Normally not passing

        6.3.2.2 Normally passing

        6.3.2.3 Typical four-way action, common inlet

    6.3.3 Double treadle, detented in both positions.

        6.3.3.1 Not passing at start of cycle

        6.3.3.2 Passing at start of cycle

        6.3.3.3 Typical four-way action, common inlet

6.3.4 Identify each foot operated valve with an FTV number, starting with FTV1 and running consecutively. There is no significance to the number other then to identify the particular foot operated valve.

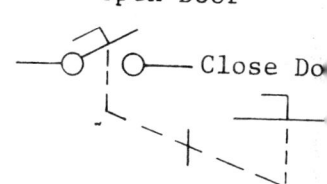

6.3.4.1 For Single Treadle Spring Return, put above the symbol the FTV number and the function which the foot operated valve performs in the cycle.

6.3.4.2 For Double Treadle Detented, put the FTV number above the symbol. Next to each treadle put a note describing what happens when that treadle takes command.

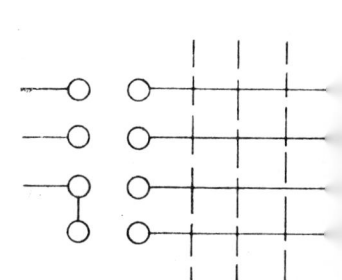

6.4 Selectors - Three-Position and Multi-Position

6.4.1 Manually operated (knob, lever, or toggle)

6.4.2 Foot operated

6.4.3 Mechanically operated

6.4.4 Flow passage designations. Represent each flow passage by a horizontal line containing a pair of terminal (port) symbols, followed by a position grid.

6.4.5 Inlet designations. Indicate a separate inlet passage with a line entering the symbol from the left. An inlet may serve only one flow passage, or it may serve several flow passages. A single line entering the symbol and connecting with several sets of terminal symbols indicates a common connection within the device.

6.4.6 Exhaust designations. Indicate special flow conditions of an exhaust path, if necessary, by a line exiting from the left side of the symbol and terminating in a dashed line. This shows that the particular connection is an exhaust to atmosphere.

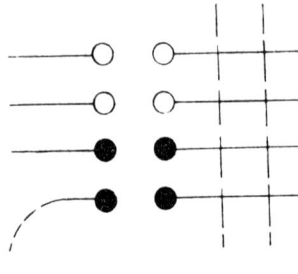

6.4.7 Actuator designations. Show the method of actuating the device by an actuator symbol placed in the top set of terminals. The position of the actuator bridge is not significant, nor is the absence of a spring symbol. The flow condition at the start of the cycle and the method of returning the actuator to the start condition are shown by the position grid.

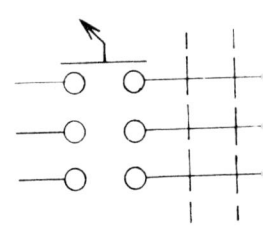

6.4.8 Actuator symbols.

   6.4.8.1 Manual

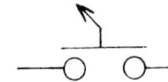

   6.4.8.2 Foot (single or double treadle)

   6.4.8.3 Mechanical

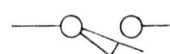

6.4.9 Position grid. Illustrate the various flow passages offered by the device by a position grid located to the right of the flow path terminals.

6.4.10 Position designations. Represent each position of the operator by a vertical dashed line, with a number above the line.

   6.4.10.1 Use the position number 0 (zero) as the starting position and number all other positions consecutively. All numbers except zero have no significance except to identify the particular position.

6.4.11 Flow passage designations. Indicate by the presence or absence of an "X" at the intersection of the horizontal line and the vertical dashed line the condition of each flow passage for each position of the operator.

"X" means that the flow passage is passing when the operator is in that position.

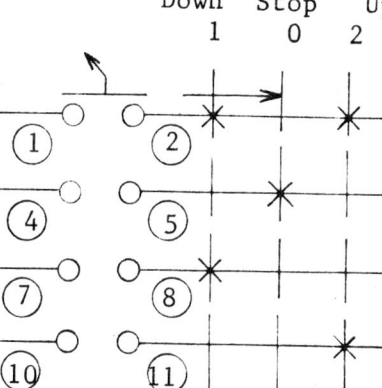

SV 6
Slide Jog

6.4.12 Type of return mechanism. Show spring return to the start position by a horizontal arrow pointing from the position shown back to the start position. (1 to zero)

6.4.12.1 Show detented action by the lack of a return arrow (2 to zero)

6.4.13 Identification. Identify each selector with a letter combination followed by a number.

The letter combination designates the type of actuator, and is the same letter combination used to identify other devices with this same type of actuation.

6.4.13.1 Manual - SV plus a number

6.4.13.2 Foot - FTV plus a number

6.4.13.3 Mechanical - LV plus a number

6.4.13.4 The numbers have no significance except to identify the particular device. In assigning these numbers, simply number consecutively with other devices using the same type of actuation.

6.4.13.5 Above each symbol put the device number and a note telling the function it performs in the cycle.

6.4.13.6   Above each position of the operator put a note telling what action occurs when the selector is in that position.

6.4.13.7   Under each inlet and outlet terminal symbol put the port identification which provides the function shown.

6.4.14   Rotary selector valves. Use either of the two approved symbols.

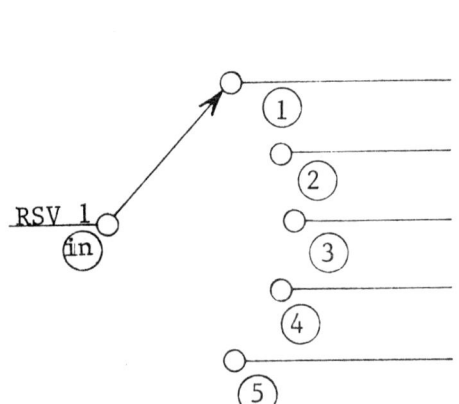
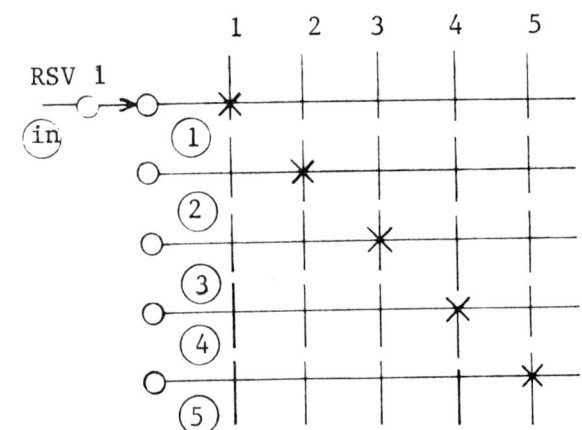

6.4.14.1   Identify each rotary selector valve with an RSV number, starting with RSV1 and numbering consecutively. The numbers have no significance other than to identify the device.

6.4.14.2   Under each inlet and outlet terminal put the port identification which provides the function shown.

# 7. SYMBOLS FOR POSITION SENSORS

## 7.1 Limit Valves

### 7.1.1
Mechanical actuator and bridge, spring return.

### 7.1.2
Mechanical actuator and bridge, detented. Indicate by the double ramp cam that this bridge is detented and is cammed in both directions.

### 7.1.3
Actuator and bridge, cable actuated, manual reset.

### 7.1.4
Condition of actuation at start of cycle. Indicate by the position of the actuator and bridge whether or not the actuator is cammed at the start of the cycle.

#### 7.1.4.1
Actuator NOT actuated at start of cycle. Indicate actuator DOWN as far as it will go.

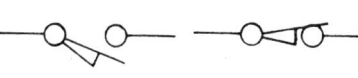

#### 7.1.4.2
Actuator ACTUATED at start of cycle. Indicate actuator UP as far as it will go.

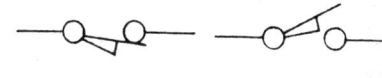

### 7.1.5
Identification. Identify each limit valve by an LV number, starting with LV1 and numbering consecutively. The number has no significance other than to identify the particular device.

#### 7.1.5.1
Place the LV number and a note telling what machine motion actuates the limit, and under what circumstances, above the limit valve symbol.

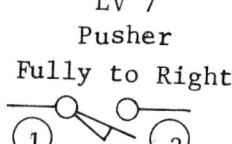

#### 7.1.5.2
If the limit is actuated on one stroke of the motion, and idles (over-rides) on the return stroke, put a brief note explaining this.

#### 7.1.5.3
Under the symbol, in small circles, put the port identification on the component which must be connected to provide the function shown.

ANSI/B93.38

7.1.6   For typical limit valve symbols see Table 2.

TABLE 2. - Typical limit valve symbols.

| Action | Symbol |
|---|---|
| 3-way, normally not passing, not actuated at start of cycle | |
| 3-way, normally not passing, actuated (held passing) at start of cycle | |
| 3-way, normally passing, not actuated at start of cycle | |
| 3-way, normally passing, actuated (held not passing) at start of cycle | |
| 4-way, spring return, not actuated at start of cycle | |
| 4-way, spring return, actuated at start of cycle | |
| Dual 3-way (2-path), spring return, not actuated at start of cycle | |
| 3-way, detented, not passing at start of cycle | |
| 3-way, detented, passing at start of cycle | |
| 3-way, 3 port used as a diverter, spring return, no exhaust in this function, alternate outlet is blocked to reverse flow | |
| 3-way, 3 port used as selector, spring return, alternate supply is blocked. | |

## 7.2 Float-Actuated Valves

**7.2.1** Normally not passing. Goes passing on rising liquid level. Resets on falling level.

**7.2.2** Normally passing. Goes not passing on rising liquid level. Resets on falling level.

**7.2.3** Identification. Identify each float operated valve by an FV number, starting with FV1, and numbering consecutively. There is no significance to the number other than to identify the particular valve. Above the symbol put a note telling what actuates the valve, and under what conditions.

FV 5
#1 Tank   500 Gallons or more

## 7.3 Air Jets

**7.3.1** Air jet. Symbol does not indicate whether or not jet is blowing at start of cycle.

**7.3.2** Air jet not blocked by target at start of cycle. The target is the device which blocks the jet, and whose position the jet is intended to sense.

**7.3.3** Air jet blocked by target at start of cycle.

**7.3.4** Identification. Identify each jet by the word "Jet" followed by a number, starting with Jet 1 and numbering consecutively. The number has no significance other than to identify the particular jet.

Jet 16
Slide Full Fwd.

**7.3.4.1** Above each jet symbol, put the jet number and a note telling what machine motion (target) blocks the jet, and under what circumstances.

**7.3.4.2** In the case of multiple jets, all acting together to actuate one receiving device, all the jets in the group may be given the same jet number, followed by dash numbers, Jet 6-1, Jet 6-2, etc.

Jet 6-1
#1 Clamp clamped

Jet 6-2
#2 Clamp clamped

Jet 6-3
#3 Clamp clamped

# 8. SYMBOLS FOR PRESSURE SENSORS

## 8.1 Pressure Sensing Valves

Use the following symbols:

### 8.1.1
Controlled flow passage which goes passing on rising pressure, and goes not passing on falling pressure. Not pressurized at start of cycle.

### 8.1.2
Controlled flow passage which goes passing on rising pressure and goes not passing on falling pressure. Pressurized at start of cycle.

### 8.1.3
Controlled flow passage which goes not passing on rising pressure and passing on falling pressure. Not pressurized at start of cycle.

### 8.1.4
Controlled flow passage which goes not passing on rising pressure and passing on falling pressure. Pressurized at start of cycle.

### 8.1.5
Pressure sensing valve having more than one controlled flow passage depicted by a composite symbol. (Example shows a 4-way pressure sensing valve.)

NOTE. - Such valves may be used to sense either pressures above atmospheric, or vacuums. In either case, the symbols are the same.

### 8.1.6
Identification. Identify each pressure sensing valve with a PSV number, starting with PSV1 and numbering consecutively. The numbers have no significance other than to identify the particular pressure sensing valve. Above each symbol, put the PSV number and a note telling what pressure on the machine actuates the pressure sensing valve, and under what circumstances.

PSV 4

Clamp pressure at or above 60 psig

## 9. SYMBOLS FOR TEMPERATURE SENSORS

### 9.1 Normally not Passing

Goes passing on increasing temperature. Resets on decreasing temperature.

### 9.2 Normally Passing

Goes not passing on increasing temperature. Resets on decreasing temperature.

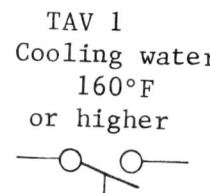

### 9.3 Identification

9.3.1 Identify each temperature actuated valve by a TAV number, starting with TAV1 and numbering consecutively. The numbers have no significance other than to identify the particular valve.

TAV 1
Cooling water
160°F
or higher

9.3.2 Put a note telling what temperature actuates the valve and under what circumstances.

## 10. SYMBOLS FOR FLOW SENSORS

### 10.1 Normally Not Passing

Goes not passing when flow reaches required conditions. Resets when flow drops below required conditions.

### 10.2 Normally Passing

Goes passing when flow is present in required quantity. Resets when flow stops.

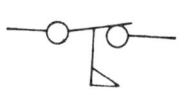

### 10.3 Identification

Identify each flow operated valve with an FLV number, starting with FLV1 and numbering consecutively. The numbers have no significance other than to identify the particular flow operated valve. Put a note telling what flow conditions actuate the valve and under what circumstances.

FLV 3

Cooling water
flowing

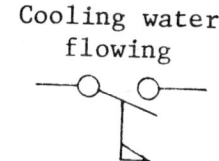

# PART III - POWER CONTROL VALVES AND INTERFACES

## 11. POWER CONTROL VALVES

### 11.1 General Principles

11.1.1    Simple and easy to draw.

11.1.2    Compact as possible.

11.1.3    Coded, if possible, by shape to indicate the type of valving action represented, and the type of control signals required to actuate the valve successfully.

NOTE. - This is particularly important in view of the international connotations of our work.

11.1.4    Using basic symbol shapes per Reference No. 2

11.1.5    Capable of representing relay valves of similar action, with suitable changes in the identification.

### 11.2 Identification of Power Control Valves

11.2.1    Supply a unique identification for each power control valve and each control point on each power control valve.

11.2.2    Use the same identification where it appears on all drawings of electrical control diagrams, pneumatic control diagrams, pneumatic fluid power diagrams or hydraulic fluid power diagrams.

11.2.3    Use the letter "H" as the basic designation for hydraulic power control.

11.2.4    Use the letter "P" as the basic designation for pneumatic power control.

11.2.5    Number each valve in clause 11.2.3 and 11.2.4 consecutively starting with the number 1. The number has no significance other than to identify the particular valve.

EXAMPLE. - H1, H2, H3, etc. for hydraulic valves and P1, P2, P3, etc. for pneumatic valves.

ANSI/B93.38

11.2.6 Further identify control points on power control valves as follows:

11.2.6.1 Identify the control point on a valve having only one control point solely by the identification of the valve.

EXAMPLE. - The control point on a single pilot spring return pneumatic valve P3 is identified as control point P3.

11.2.6.2 Identify by the number of the valve and the letters "A" and "B" the control points on a power control valve which has two control points, one acting to CHANGE the starting condition of the valve, and the other acting to RESTORE the starting condition of the valve.

11.2.6.2.1 "A" actuator indicates the control point which, when it takes command of the valve, causes the machine element to move AWAY from its starting position.

11.2.6.2.2 "B" actuator indicates the control point which, when it takes command of the valve, causes the machine element to move BACK to its starting position.

NOTE. - The complete identification of the control point is the identification of the valve followed by the "A" or "B" such as H1A, H1B, etc. for hydraulic valves, P1A, P1B, P2A, P2B, etc. for pneumatic valves.

11.2.7 Use the above noted identification for multiple control points which perform identical functions on a power control valve. In addition, identify by dash numbers added after the usual control point identification, starting with 1, and numbering consecutively. The dash numbers have no significance other than to identify the particular control point.

EXAMPLE. - A single pilot spring return valve with multiple control points serving the "A" actuator, any one control point capable of actuating the valve. If the valve identifier is P4, the control points would be identified as P4-1, P4-2, P4-3, etc.

11.2.8 In the case of multiple control points on the "A" actuator of a double pilot detented power control valve, the valve identifier being P6, identify the control points as P6A-1, P6A-2, P6A-3, etc.

NOTE. - Clause 11.2.7 anticipates that in the future, developments in pneumatically actuated power control valving will build multiple actuators into the valve. Present standards make no provision for such future developments.

11.2.9 Arrange the power control points in the circuit diagram in a vertical column to the right of all relay valve control points, and use dashed lines leading from the control system to the power control points.

11.2.10 Place a note to the right of each power valve control point telling what machine motion takes place when that control point takes command of its power control valve.

11.2.11 In the case of a control point on a power control valve having control points on both "A" and "B" actuators, place a suitable cross-reference notation on the drawing adjacent to each control point, giving the location on the drawing of all the opposing control points.

11.2.11.1 Put the notation per clause 11.2.11 in the form of a set of brackets to the right of the control point symbol, under the machine motion note, the brackets containing the guide line numbers of the circuit lines containing the opposing control points.

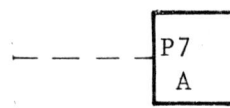

11.3 Symbols for Control Points on Power Control Valves

11.3.1 Control point on a valve having one actuator and spring return or automatic return, the automatic return accomplished entirely within the valve assembly, and requiring no connections from the control system to make it operational.

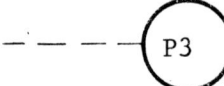

When this control point is pressurized, it takes command of the valve and keeps command as long as it is pressurized. When it is exhausted, the spring (or automatic return) returns the valve to the starting position.

11.3.2  Control points on a double control point detented (or equivalent) power control valve.

When either control point is exhausted, a momentary fluid signal into the opposite control point will cause that control point to take command and shift the valve action. When the fluid signal is exhausted, the detent holds the valve action in the new position until the second control point will cause it to take command, and shift the valve action back to the start position.

If control pressure is held in either control point, a fluid signal of equal pressure into the opposing control point will not shift the valve until the original control point is exhausted.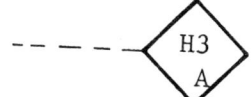

11.3.3  Control points on a double control point 3-position spring centered (or power centered) power control valve.

When both control points are exhausted or at the same pressure, the springs center the valve.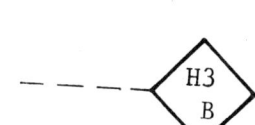

When either control point is pressurized and the opposing control point exhausted, the control point which is pressurized takes command of the valve and shifts it towards the exhausted control point.

When the opposing control point is pressurized, or the pressurized control point is exhausted again, the springs again center the valve action.

11.3.4  Control points on a double control point spring offset power control valve.

Control point on end opposite the spring.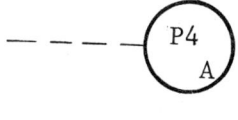

Control point on the spring end.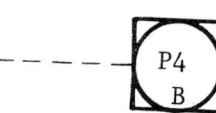

-409-

With the spring end control point exhausted, the "A" control point acts like the control point on a single control point spring return valve.

With pressure in the spring end ("B") control point, an equal pressure in the "A" control point will not shift the valve until the "B" control point is exhausted.

If the "B" control point is exhausted, a fluid signal into the "A" control point will shift the valve and it will stay shifted as long as the fluid signal into "A" is maintained.

However, if, while the "A" control point is pressurized and in command of the valve, an equal pressure is introduced into the "B" control point, the two fluid signals will cancel each other, and the spring will return the valve to the starting position EVEN THOUGH THE "A" CONTROL POINT IS STILL PRESSURIZED.

Finally, if the "A" control point is pressurized by a maintained fluid signal, it will take command of the valve and hold the valve shifted against the spring. The "B" control point can then be pressurized and exhausted, and will act like the actuator on a single control point spring return valve.

11.3.5 Cross-hatch all control points which are pressurized at the start of a normal automatic cycle. This cross-hatching is a major aid in understanding the operation of the circuit.

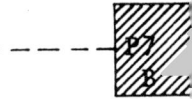

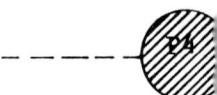

ANSI/B93.38

## 12. SYMBOLS FOR PRESSURE SWITCHES

### 12.1 Identification of Pressure Actuated Switches

Identify each switch by a PS number, starting with PS1 and numbering consecutively. The number has no significance other than to identify the particular switch.

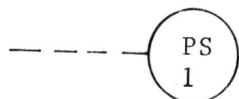

### 12.2 Location on the Diagram

#### 12.2.1
Arrange symbols for pressure actuated switches on the circuit diagram in a vertical column to the right of all relay valve control point symbols, and in line with the power control valve control symbols.

#### 12.2.2
Use dashed lines leading from the control system to the pressure actuated switch symbols.

# PART IV - ATTACHED METHOD OF DIAGRAMMING (LOGIC DEVICES)

13. GRAPHIC SYMBOLS, ATTACHED METHOD

   13.1   Use graphic symbols to express the logic functions of a control circuit.

   13.2   Establish symbols that are easy to draw and represent an obvious function.

   NOTE. - Since all logic functions can be described using standard valve symbols or combinations of valve symbols, use Reference No. 1 to describe each of the basic logic functions.

   13.3   Show the descriptive shaped logic symbols with a Truth Table indicating the output state (1 = pressurized, 0 = exhausted) for all possible combinations of input states.

   13.4   Use the AND output to assume the 1-state, if and only if, all of the inputs assume the 1-state. The AND is a passive device (See Table 3).

TABLE 3. - Truth table for AND output.

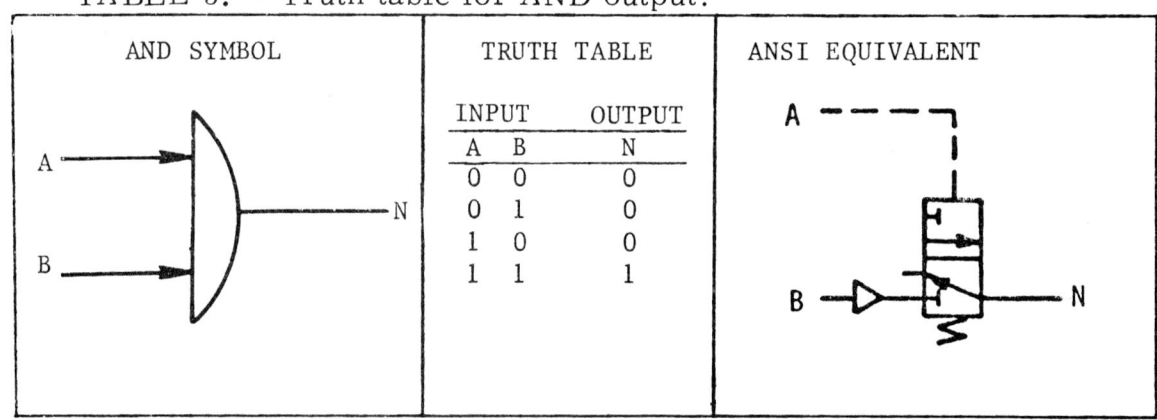

   13.5   Use the OR output to assume the 1-state, if any or all, of the inputs assume the 1-state. The OR is a passive device. (See Table 4)

# Controls Diagramming

ANSI/B93.38

TABLE 4. - Truth table for OR output.

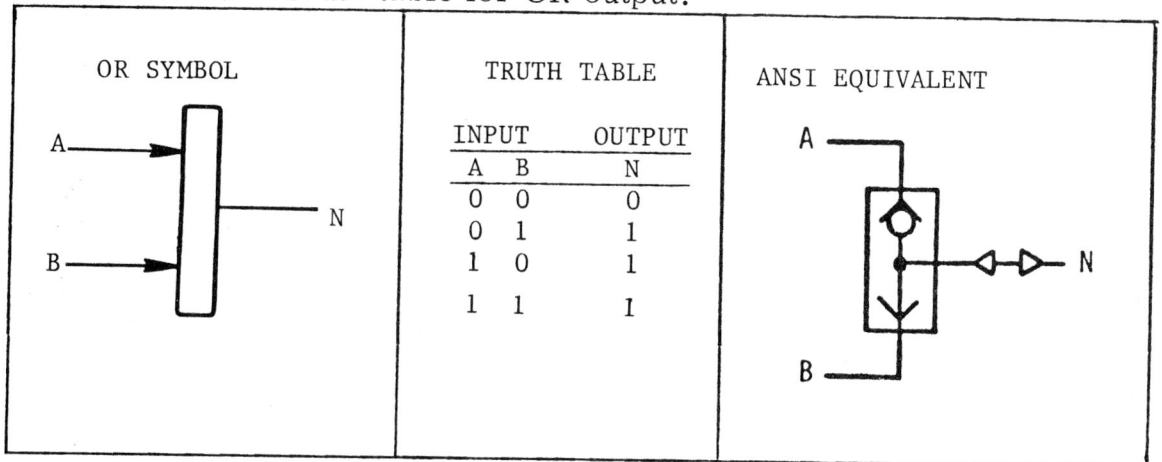

| OR SYMBOL | TRUTH TABLE | ANSI EQUIVALENT |
|---|---|---|
| | INPUT / OUTPUT | |
| | A B / N | |
| | 0 0 / 0 | |
| | 0 1 / 1 | |
| | 1 0 / 1 | |
| | 1 1 / 1 | |

13.6  Use the NOT output to assume the 1-state only when the input assumes the 0-state. The NOT is an active device. (See Table 5)

TABLE 5. - Truth table for NOT output.

| NOT SYMBOL | TRUTH TABLE | ANSI EQUIVALENT |
|---|---|---|
| | INPUT / OUTPUT | |
| | A / N | |
| | 0 / 1 | |
| | 1 / 0 | |

13.7  Use the Inhibitor output to assume the 1-state when input "B" is in the 1-state and the input "A" is in the 0-state. The inhibitor is a passive device. (See Table 6)

TABLE 6. - Truth table for inhibitor (nonimplication) output.

| INHIBITOR SYMBOL | TRUTH TABLE | ANSI EQUIVALENT |
|---|---|---|
| | INPUT / OUTPUT | |
| | A B / N | |
| | 0 0 / 0 | |
| | 0 1 / 1 | |
| | 1 0 / 0 | |
| | 1 1 / 0 | |

13.8   Use the Fluid Memory as an active device to store a single bit of information. It has two inputs, set (S) and reset (R) and may have one or two outputs. The storage of the single bit of information is maintained only as long as power is maintained. (See Table 7)

TABLE 7. - Truth table for Fluid Memory device.

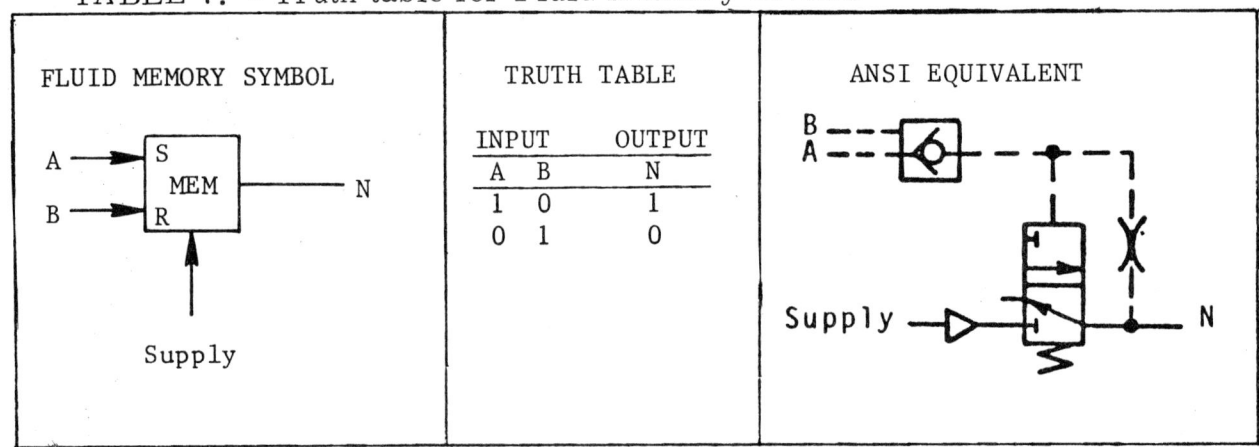

13.9   Use the Flip-Flop as a sequential device whose output reflects not only the present input condition but the previous input history. (See Table 8). Its storage capability is not affected by power interruptions.

TABLE 8. - Truth table for Flip-Flop device.

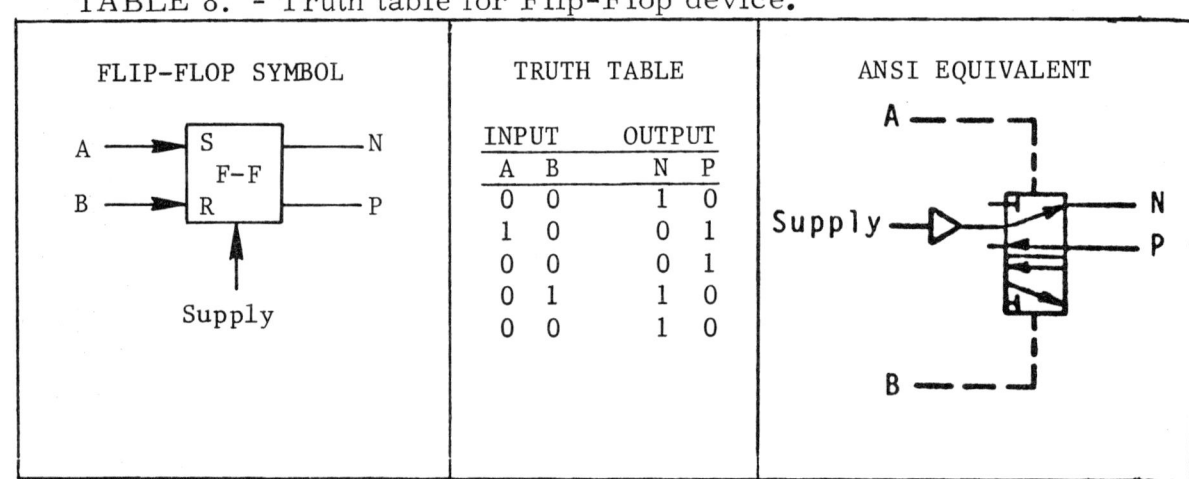

13.10 Use the One-Shot as a passive single shot device to provide a timed single output. When the input assumes the 1-state the output also assumes the 1-state for a predetermined period of time and then returns to the 0-state. (See Table 9.)

TABLE 9. - Truth table for one-shot device.

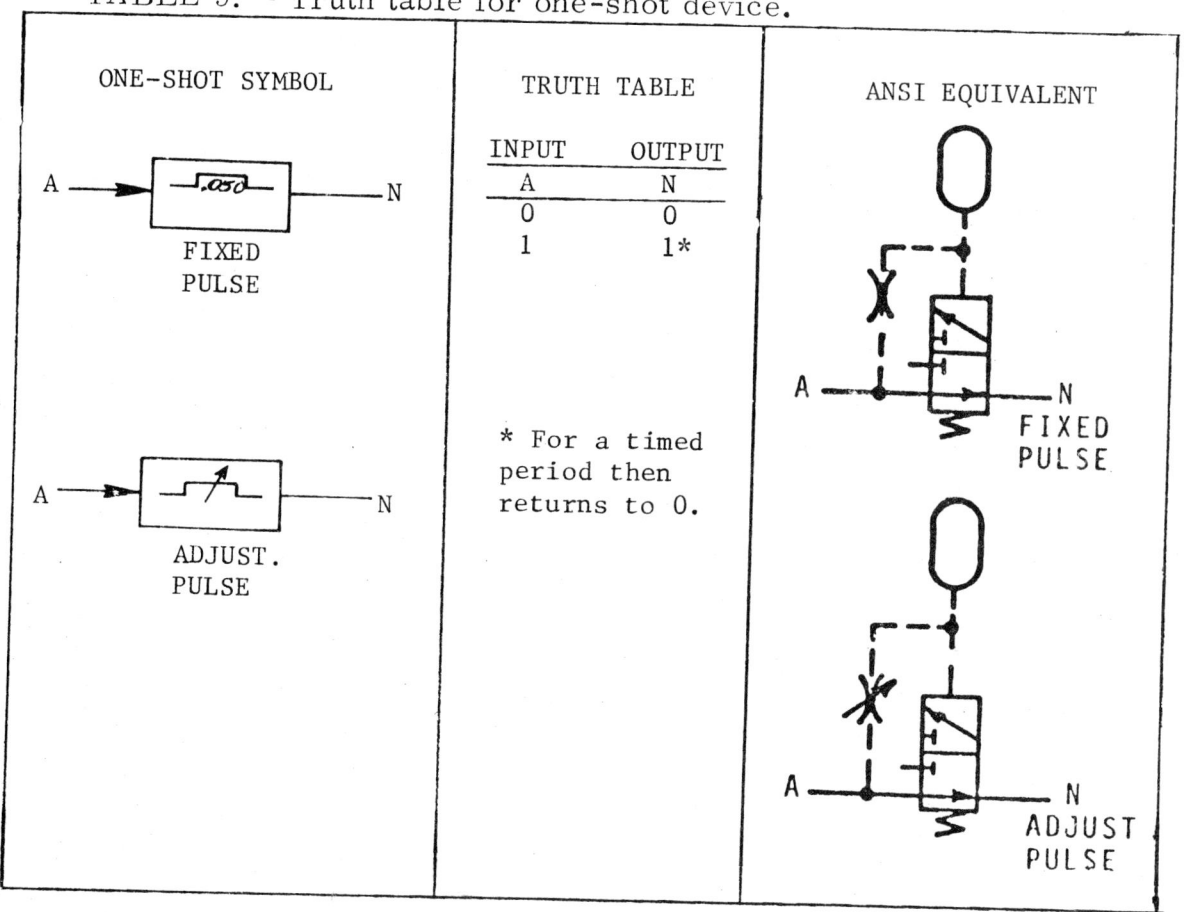

13.11 Use the delay, Timing In output to assume the 1-state after a controlled period of time after the input assumes the 1-state. (See Table 10)

TABLE 10. - Truth table for delay, timing in.

13.12   Use the delay, timing out output (which is in 1-state) to assume the 0-state after a controlled period of time after the input assumes the 0-state. (See Table 11)

TABLE 11. - Truth table for delay, timing out.

| DELAY, TIMING OUT SYMBOL | TRUTH TABLE | ANSI EQUIVALENT |
|---|---|---|
| A → [DEL] → N | INPUT   OUTPUT<br>A          N<br>1          1<br>0          0 Delayed | |

13.13   Use the Amplifier as an active device to allow an input signal of low energy to control an output fluid signal of a high energy level.

13.14   Show the Amplifier, consisting of one or more stages, connected to the AND or NOT symbol to designate a normally 1-state (NOT) or a normally 0-state (AND).

13.15   Use the General Logic Symbol for functions not elsewhere specified.

13.15.1   Label this symbol adequately to identify the function performed.

NOTE. - It is not intended that this symbol be used for functions which can be logically expressed by a single symbol established in this document.

TABLE 12. – Truth table for amplifier device.

| AMPLIFIER SYMBOL | TRUTH TABLE | ANSI EQUIVALENT |
|---|---|---|
| (amplifier symbol with inputs A and Supply, output N) | INPUT / OUTPUT<br>A / N<br>0 / 0<br>1 / 1 | (ANSI equivalent schematic) |
| (amplifier symbol with N block, inputs A and Supply, output N) | INPUT / OUTPUT<br>A / N<br>0 / 1<br>1 / 0 | (ANSI equivalent schematic) |

14. PRACTICE, ATTACHED METHOD

   14.1   Multiple Inputs to a Single Function

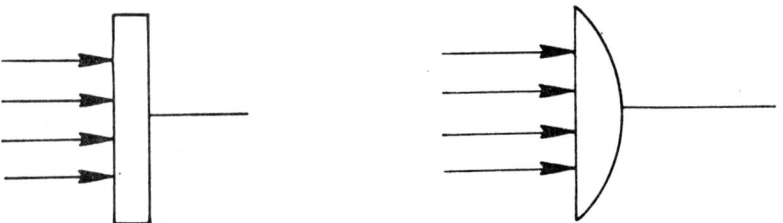

   14.2   Multiple Inputs to Physically Separate Functions with Common Outputs

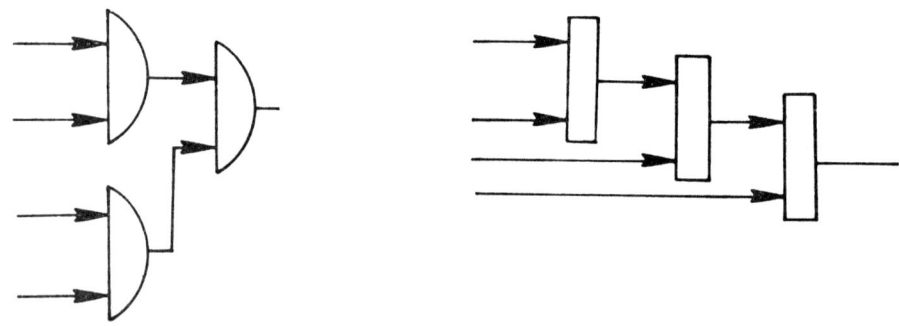

15. DIAGRAM RULES, ATTACHED METHOD

   15.1   Arrangement of Symbols

      15.1.1   Arrange symbols in the diagram to facilitate the use of direct and straight interconnecting lines.

      15.1.2   Prepare lines between symbol inputs and outputs horizontal or vertical with a minimum of line crossing and with spacing to avoid crossing.

   15.2   Circuit Arrangement

      15.2.1   Arrange the circuit in functional sequence, left to right and top to bottom. Follow this rule rigidly where excessive line crossing impairs the clarity of the diagram.

      15.2.2   Show all inputs to the control circuit on the left and all outputs on the right.

         NOTE. - For clarity of the total system, the operational diagram may be combined and shown connected to the logic diagram.

      15.2.3   Use solid lines to represent interconnecting flow paths between logic devices within the control circuit.

## 15.3 Designations

### 15.3.1
Use designations of input(s), output(s), flow paths and logic devices to facilitate the checking of circuits. The use of designation is determined by the overall adaptability of the design, assembly, installation and maintenance of the equipment.

### 15.3.2
Assign all input(s) and output(s) of the control circuit arbitrary designations. These are comprised of letters and combinations of letters and numbers (A, B, C, X, Y, Z, etc. or A1, A2, A3, Z7, Z10, etc.) To further identify the inputs from the outputs, the inputs are assigned a letter from the first half of the alphabet with the outputs assigned a letter from the last half of the alphabet (O and I are omitted).

### 15.3.3
To identify the flow path, use detailed logic diagrams to show each interconnecting flow path (between input(s) or output(s) and logic devices and between logic devices) and number consecutively.

### 15.3.4
For logic devices, use detailed logic diagrams to identify each logic device with a numerical designation starting with the number 1 and numbering consecutively.

## 15.4 Explanatory Notes

Add explanatory notes for clarification of function description where the function is not clear.

NOTE. - Where notes are lengthy or must be repeated, they may be shown at a common location with proper reference at the point of application.

## 15.5 Sequence of Operation

### 15.5.1
When applicable, list in the order in which they occur, the sequence of operation of the control circuit with reference to input conditions and resulting output conditions.

### 15.5.2
Number each phase followed by a brief description of the input conditions and the resulting outputs. These sequence statements always supercede a statement concerning the condition of the inputs and outputs as they appear at the start of the sequence.

## 15.6 Installation

Show data pertinent to installation on the detail logic diagrams. Items to be included are tubing size, operating pressure, degree of filtration required and enclosure (if required).

# 16. LOGIC DIAGRAMS, ATTACHED METHOD

## 16.1 Typical Logic Diagrams Showing Logic Function in a Typical Circuit

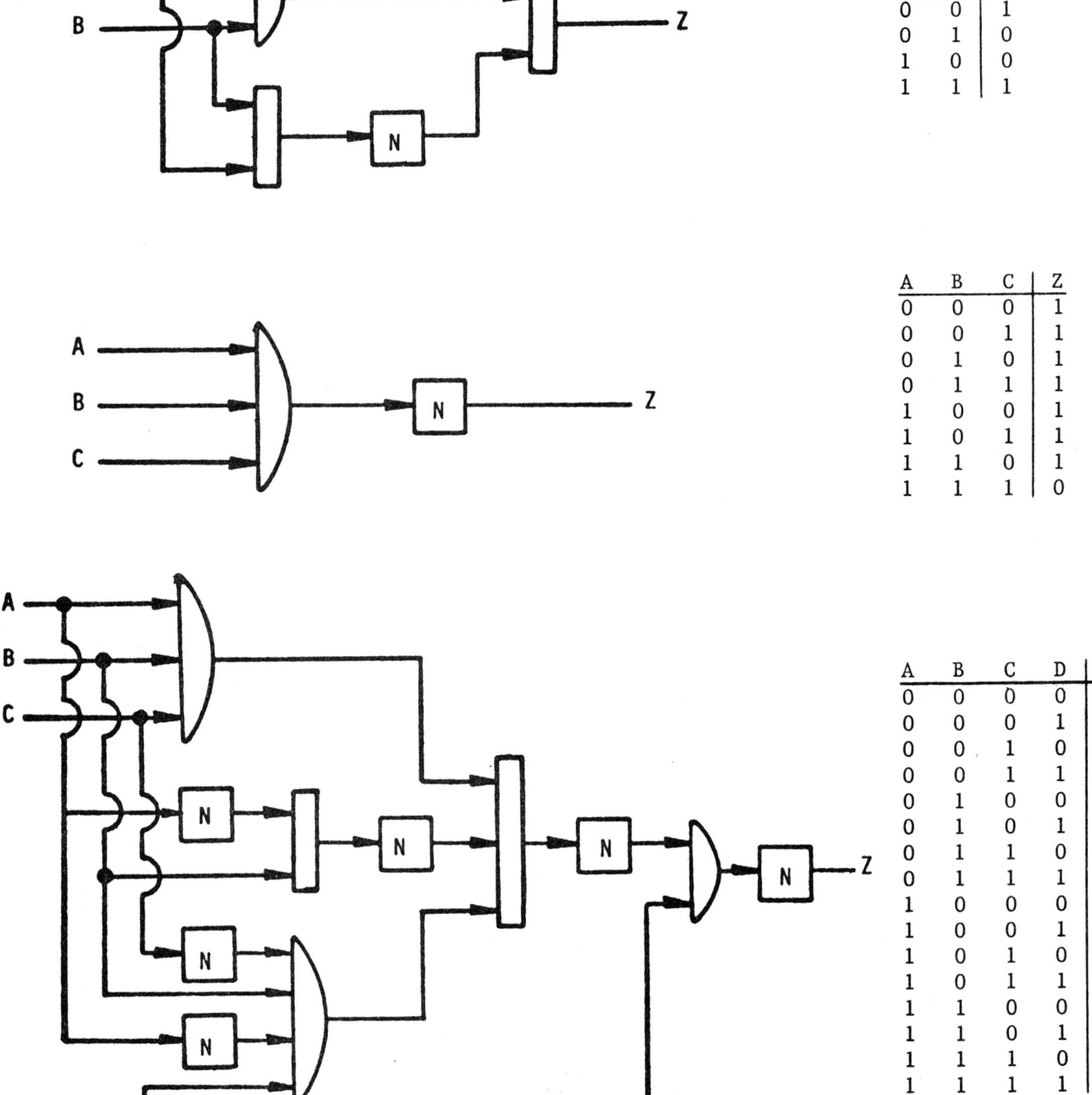

# PART V - DETACHED METHOD OF DIAGRAMMING
## (RELAY VALVES AND LOGIC DEVICES)

### 17. SYMBOLS FOR RELAY VALVES

**17.1** Identification of Relay Valves

    **17.1.1** Identify each relay valve by the letters "RV" followed by a number, starting with the number 1 and numbering consecutively. There is no significance to the numbers other than to identify the particular relay valve.

    **17.1.2** Identify one relay valve in each circuit as a master relay valve, RVM

**17.2** Identification of Special Purpose Relay Valves

Identify either by standard RV numbers, or by special identification which clearly indicates the function when a relay valve is used in conjunction with other devices to provide a special-purpose function (jet position-sensing, pressure-sensing, timing, etc.)

**17.3** Special Identification

Use the following rules:

    **17.3.1** Identify by a JRV number (JRV1, JRV2, etc.) a relay valve which cooperates with a jet in a jet position-sensing function.

    **17.3.2** Identify by a PSRV number (PSRV1, PSRV2, etc.) a relay valve used as the sensing element in a pressure-sensing function.

    **17.3.3** Identify by a TDRV number (TDRV1, TDRV2, etc.) a relay valve used in a time delay function.

    **17.3.4** Explain identifications not covered by the above by notes on the drawing.

**17.4** Identification of Relay Valve Control Points

Use the following rules for each control point on each relay valve having a unique identification:

17.4.1 Identify the control point on a relay valve having only one control point with the number of that relay only.

> EXAMPLE. - The control point on a single control point spring return relay valve RV3 carries the identification "RV3".

17.4.2 Identify the control points on a relay valve having two opposing control points as "A" and "B". The complete identification is by the number of the relay valve followed by the "A" or "B".

> EXAMPLE. - RV4A, RV4B, etc.

> NOTES. -
>
> 1. It is anticipated that new styles of relay valves will make their appearance in the future, having more than one control point in each end of the relay valve. For example, a relay valve similar to a single control point spring return relay valve, except that instead of having just one control point capable of opposing the spring, this relay valve may have several control points on the "A" end, each control point independent of the others and capable of opposing the spring.
>
> 2. In the case of a relay valve having multiple control points, all serving the same basic function of actuating the relay valve, these control points would be identified by the usual designation of the control point (e.g., RV6A or RV6B) followed by dash numbers, -1, -2, -3, etc.
>
> 3. Thus, a single pilot spring return relay valve with multiple control points opposing the spring would identify these points as RV5-1, RV5-2, RV5-3, etc.
>
> 4. In the case of a double pilot relay valve the control points would be identified RV6A-1, RV6A-2, RV6A-3, etc., and RV6B-1, RV6B-2, RV6B-3, etc.

## 17.5 Symbols for Control Points on Relay Valves

### 17.5.1
Control point on a relay valve having one actuator and spring return or automatic return. The automatic return is accomplished entirely within the relay valve assembly and requires no connection from the control system.

⎯⎯( RV 2 )

When this control point is pressurized, it takes command of the relay valve and keeps command as long as it is pressurized. When it is exhausted, the spring (or automatic return) returns the relay valve to the starting position.

### 17.5.2
Control points on a double control point detented (or equivalent) relay valve. When either control point is exhausted, a momentary fluid signal into the opposite control point will cause that control point to take command and shift the relay valve.

⎯⎯[ RV 5A ]

When that fluid signal is exhausted, the detent holds the relay valve in the new position until the second control point is pressurized.

⎯⎯[ RV 5B ]

A momentary fluid signal into the second control point will cause it to assume command and shift the relay valve back to the start position.

If control pressure is held in either control point, a fluid signal of equal pressure into the opposite control point will not shift the relay valve until the original control point is exhausted.

NOTE. - These are the symbols for the control points on a true "retentive fluid memory" device, where the relay valve is physically held in position by mechanical or magnetic means.

Thus, once the relay valve has been shifted to a specific position, the control air to the system may be turned on and off any number of times without affecting the position of the relay valve. The only thing which will change its position is a fluid signal into the opposite control point.

17.5.3 Control points on a double control point spring offset relay valve. Control point on the end opposite the spring and control point on the spring end.

When the spring end control point is exhausted, the "A" control point acts exactly like the control point on a single control point spring return relay valve.

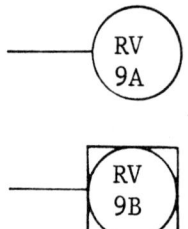

With pressure in the "B" control point, an equal pressure in the "A" control point will not shift the relay valve until the "B" control point is exhausted.

If the "B" control point is exhausted, a fluid signal into the "A" control point will shift the relay valve and it will stay shifted as long as the fluid signal into "A" is maintained. If, however, while the "A" control point is pressurized, and in command of the relay valve, an equal pressure is introduced into the "B" control point, the two fluid signals will cancel each other, and the spring will return the relay valve to the starting position, EVEN THOUGH THE "A" CONTROL POINT IS STILL PRESSURIZED.

Finally, if the "A" control point is pressurized by a maintained fluid signal, it will take command of the relay valve and hold the relay valve shifted against the spring. The "B" control point can then be pressurized and exhausted, and will act like the actuator on a single control point spring return valve.

17.5.4 Control Points on a Free-Floating Relay Valve. These symbols represent the control points on a double-control point relay valve wherein the valving action moves so freely that it has no inherent mechanical restraint, and is normally used with bias pressure in one control point. The "B" control point will normally be the control point connected to the bias pressure.

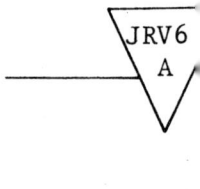

## 17.6 Symbols for Controlled Flow Passages of Relay Valves

**17.6.1** Three-way function, not passing at the start of the cycle. This may be a controlled flow passage on a detented relay valve, which is not passing at the start of the cycle, or it may be a normally not passing passage on a spring return relay valve with the relay valve in its normal position at the start of the cycle. Exhaust is assumed but not shown.

**17.6.2** Three-way function, passing at the start of the cycle. This may be a controlled flow passage on a detented relay valve, or it may be a normally passing passage on a spring return relay valve with the relay valve in its normal condition at the start of the cycle. Exhaust is assumed but not shown.

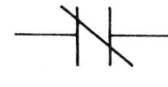

**17.6.3** Three-way function, normally passing but held not passing at the start of the cycle because the relay valve control point is pressurized. Used only in the case of spring return or automatic return relay valves. Exhaust is assumed but not shown.

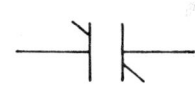

**17.6.4** Three-way function, normally not passing but held passing at the start of the cycle because the relay valve control point is pressurized. Used only in the case of spring return or automatic return relay valves. Exhaust is assumed but not shown.

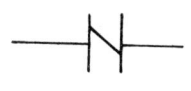

**17.6.5** Non-exhausting function. The solid dots indicate that the function is a non-exhausting function.

NOTE. - All the three-way symbols under clause 17.6 may be used with the dots to indicate a non-exhausting function.

**ANSI/B93.38**

### 17.7 Relay Valve Symbol Cross-Reference System

Prepare an identifying and cross-reference system in conjunction with all relay valve control points and controlled flow passage symbols, so that the control points and their associated controlled flow passages may be easily located on the diagram.

**17.7.1** Number each horizontal flow path in the diagram, starting with number 1 and numbering consecutively from the top of the diagram downward. Arrange these guide numbers (or line numbers) in a vertical row down the left side of the diagram.

**17.7.2** Where the size of the diagram requires it to be separated into several columns on one sheet or continued over from one sheet to another, carry over the guide numbers and continue in sequence on the top of the next column or on the top of the next sheet.

**17.7.3** In the case of relay valves having two or more opposing control points, enter a set of brackets to the right of each control point, containing the guide numbers of the lines in the diagram which contain the opposing control points.

EXAMPLE. - A double control point detented relay valve with the "A" control point in line 12 and the "B" control point in line 26.

**17.7.4** Enter, in parentheses, to the right of each control point, the numbers of the lines where the controlled flow passages of the relay valve are located.

**17.7.4.1** Underline controlled flow passages which a control point causes to go not passing.

**17.7.4.2** DO NOT underline controlled flow passages which a control point causes to go passing.

ANSI/B93.38

## 17.8 Relay Valve Controlled Flow Passage Cross-Referencing

17.8.1 Note above each controlled flow passage symbol the identification of the relay valve control point which, when it takes command of the relay valve, causes the flow condition to change from the condition shown.

17.8.2 List only the basic identification of the control points in the case of a relay valve with multiple control points all performing the same function.

EXAMPLE. - A relay valve controlled flow passage where any of the "A" control points, RV6A-1, RV6A-2, or RV6A-3 can change the flow condition. Show only the identification RV6A.

17.8.3 Where the complexity of the drawing warrants, enter underneath the controlled flow passage symbol the guide numbers of the lines where the control points of the relay valve are located.

List the locations of the control points which CHANGE the flow condition first.

## 18. SYMBOLS FOR OFF-RETURN FLUID MEMORY RELAY VALVES

### 18.1 Three Versions of the Off-Return Relay Valve

18.1.1 Single control point. Reset by exhausting the input to the controlled flow passages.

18.1.2 Double control point. Reset either by exhausting the input to the controlled flow passages, or by applying a momentary control fluid signal to the reset control point.

18.1.3 Double control point with an external seal-in circuit. Reset by either:

18.1.3.1 Exhausting the input to the controlled flow passages.

18.1.3.2 Applying a momentary control fluid signal to the reset control point.

18.1.3.3 Interrupting the external seal-in circuit by means of a separate, external control device.

18.2 Symbols for Control Points on Off-Return Relay Valve

Control point on a single control point off-return relay valve.

18.3 Symbols for Control Points on a Double Control Point Off-Retur

18.3.1 Control point which causes actuation.

18.3.2 Control point which causes reset.

18.4 Symbols for Controlled Flow Passages of an Off-Return Relay Valve

18.4.1 Controlled flow passage which controls the seal-in circuit (internal seal-in circuit).

18.4.2 Controlled flow passage which controls the seal-in circuit (external seal-in circuit) showing a typical interrupter passage in the seal-in circuit (in this case a relay valve RV16).

18.4.3 Controlled flow passage which does not control the seal-in circuit.

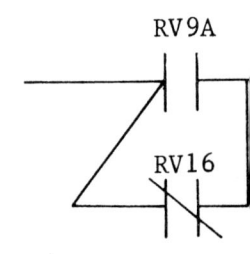

18.5 Identification

Identify each off-return relay valve in the same manner as other relay valves. Use normal relay valve control point identification and location cross-references.

ANSI/B93.38

19. SYMBOLS FOR TIME DELAY RELAY VALVES

    19.1    Time Delay Relay Valves - Control Points

For control points on time delay relay valves, use the same symbols as those used for the corresponding action of a relay valve.

19.1.1    Control point on a single control point time delay relay valve. Use this symbol for both time delay after pressurizing and time delay after exhausting. The symbols for the controlled flow passages indicate whether the action is time delay after pressurizing the control point or time delay after exhausting the control point.

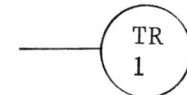

19.1.2    Control points on a double control point detented time delay relay valve. Use these symbols for both delayed actuation and delayed reset relay valves. The symbols for the controlled flow passages indicate whether the action is that of a delayed actuation or a delayed reset.

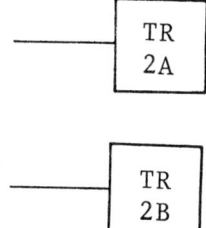

    19.2    Identification - Time Delay Relay Valves

Identify by a TR number starting with TR1 and numbering consecutively. The number has no significance other than to identify the particular time delay relay valve.

19.2.1    Identify the control point on a single point time delay by the TR number of the time delay relay valve.

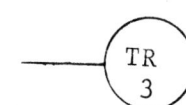

19.2.2    Identify the control points on a double control point detented time delay relay valve by the TR number and the letter "A" or "B".

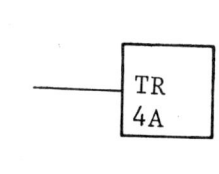

19.2.2.1    Use "A" control point, when it takes command, to start the actuation process.

19.2.2.2    Use "B" control point, when it takes command, to start the reset process.

## 19.3 Symbols for Time Delay Relay Valves Controlled Flow Passages

Depict each controlled flow passage by a set of terminals and a swinging gate, similar to those used to depict limit valves. Under each swinging gate, use a symbol which indicates the details of the timing action, as follows:

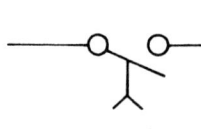

### 19.3.1
Controlled flow passage on a time delay after pressurization relay valve. Normally not passing. After receipt of a control fluid signal there is a delay before this passage goes passing. It resets immediately when the control fluid signal is removed.

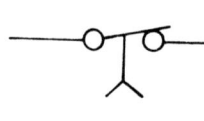

### 19.3.2
Controlled flow passage on a time delay after pressurization relay valve. Normally passing. After receipt of a control fluid signal there is a delay before this passage goes not passing. It resets immediately when the control fluid signal is removed.

### 19.3.3
Controlled flow passage on a time delay after exhausting relay valve. Normally not passing. Goes passing immediately upon receipt of a control fluid signal. After the control fluid signal is removed, there is a delay before this passage resets.

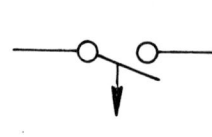

### 19.3.4
Controlled flow passage on a time delay after exhausting relay valve. Normally passing. Goes not passing immediately upon receipt of a control fluid signal. After the control fluid signal is removed, there is a delay before this passage resets.

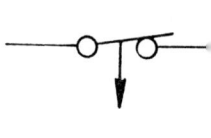

### 19.3.5
Controlled flow passage on a double control point detented time delay relay valve with delayed actuation.

Not passing at start of cycle.

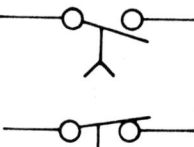

Passing at start of cycle.

19.3.6 Controlled flow passages on a double control point detented time delay relay valve with delayed reset.

Not passing at start of cycle.

Passing at start of cycle.

19.4 Identification of Controlled Flow Passages

19.4.1 Above each controlled flow passage symbol put the identification of the time delay relay valve.

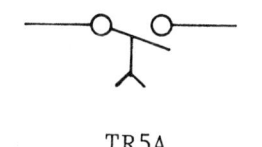

19.4.2 Put the identification of the "A" control point for a double control point detented relay valve.

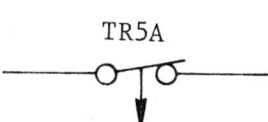

19.5 Relay Symbol Cross-Reference System

Cross-reference time delay relay valve symbols in a manner similar to that used for relay valves, so that control points and their associated passages may be easily located on the diagram.

20. SYMBOLS FOR RESISTANCE DEVICES

20.1 Fixed Resistance

Resists flow in both directions.

20.2 Adjustable Resistance

Resists flow in both directions.

20.3 Fixed Resistance with Free Return Check

20.3.1 Choked flow left to right; free flow right to left.

20.3.2 Choked flow right to left; free flow left to right.

## 20.4 Adjustable Resistance with Free Return Check

20.4.1 Choked flow left to right; free flow right to left.

20.4.2 Choked flow right to left; free flow left to right.

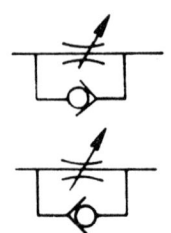

## 21. SYMBOLS FOR RELAY VALVES IN A TIMING CIRCUIT

Use relay valves in conjunction with resistance devices to perform timing functions.

21.1 When the resistance device and the relay valve are separate devices, and external connections are made to make them an operational unit, diagram as separate units, with a tube connection between them.

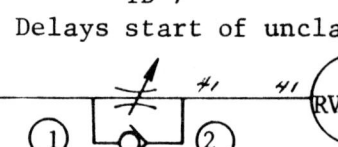

TD 7
Delays start of uncla

21.2 Identify each resistance device by a TD number, starting with TD1 and numbering consecutively. The numbers have no significance other than to identify the particular component.

Under the TD number put a note telling what action is delayed by this resistance device.

21.3 Depict the relay valve by relay valve symbols.

21.4 Identify the relay valve by standard RV numbers, or by special identification which indicates that it is performing a timing function.

If special identification is used, identify the relay valve by a TDRV number, the number to correspond to the TD number assigned to its resistance device.

21.5 Diagram the relay valve controlled flow passages by relay valve symbols, identified by the RV number or TDRV number, as the case may be.

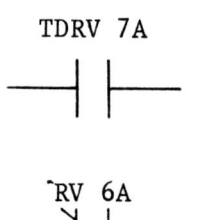

ANSI/B93.38

## 22. SYMBOLS FOR ONE-SHOT RELAY VALVES

NOTE. - In certain versions of a one-shot relay valve the device may have two control points. Its action is similar to that of the single control point device, except that when the initiating control fluid signal is removed, the device does not reset. To reset the device the second control point must receive a control fluid signal.

### 22.1 Control Point Symbols for One-Shot Relay Valves

22.1.1 Single control point - automatic reset when control point is exhausted.

22.1.2 Double control point - requiring a pneumatic reset fluid signal. Identify by "A" the initiating control point, and by "B" the resetting control point.

22.1.3 Use above control point symbols for both fixed pulse length and adjustable pulse length devices.

### 22.2 Identification

Identify each one-shot relay valve by an OSR number, starting with OSR-1 and numbering consecutively. The numbers have no significance other than to identify the particular one-shot.

### 22.3 Controlled Flow Passage Symbols for One-Shot Relay Valves

22.3.1 Three-way, normally not passing. Goes passing immediately upon receipt of initiating control fluid signal, returns to not passing after a non-adjustable length of time. Numeral indicates length of fixed pulse in seconds.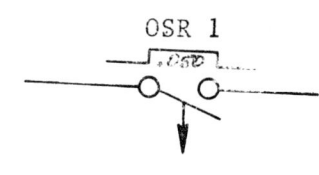

22.3.2 Three-way, normally passing. Goes not passing immediately upon receipt of control fluid signal. Returns to passing after a non-adjustable length of time. Numeral indicates length of fixed pulse in seconds.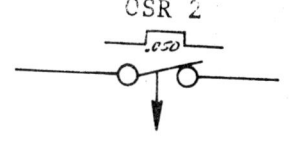

22.3.3 Three-way, normally not passing. Goes passing immediately upon receipt of a control fluid signal. Returns to not passing after an adjustable length of time.

22.3.4 Three-way, normally passing. Goes not passing immediately upon receipt of a control fluid signal. Returns to passing after an adjustable length of time.

22.3.5 Identification. Identify the one-shot relay valve controlled flow passages by placing the OSR number above the controlled flow passage symbol.

22.4 One-Shot Relay Valve Location Cross-References

Use a cross-reference for control points and controlled flow passages to facilitate location on the diagram, using standard relay valve cross-references.

23. SYMBOLS FOR ELECTRIC-TO-AIR RELAY VALVES

23.1 Identification

Identify each electrical-to-air interface with an ERV (electrical relay valve) number, starting with ERV1 and numbering consecutively. The numbers have no significance other than to identify the particular component.

23.2 Symbols - Actuating Devices

23.2.1 Single Solenoid, Spring or Automatic Return. When the solenoid is energized, one or more controlled flow passages change condition. When the solenoid is de-energized, the controlled flow passages return to the original condition.

23.2.2 Double Solenoid, Detented. When the "A" solenoid is energized momentarily, the controlled flow passages change condition. When the "B" solenoid is energized momentarily, the controlled flow passages return to the original condition.

### 23.2.3 Symbols for a Combination of Actuators.

When both an air signal AND an electrical signal must be present to cause energization, show the solenoid and the air actuators end to end. When either the air signal OR the electrical signal can cause energization, show the symbols for the solenoid and air actuator side by side.

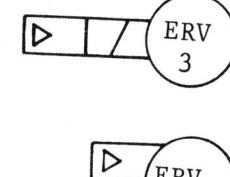

## 23.3 Location of Actuator Symbols

### 23.3.1
Locate the symbols for the ERV actuators in a vertical column on the circuit diagram, in line with the control point symbols for the power control valves.

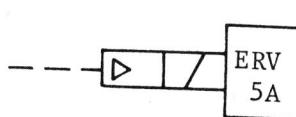

### 23.3.2
In the case of combination actuators requiring an air signal from the control system, use a dashed line from the control system to the air actuator.

## 23.4 Symbols for Controlled Flow Passages

Use standard symbols for relay valve controlled flow passages.

## 23.5 Identification of Controlled Flow Passages

Identify each controlled flow passage by the ERV number located above the controlled passage symbol. In the case of double solenoid detented relay valves add the letter of the solenoid acutator which causes the controlled flow passage to change from the condition shown.

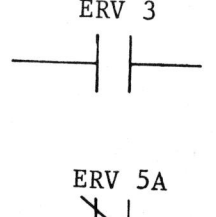

## 23.6 Symbol Location Cross-Reference

Cross-reference solenoid actuators and controlled flow passages to facilitate location on the diagram, using standard relay valve cross-references.

ANSI/B93.38

## PART VI - MISCELLANEOUS DEVICES

24. VISUAL INDICATORS

   24.1   Spring Return Type, Single Control Point

   When control point is pressurized, indicator shows color. When control fluid signal is removed, indicator returns to start conditions.

   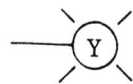

   NOTE. - Letter in symbol denotes color.

   24.2   Two-Position Detented, Double Control Point

   When "A" control point is pressurized by a momentary pulse, indicator changes color.

   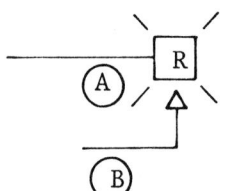

   When "B" control point is pulsed, indicator returns to starting condition.

   24.3   If necessary, put a note giving details of indicator operation or legend.

   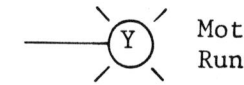 Motors Running

   24.4   Identification

   Give each visual indicator a VI number, starting with VI-1 and numbering consecutively. The numbers have no significance other than to identify the particular visual indicator.

25. TEST POINT

   Use a test point symbol as a special fitting allowing the ready application of a pressure gage to the circuit for troubleshooting purposes.

26. PRESSURE INDICATOR

   Use a pressure indicator symbol as a visual indicator installed permanently in the circuit, for the purpose of indicating visually the presence of pressure at that point in the circuit.

27. PRESSURE GAGE

Use a pressure gage symbol as a pressure indicator calibrated in such a way as to indicate not only the presence of pressure, but the exact pressure level.

28. PRESSURE CONTROL VALVE (PRESSURE REGULATOR)

   28.1  Adjustable, Relieving Type

   28.1.1  Identify each pressure control valve with an "R" number, starting with R1 and numbering consecutively. The number has no significance other than to identify the particular pressure control valve.

   28.1.2  Put a note telling to what pressure the pressure control valve is set.

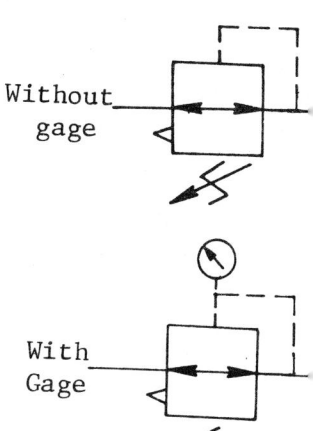

Without gage

With Gage

29. FLUID CONDITIONERS

   29.1  Filters

   Identify each filter with an "F" number, starting with F1 and numbering consecutively. The numbers have no significance other than to identify the particular filter.

   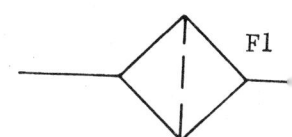

   29.2  Lubricators

   Identify each lubricator with an "L" number, starting with L1 and numbering consecutively. The numbers have no significance other than to identify the particular lubricator.

   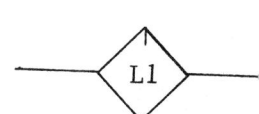

   29.3  Air Dryers

   Identify each air dryer with a "D" number, starting with D1 and numbering consecutively. The numbers have no significance other than to identify the particular air dryer.

   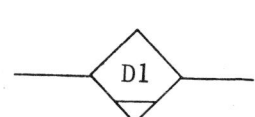

30. CHECK VALVE

    30.1    Check Valve, Pilot-Operated

        30.1.1    Pilot-operated to open.

        30.1.2    Pilot-operated to close.

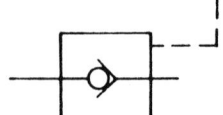
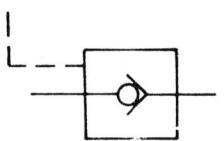

31. SHUTTLE VALVE (TWO-INPUT PASSIVE OR)

Identify each shuttle valve with an "S" number, starting with S1 and numbering consecutively. The number has no significance other than to identify the particular component.

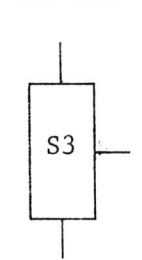

32. IDENTIFICATION STATEMENT

Use the following statement in catalogs and sales literature when electing to comply with this voluntary standard:

"Method of diagramming conforms to American National Standard, ANSI/B93.38-1976."

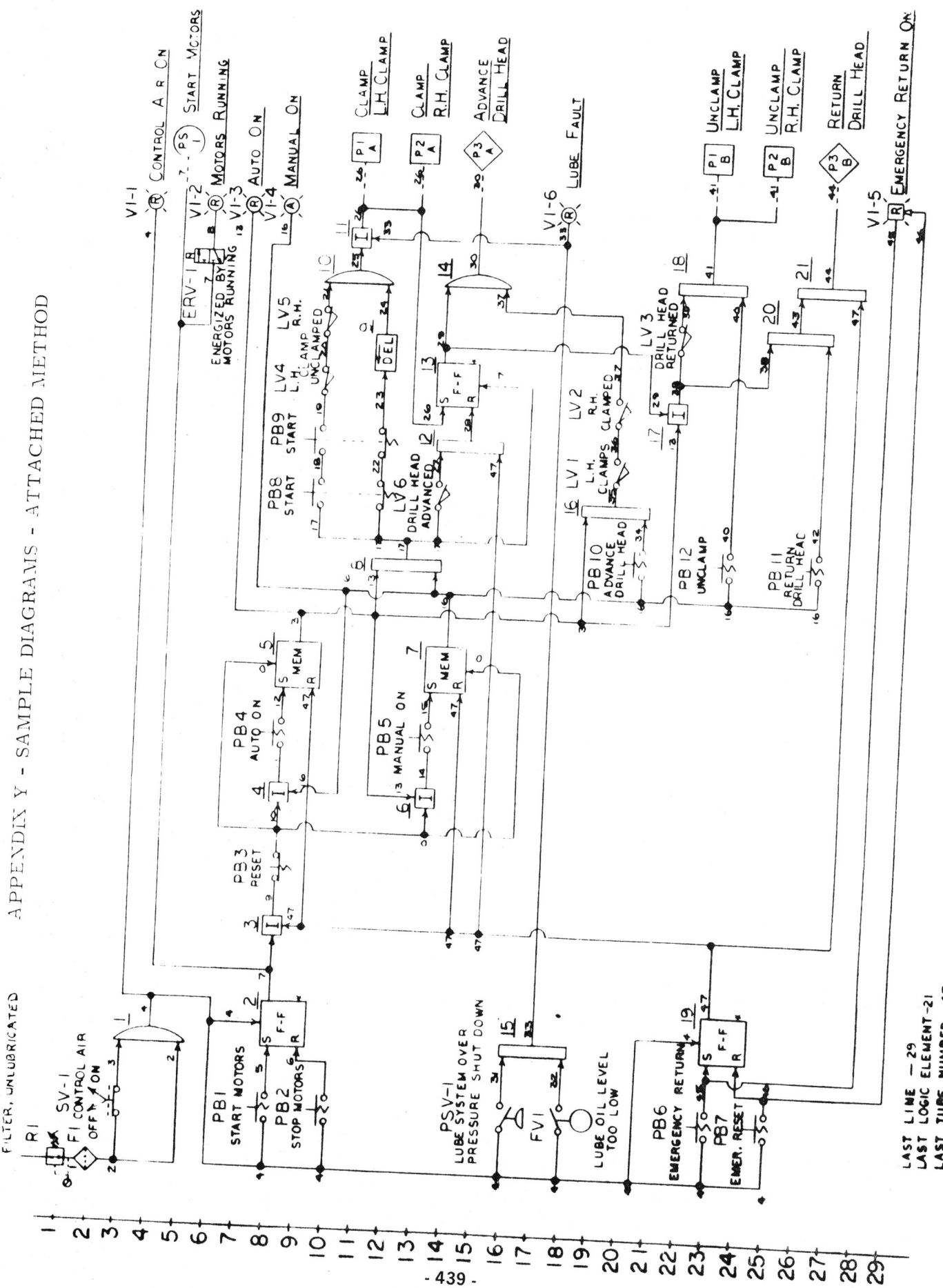

APPENDIX Y - SAMPLE DIAGRAMS - ATTACHED METHOD

ANSI/B93.38

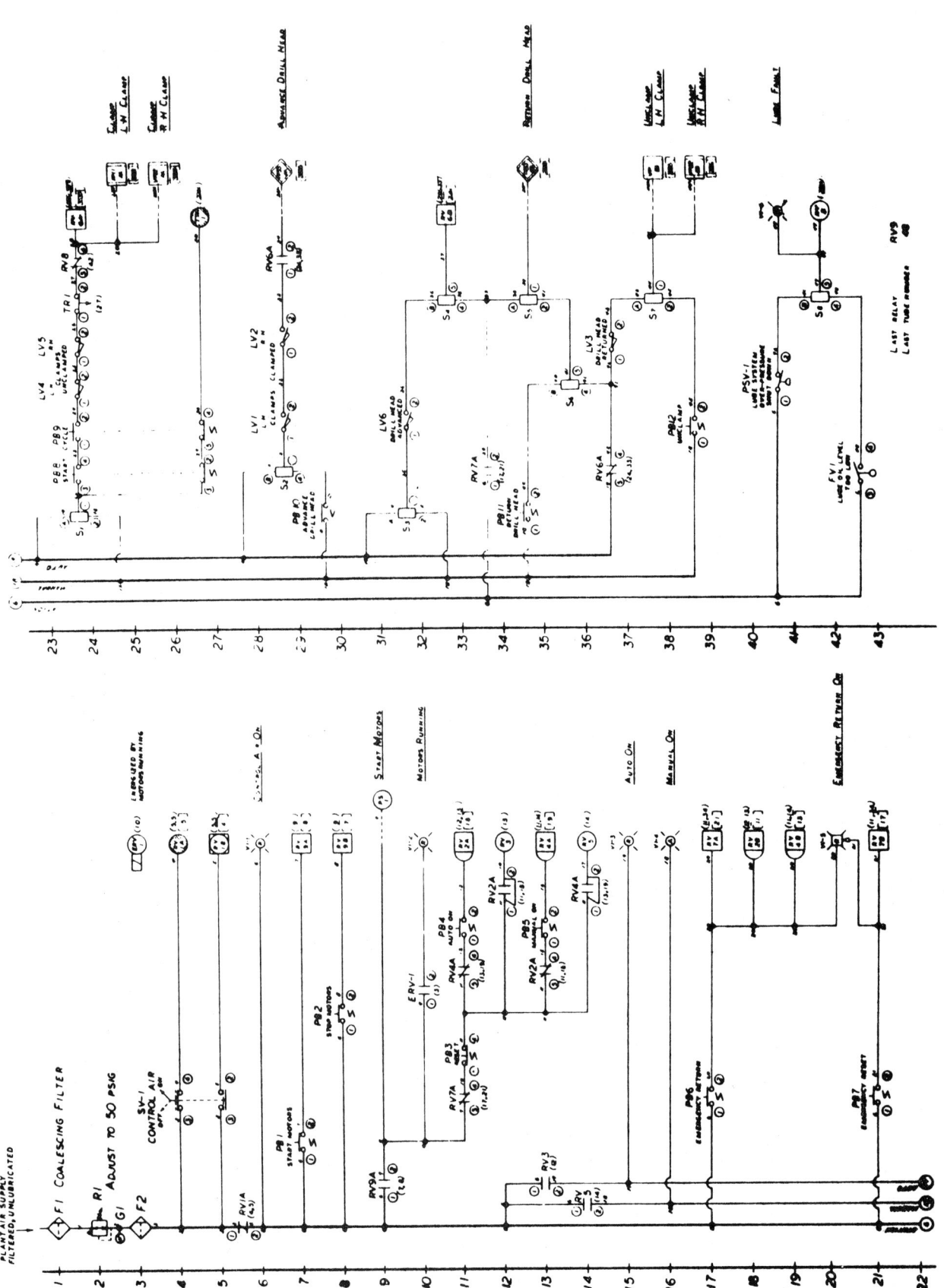

# APPENDIX Z - SAMPLE POWER CONTROL DIAGRAM

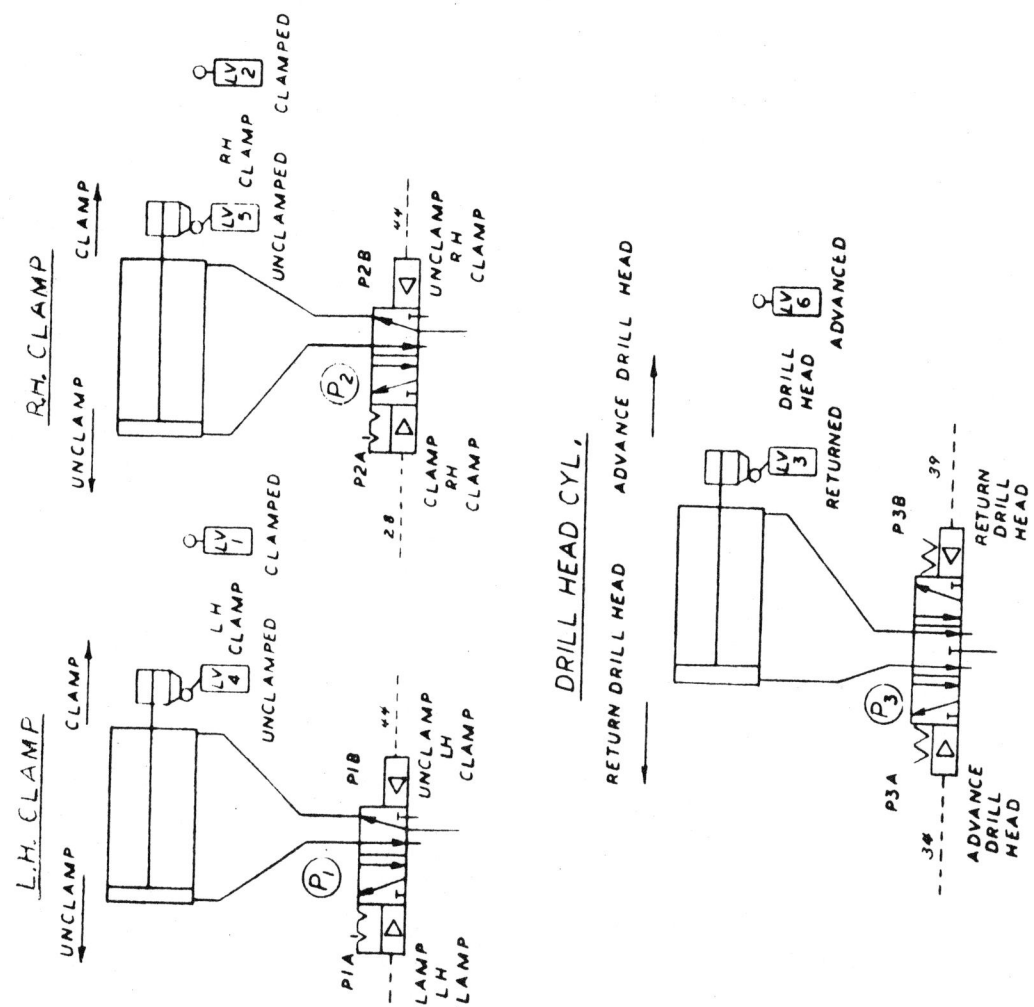

ANSI/B93.38

# APPENDIX Z - SAMPLE POWER CONTROL DIAGRAM

SEQUENCE OF CONTROL OPERATIONS:

Initial Conditions:

    (1)    Electrical power on. Lube oil pump motor not running. Spindle motor not running.

    (2)    Control air on. "Control Air On" visual indicator showing red.

    (3)    L.H. and R.H. fixture clamps unclamped. Fixtures empty. Drill head returned.

    (4)    Ready to start.

Automatic Cycle:

1. Operator momentarily depresses "Start Motors" button.

    a.    Lube oil pump motor starts.

    b.    Spindle motor starts.

2. As motors start, electrical power to motors energizes ERV-1.

    a.    "Motors Running" indicator turns red.

3. Operator momentarily depresses "Auto On" button.

    a.    Controls switch into "Auto" mode.

    b.    "Auto On" indicator turns red.

4. Operator manually loads two parts into the L.H. and R.H. fixtures.

ANSI/B93.38

5. Operator simultaneously depresses both "Start Cycle" buttons momentarily.

   a.   L.H. Clamp clamps part.

   b.   R.H. Clamp clamps part.

6. As L.H. Clamp starts to clamp, it releases LV4, and as R.H. Clamp starts to clamp, it releases LV5.

   a.   No action.

7. When clamps are fully clamped, L.H. Clamp makes LV1, and R.H. Clamp makes LV2.

   a.   Drill head advances toward work.

8. As drill head starts to advance, it releases LV3.

   a.   No action.

9. When drills reach full depth, drill head makes LV6.

   a.   Drill head starts to return.

10. As drill head starts to return, it releases LV6.

    a.   No action.

11. When drill head is fully returned, it makes LV3.

    a.   L.H. Clamp unclamps.

    b.   R.H. Clamp unclamps.

12. As L.H. Clamp starts to unclamp, it releases LV1, and as R.H. clamp starts to unclamp, it releases LV2.

    a.   No action.

13. When L.H. Clamp is unclamped, it makes LV4, and when R.H. Clamp is unclamped, it makes LV5.

    a.   No action.

14. Operator manually removes both finished parts from fixtures.

End of cycle. Cycle repeats from step 4 above.

## SAFETIES AND INTERLOCKS:

1. Drill head cannot advance unless both L.H. and R.H. clamps are clamped, and limits LV4 and LV5 are actuated. Effective in both Manual and Automatic modes.

2. Lubrication fault:

   Effective in both Manual and Auto modes. If lube oil pressure rises too high, or if lube oil level drops too low, the "Lube Fault" indicator will turn red, and the machine will finish the cycle it is in, but the "Start Cycle" buttons will be unable to start the next cycle until the fault is corrected.

3. Two-Hand Non-Tiedown:

   The "Start Cycle" buttons are so arranged so that when either button is depressed, it starts timing relay TR1. If the second button is not depressed before TR1 times out, then the buttons become inoperative, and both must be released and be up at the same time to reset TR1 for another attempt.

4. Emergency Return:

   If the "Emergency Return" button is depressed momentarily at any point in the cycle, the following happens:

   a. The drill head immediately returns.

   b. The clamps stay clamped.

   c. The "Emergency Return On" indicator turns red.

   d. The controls drop out of the mode in which they were operating, and stay in neutral, neither in Manual nor in Auto.

   e. The "Manual On" and "Auto On" buttons become inoperative.

   To clear the machine and start another cycle, the operator must first momentarily depress the "Emergency Reset" button, then select "Manual On" and clear the machine manually. He can then select "Auto On" mode and start another cycle.

NFPA Bibliography
T3.5.27-1976 (R1982)

## AN INDUSTRY STANDARD FOR FLUID POWER

A Bibliography of

Hydraulic Valve Standards and Test Procedures

published by
## NATIONAL FLUID POWER ASSOCIATION, INC.

3333 N. Mayfair Road  /  Milwaukee, WI 53222  /  414-778-3344  /  TLX 26898

NFPA/T3.5.27

## SOURCE OF DOCUMENTS

CSA — Canadian Standards Association, 178 Rexdale Boulevard., Rexdale 603, Ontario, Canada

DOT — Department of Transportation, Distribution Unit, TAD-434.3, Washington, DC 20590

FCI — Fluid Controls Institute, Inc., 12 Bank Street, Summit, NJ 07901

ISA — Instrument Society of America, 400 Stanwix Street, Pittsburgh, PA 15222

JIC — Joint Industrial Council, 7901 Westpark Drive, McLean, VA 22101

MSS — Manufacturers Standardization Society of the Valve and Fittings Industry, 1815 North Fort Myer Drive, Arlington, VA 22209

NPFC — Commanding Officer, Naval Publications and Forms Center, 5801 Tabor Avenue, Philadelphia, PA 19120, Attn: NPFC 105

OSU — Oklahoma State University, Fluid Power Research Center, Stillwater, OK 74074

SAE — Society of Automotive Engineers, Inc., 400 Commonwealth Drive, Warrendale, PA 15096

UL — Underwriters Laboratories, Publications Stock, 333 Pfingsten Road, Northbrook, IL 60062

---

NOTE: This bibliography provides a list of standards related to valves which have been issued by standards-writing organizations other than NFPA, ANSI/B93 and ISO/TC 131. Listing in this bibliography is only for reference and does not imply Association endorsement.

NFPA/T3.5.27

| Identification Number | Title | Source | Date of Issue |
|---|---|---|---|
| AN6247 REV 2 | Valve - Controllable Hydraulic Check (Rotating Action) | NPFC | 1967 |
| AN6277 REV 3 | Valve - Hydraulic Shuttle, 3000 psi (Internal Thread Tube Fitting Outlet) | NPFC | 1971 |
| AN6245 REV 4 | Valve - Hydraulic Thermal Expansion Relief | NPFC | 1953 |
| AN5830 REV 5 | Valve - Low Pressure Check, Internal Pipe Thread (Use MS 28050) | NPFC | 1955 |
| AN6215 REV 3 | Valve - Radial Type 16 GPM Hydraulic Directional Control | NPFC | 1950 |
| AN6214 REV 3 | Valve - Radial Type 6 GPM Hydraulic Directional Control | NPFC | 1950 |
| AN5831 REV 2 | Valve - Static Pressure Selector (Use MS 28051) | NPFC | 1955 |
| AN6207 REV 4 | Valve - 1500 psi Hydraulic Check | NPFC | 1945 |
| AN6280 REV 1 | Valve - 3000 psi Hydraulic Check Internal Parts | NPFC | 1949 |
| AN6210 REV 3 | Valve - In Line Type 3 1/2 GPM Hydraulic Directional Control | NPFC | 1950 |
| AN6294 REV 3 | Valve - 4-Way Rotary Selector, 3000 psi, Non-Interflow | NPFC | 1955 |
| AN6293 REV 3 | Valve - 4-Way Rotary Selector, 3000 psi, Non-Interflow | NPFC | 1955 |

NFPA/T3.5.27

| Identification Number | Title | Source | Date of Issue |
|---|---|---|---|
| AN6211 REV 3 | Valve - In Line Type 6 GPM Directional Control | NPFC | 1950 |
| AN6279 REV 5 | Valve - Hydraulic Pressure Relief Two Port (Inactive) (Use MS 28893) | NPFC | 1961 |
| AN6213 REV 3 | Valve - Radial Type 3 1/3 GPM Hydraulic Directional Control | NPFC | 1950 |
| AN6204 REV 5 | Valve - Hydraulic Bleeder | NPFC | 1956 |
| AN6200 REV 4 | Valve - Hydraulic Pressure Relief, Class Ab (Inactive) (Use AN6279) | NPFC | 1955 |
| AN6278 REV 3 | Valve - Hydraulic Shuttle, 3000 psi (Universal and Internal Fitting Outlet) (Inactive 5-26-70) | NPFC | 1955 |
| AN6249 REV 2 | Valve - 3000 psi Hydraulic Check | NPFC | 1959 |
| AND 10067 REV 3 | Valve Installation - Hydraulic Bleeder (Standard Dimensions for) | NPFC | 1954 |
| CG-115 | Marine Engineering Regulations Subcharter F | DOT | 1973 |
| C22.2 No. 139 | Electrically Operated Valves | CSA | 1973 |
| FCI 58-2 | Measurement Procedure for Determining Control Valve Flow Capacity | FCI | 1963 |

NFPA/T3.5.27

| Identification Number | Title | Source | Date of Issue |
|---|---|---|---|
| FCI 68-1 | Recommended Procedure in Rating Flow and Pressure Characteristics of Solenoid Valves for Gas Service | FCI | 1968 |
| FCI 62-1 | Recommended Voluntary Standard Formulas for Sizing Control Valves | FCI | 1962 |
| FCI 75-1 | Test Conditions and Procedures for Measuring Electrical Characteristics of Solenoid Valves | FCI | 1975 |
| FCI 65-2 | Face-to-Face Dimensions of Control Valves | FCI | 1965 |
| FCI 68-2 | Recommended Procedure in Rating Flow and Pressure Characteristics of Solenoid Valves for Liquid Service | FCI | 1968 |
| FCI 70-2 | Quality Control Standards for Control Valve Seal Leakage | FCI | 1970 |
| H-1-1973 | Hydraulic Standards for Industrial Equipment and General Purpose Machine Tools | JIC | 1973 |
| ISA-S39.1 | Standard Control Valve Sizing Equations for Incompressible Fluids | ISA | 1972 |
| ISA-S39.2 | Standard Control Valve Capacity Test Procedure for Incompressible Fluids | ISA | 1972 |
| ISA-RP4.2 | Standard Control Valve Manifold Designs (Carbon Steel Valves) | ISA | 1956 |

NFPA/T3.5.27

| Identification Number | Title | Source | Date of Issue |
|---|---|---|---|
| MIL-V-82444(1) | Valve | NPFC | 1968 |
| MIL-V-62010 | Valve Assembly, Hydraulic Locking | NPFC | 1964 |
| MIL-V-25517A | Valve, Aircraft Hydraulic Restrictor | NPFC | 1959 |
| MIL-V-5524B(2) | Valve, Check, Hydraulic, Aircraft, Type I Systems (Asg) | NPFC | 1959 |
| MIL-V-19069A(1) | Valve, Check, Hydraulic, Aircraft, Type II Systems (Asg) | NPFC | 1959 |
| MIL-V-25675B(4) | Valve, Check, Miniature, Hydraulic, Aircraft and Missile | NPFC | 1972 |
| MIL-V-23254A | Valve, Control | NPFC | 1970 |
| MIL-V-26880B | Valve, Diaphragm Actuated Hydraulic Operated, Pressure Reducing and Check | NPFC | 1971 |
| MIL-V-5529A(2) | Valve, Hydraulic Directional Control | NPFC | 1956 |
| MIL-V-52688 | Valve, Hydraulic, Directional Control, Spool Type, Manually Operated | NPFC | 1970 |
| MIL-V-81073(7) | Valve, Hydraulic, Four Way Electrically Actuated (For Torpedo MA 46 Mod 0) | NPFC | 1967 |
| MIL-V-23255(1) | Valve, Pressure Relief | NPFC | 1965 |
| MIL-V-55341 | Valve, Pressure Relief | NPFC | 1971 |
| MIL-V-52687(1) | Valve, Relief, Hydraulic Pressure | NPFC | 1972 |

NFPA/T3.5.27

| Identification Number | Title | Source | Date of Issue |
|---|---|---|---|
| MIL-V-55230(1) | Valve, Relief, Hydraulic Pressure | NPFC | 1971 |
| MIL-V-27162 | Valve, Servo Control, Electro-Hydraulic, General Specification for | NPFC | 1959 |
| MIL-V-81072(5) | Valve, Servo, Electro-Hydraulic, Four Way (For Torpedo Mk 46 Mod 0) | NPFC | 1969 |
| MIL-V-46720(1) | Valve, Solenoid, Hydraulic | NPFC | 1963 |
| MIL-V-46727A | Valve, Three-Position, Four-Way, Hydraulic, Manual | NPFC | 1969 |
| MIL-V-46725A | Valve, Two-Position, Four-Way, Hydraulic, Manual | NPFC | 1969 |
| MIL-V-8566A | Valves, Aircraft Hydraulic Flow Regulator | NPFC | 1964 |
| MIL-V-5580B(4) | Valves, Aircraft Hydraulic Shuttle | NPFC | 1970 |
| MIL-V-5519C(1) | Valves, Aircraft Hydraulic Unloading | NPFC | 1971 |
| MIL-V-5525C | Valves, Aircraft Power Brake | NPFC | 1959 |
| MIL-V-8313 | Valves, Aircraft, Hydraulic Pressure Relief Type II Systems (Asg) | NPFC | 1957 |
| MIL-V-5527A | Valves, Aircraft, Hydraulic Thermal Expansion Relief (Supersedes AN-V-28) | NPFC | 1951 |
| MIL-V-19067A | Valves, Check, Controllable, Hydraulic Aircraft Type II Systems (Asg) | NPFC | 1957 |

NFPA/T3.5.27

| Identification Number | Title | Source | Date of Issue |
|---|---|---|---|
| MIL-V-5528A | Valves, Hydraulic Controllable Check | NPFC | 1951 |
| MIL-V-7915(2) | Valves, Hydraulic, Directional Control, Slide Selector | NPFC | 1956 |
| MIL-V-46726A | Valves, Restrictor, Check, Hydraulic | NPFC | 1969 |
| MIL-V-19068A | Valves, Shuttle, Hydraulic, Aircraft, Type II Systems (Asg) | NPFC | 1957 |
| MS-28880 | Valve - Controllable Hydraulic Check, Rotating Action, 3000 psi | NPFC | 1951 |
| MS-28016B | Valve - Hydraulic Dual Thermal Expansion Relief | NPFC | 1953 |
| MS-28881A | Valve - Hydraulic, Directional Control, Slide Selector, 4-Way, 3000 psi | NPFC | 1953 |
| MS-28886C | Valve, Aircraft Hydraulic Flow Regulator | NPFC | 1961 |
| MS-24352 | Valve, Aircraft Hydraulic Restrictor, One-Way, Unfiltered | NPFC | 1957 |
| MS-24418 | Valve, Aircraft Hydraulic Restrictor, One-Way, Unfiltered | NPFC | 1957 |
| MS-24353 | Valve, Aircraft Hydraulic Restrictor, Two-Way, Filtered | NPFC | 1957 |
| MS-24419A | Valve, Aircraft Hydraulic Restrictor, Two-Way, Unfiltered | NPFC | 1957 |

NFPA/T3.5.27

| Identification Number | Title | Source | Date of Issue |
|---|---|---|---|
| MS-28764A | Valve, Check, Controllable, Rotary-Action, Hydraulic, 3000 psi, Type II Systems (Asg) | NPFC | 1957 |
| MS-28771A | Valve, Check, Hydraulic, Flared, 3000 psi, Type II Systems (Asg) | NPFC | 1957 |
| MS-28765A | Valve, Check, Hydraulic, Internal Ports, 3000 psi, Type II Systems (Asg) | NPFC | 1957 |
| MS-28050 | Valve, Check, Low Pressure, Internal Threads (Asg) | NPFC | 1955 |
| MS-24423B | Valve, Check, Miniature Hydraulic, Aircraft and Missile, Minus 65 Deg F to Plus 450 Deg F, 3000 psi, Flareless Tube | NPFC | 1968 |
| MS-28052B | Valve, Check, Pressure Gage Lines Bleeder and Filler (Asg) | NPFC | 1956 |
| MS-27611A | Valve, Hydraulic Aircraft Wheel Brake | NPFC | 1971 |
| MS-28890B | Valve, Hydraulic Check, 3000 psi, Flareless, Type I Systems | NPFC | 1956 |
| MS-28893 | Valve, Hydraulic Relief, Cylindrical, Type II Systems (Asg) | NPFC | 1957 |
| MS-28766A | Valve, Shuttle, Hydraulic, Direct Mounting, 3000 psi, Type II Systems (Asg) | NPFC | 1957 |

NFPA/T3.5.27

| Identification Number | Title | Source | Date of Issue |
|---|---|---|---|
| MS-28767A | Valve, Shuttle, Hydraulic, Internal Thread, Tube Fitting Outlet, 3000 psi, Type II Systems (Asg) | NPFC | 1957 |
| MS-28891A | Valve, Shuttle, Hydraulic, 3000 psi, Direct Mounting | NPFC | 1956 |
| MS-28892 | Valve, 3000 psi Hydraulic Check, Flareless, Type II Systems (Asg) | NPFC | 1956 |
| OSU-V-1 | Method of Determining and Reporting a Pressure Differential-Flow Characteristic of a Fluid Power Valve | OSU | 1972 |
| OSU-V-2 | Method of Determining and Reporting the External Control Characteristics of a Fluid Power Valve | OSU | 1972 |
| OSU-V-3 | Method for Determining the Metering Characteristics of a Fluid Power Directional Control Valve | OSU | 1972 |
| OSU-V-4 | Method for Determining a Leakage Characteristic of a Fluid Power Valve | OSU | 1972 |
| OSU-V-5 | Method for Determining the Steady State Output Flow - Input Flow Characteristics of a Fluid Power Control Valve | OSU | 1972 |
| OSU-V-6 | Method of Determining the Dynamic Pressure Response Associated with a Fluid Power Valve Subjected to a Specified Input of a Control Variable | OSU | 1972 |

NFPA/T3.5.27

| Identification Number | Title | Source | Date of Issue |
|---|---|---|---|
| OSU-V-7 | Method for Determining the Dynamic Pressure Response Associated with a Fluid Power Valve Subjected to a Specified Input of Flow | OSU | 1972 |
| OSU-V-8 | Method for Determining the Dynamic Flow Response Associated with a Fluid Power Valve Subjected to a Specified Flow Input | OSU | 1972 |
| OSU-V-9 | Method for Determining the Durability of a Fluid Power Valve | OSU | 1972 |
| OSU-V-10 | Method for Determining the Low Temperature Performance of a Fluid Power Valve | OSU | 1972 |
| OSU-V-11 | Method for Determining the Contaminant Sensitivity of a Fluid Power Valve | OSU | 1972 |
| OSU-V-12 | Method for Comparing the Mathematically Predicted Response of a Fluid Power Valve with its Actual Response | OSU | 1972 |
| OSU-V-13 | Method for Determining the Regulation Characteristics of a Fluid Power Pressure Control Valve | OSU | 1972 |
| OSU-V-14 | Method for Evaluating the Structural Integrity of a Fluid Power Control Valve | OSU | 1972 |
| SAE J117 | Method of Measuring and Reporting the Pressure Differential Flow Characteristics of a Hydraulic Fluid Power Valve | SAE | 1975 |

| Identification Number | Title | Source | Date of Issue |
|---|---|---|---|
| SAE J747b | Hydraulic Equipment Test Code for Directional Control Valves | SAE | 1975 |
| SAE J748 | Hydraulic Directional Control Valves for 3000 psi (Maximum) | SAE | 1960 |
| SP-6 | Finishes for Contact Faces of Pipe Flanges and Connecting End Flanges of Valves and Fittings | MSS | 1963 |
| SP-61 | Hydrostatic Testing of Steel Valves | MSS | 1961 |
| UL-429 | Standard for Electrically Operated Valves | UL | 1973 |
| UL 1002 | Standard for Electrically Operated Valves for Use in Hazardous Locations, Class I, Groups A, B, C and D; Class II, Groups E, F and G | UL | 1972 |

NFPA/T3.5.27

# NFPA AND ANSI/B93 HYDRAULIC VALVE STANDARDS

In addition to the standards listed in this bibliography, the following NFPA and ANSI/B93 hydraulic valve standards have been published:

- Symbols for Marking Electrical Leads and Ports on Fluid Power Valves, ANSI/B93.9-1969 (R1975)

- Dimensions for Mounting Surfaces of Sub-Plate Type Hydraulic Fluid Power Valves, ANSI/B93.7-1968

- Series of Mounting Interfaces for 4567 Maximum psi (315 bar) Four Port Hydraulic Fluid Power Directional Valves, ANSI/B93.40-1976

These standards are available from the National Fluid Power Association, Inc., 3333 North Mayfair Road, Milwaukee, Wisconsin 53222.

# NOTES

NFPA Bibliography
T3.12.9-1977

---

**AN INDUSTRY STANDARD FOR FLUID POWER**

A Bibliography of

Fluid Power Pneumatic FRL Standards

---

published by
## NATIONAL FLUID POWER ASSOCIATION, INC.

3333 N. Mayfair Road  /  Milwaukee, WI 53222  /  414-778-3344  /  TLX 26898

## SOURCE OF DOCUMENTS

| | |
|---|---|
| AAR | Association of American Railroads<br>1920 L Street, N.W.<br>Washington, D.C. 20036 |
| AGA | American Gas Association<br>1515 Wilson Blvd.<br>Arlington, Virginia 22209 |
| ANSI | American National Standards Institute<br>1430 Broadway<br>New York, New York 10018 |
| FCI | Fluid Controls Institute, Inc.<br>P.O. Box 3854<br>Tequesta, Florida 33458 |
| ISO | International Standards Organization<br>1, rue de Varembe<br>1211 Geneva 20, SWITZERLAND |
| JIC | Joint Industrial Council<br>7901 Westpark Drive<br>McLean, Virginia 22101 |
| NFPA | National Fluid Power Association<br>3333 North Mayfair Road<br>Milwaukee, Wisconsin 53222 |
| NPFC | Commanding Officer, Naval Publications and Forms Center<br>5801 Tabor Avenue<br>Philadelphia, Pennsylvania 19120<br>Attn: NPFC 105 |
| SAE | Society of Automotive Engineers<br>400 Commonwealth Drive<br>Warrendale, Pennsylvania 15096 |
| UL | Underwriters Laboratories, Inc.<br>333 Pfingsten Road<br>Northbrook, Illinois 60062 |

NFPA/T3.12.9

USGPO    United States Government
U.S. Government Printing Office
Washington, DC 20402

Note: This bibliography provides a list of standards related to fluid power pneumatic FRLs which have been issued by NFPA and other standards-writing organizations. Listing in this bibliography is only for reference and does not imply Association endorsement.

NFPA/T3.12.9

| Identification Number | Title | Source | Date of Issue |
|---|---|---|---|
| AAR CSLT | Standard for High Efficiency Particulate Air Filter Units | AAR | 1968 |
| AGA X-50865 | Service-Type Regulator Specifications | AGA | |
| ANSI/B93.13[1] | Fluid Power Industrial Type Air Line Pressure Regulators | ANSI NFPA | 1971 |
| 46 CFR 58.30 | Code of Federal Regulations, Title 46 - Shipping, Chapter 1 - Coast Guard, Department of Transportation, Subchapter F - Marine Engineering, Subpart 58.30 - Fluid Power and Control Systems | USGPO | 1975 |
| FCI 58-1 | Definitions of Regulator Capacities | FCI | |
| FCI 65-1 | Guide to Material Selection for Industrial Regulators | FCI | 1964 |
| FCI 71-1 | Standard Terminology for Regulators | FCI | 1971 |
| JIC P-1 | Pneumatic Standards for Industrial Equipment and General Purpose Machine Tools | JIC | 1975 |
| MIL F 82042A | Filter and Oiler, Airline | NPFC | 1970 |
| MIL F 83857 | Filter Element, Aircraft Bleed Air | NPFC | 1971 |
| MIL R 3295B | Regulator, Air Pressure, Protective Shelter | NPFC | 1958 |
| MIL R 9345 | Regulator, Air Pressure, Aircraft Cabin, Specifications | NPFC | 1957 |
| MIL R 13877B | Regulator, Pressure, Compressed Gas | NPFC | 1968 |

- 462 -

NFPA/T3.12.9

| Identification Number | Title | Source | Date of Issue |
|---|---|---|---|
| MIL R 19180 | Regulators, Pressure, Compressed Gas | NPFC | 1963 |
| MIL R 23057 | Regulator, Compressed Air, High Pressure (3500 PSI) | NPFC | 1962 |
| MS 18284 | Regulating and Control Valves, Accuracy of Regulation Requirement | NPFC | 1968 |
| NFPA/T3.12.2 | Requirements for Presenttation of Catalog Data, Fluid Compatibility, Cleaning Media, Markings and Dimensional Identification Codes for Fluid Power Air Line Filters | NFPA | 1975 |
| NFPA/T3.12.2.6 | Pressure Drop Characteristics, Test Procedure and Data Presentation for Fluid Power Air Line Filters | NFPA | 1975 |
| NFPA/T3.12.10 | Air Line Filter, Regulator and/or Lubricator Pressure Rating | NFPA | 1976 |
| SAE AS1297 | Reducers, Pneumatic Pressure, Missile | SAE | |
| UL 252 [1] | Gas, Pressure Regulators | UL | 1973 |

---

[1] Under Revision

NOTES

NFPA Bibliography
T3.21.5-1978

**AN INDUSTRY STANDARD FOR FLUID POWER**

A Bibliography of

Fluid Power Pneumatic Valve Standards

published by
# NATIONAL FLUID POWER ASSOCIATION, INC.

3333 N. Mayfair Road / Milwaukee, WI 53222 / 414-778-3344 / TLX 26898

## SOURCE OF DOCUMENTS

| | |
|---|---|
| AAR | Association of American Railroads<br>1920 L Street, N.W.<br>Washington, DC  20036 |
| ANSI | American National Standards Institute<br>1430 Broadway<br>New York, NY  10018 |
| CETOP | European Committee for Oilhydraulics and Pneumatic Cont<br>c/o S.C.T.H.P.<br>10 Avenue Hoch<br>F 75382 Paris  Cedex 08<br>FRANCE |
| CSA | Canadian Standards Association<br>178 Rexdale Boulevard<br>Rexdale, Ontario  M9W  1R3<br>CANADA |
| FCI | Fluid Controls Institute, Inc.<br>Plaza 222   U.S. Highway 1<br>P.O. Box 3854<br>Tequesta, FL  33458 |
| ISA | Instrument Society of America<br>400 Stanwick Street<br>Pittsburg, PA  15222 |
| JIC | Joint Industrial Council<br>7901 Westpark Drive<br>McLean, VA  22101 |
| NFPA | National Fluid Power Association, Inc.<br>3333 North Mayfair Road<br>Milwaukee, WI  53222 |
| NPFC | Commanding Officer, Naval Publications and Forms Center<br>5801 Tabor Avenue<br>Philadelphia, PA  19120<br>Attn: NPFC 105 |

NFPA/T3.21.5

## SOURCE OF DOCUMENTS (continued)

SAE        Society of Automotive Engineers
             400 Commonwealth Drive
             Warrendale, PA    15096

UL          Underwriters Laboratories, Inc.
             333 Pfingsten Road
             Northbrook, IL    60062

USGPO      United States Government
             U.S. Government Printing Office
             Washington, D.C.    20402

---

Note: This bibliography provides a list of standards related to fluid power pneumatic valves which have been issued by NFPA and other standards-writing organizations. Listing in this bibliography is only for reference and does not imply Association endorsement.

---

NFPA/T3.21.5

| Identification Number | Title | Source | Date of Issue |
|---|---|---|---|
| AAR/SM247 | Lubricants to Use in Electro-Pneumatic Valves and Cylinders | AAR | 1949 |
| ANSI/B16.104 (FCI 70-2) | Control Valve Seat Leakage | ANSI | 1976 |
| ANSI/B93.9 | Symbols for Marking Electrical Leads and Ports on Fluid Power Valves | NFPA (ANSI) | 1969 (R 1975) |
| ANSI/B93.33 | Interfaces for 4-Way General Purpose Industrial Pneumatic Directional Control Valves | NFPA (ANSI) | 1974 |
| CETOP/RP 19P | Pneumatic Directional Control Valves | CETOP | 1965 |
| CETOP/RP 20P | Pneumatic Flow Control Valves | CETOP | 1965 |
| CETOP/RP 21P | Pneumatic Pressure Control Valves | CETOP | 1965 |
| CETOP/RP 22P | Pneumatic Shuttle, Non-Return and Quick-Exhaust Valves | CETOP | 1965 |
| CETOP/RP 23P | Pneumatic Pressure Intensifiers | CETOP | 1965 |
| CETOP/RP 32P | Sub-Plates for Directional Control Valves | CETOP | 1969 |
| CSA/C22.2 No. 139 | Electrically Operated Valves | CSA | 1973 |
| FCI 68-1 | Procedure in Rating Flow and Pressure Characteristics of Solenoid Valves For Gas Service | FCI | 1968 |

| Identification Number | Title | Source | Date of Issue |
|---|---|---|---|
| FCI 68-4 | Nomenclature and Definition for Space Heating Manual Supply Valves | FCI | 1968 |
| FCI 74-1 | Silent Check Valve Standard | FCI | 1974 |
| FCI 75-1 | Test Conditions and Procedure for Measuring Electrical Characteristics of Solenoid Valves | FCI | 1975 |
| ISA-S39.3 | Control Valve Sizing Equations for Compressible Fluids | ISA | 1973 |
| ISA-S39.4 | Control Valve Capacity Test Procedure for Compressible Fluids | ISA | 1974 |
| JIC P-1 | Pneumatic Standards for Industrial Equipment and General Purpose Machine Tools | JIC | 1975 |
| MIL-8712A | Valve, Air Relief, Low Pressure | NPFC | 1965 |
| MIL-V-19593 | Valve Control, Aircraft Pneumatic 28 V DC Solenoid Operated | NPFC | 1956 |
| MIL-V-6164C | Valve, Aircraft, Air, High Pressure | NPFC | 1975 |
| MS-28889H | Valve, Air, High Pressure Charging | NPFC | 1976 |

NFPA/T3.21.5

| Identification Number | Title | Source | Date of Issue |
|---|---|---|---|
| NFPA/T3.21.4 | Pneumatic Valve Pressure Rating Supplement No. 2 to NFPA Recommended Standard for Verifying the Fatigue and Static Pressure Ratings of the Pressure Containing Envelope of a Metal Fluid Power Component | NFPA | 1977 |
| NFPA/T3.21.7 | Defining Interface Surfaces for each Pneumatic Valve Interface in ANSI/B93.33-1974 | NFPA | 1976 |
| NFPA/T3.21.9 | Definition of Port Communication for the Fluid Power Valve Interface to NFPA Recommended Standard T3.21.1 With the Valve in Position in Response to a Remote Pilot Signal or Electrical Energization | NFPA | 1976 |
| SAE 513AS | Qualification Test for Aircraft Air Valves | SAE | 1960 |
| UL 429 | Electrically Operated Valves | UL | 1955 |
| UL 1002 | Electrically Operated Valves for Use in Hazardous Locations | UL | 1976 |
| 46 CFR 58.30 | Code of Federal Regulations, Title 46 - Shipping. Chapter 1 - Coast Guard Department of Transportation, Subchapter F - Marine Engineering, Subpart 58.30 - Fluid Power and Control Systems. | USGPO | 1975 |

NFPA Bibliography
T3.27.4-1979

## AN INDUSTRY STANDARD FOR FLUID POWER

A Bibliography of

Compressed Air Dryers Standards

published by
**NATIONAL FLUID POWER ASSOCIATION, INC.**

3333 N. Mayfair Road  /  Milwaukee, WI 53222  /  414-778-3344  /  TLX 26898

## SOURCE OF DOCUMENTS

| | |
|---|---|
| AAR | Association of American Railroads<br>1920 L Street, N.W.<br>Washington, DC 20006 |
| AMCA | Air Moving and Conditioning Association<br>30 West University Drive<br>Arlington Heights, IL 60004 |
| ANSI | American National Standards Institute<br>1430 Broadway<br>New York, NY 10018 |
| ARI | Air Conditioning and Refrigeration Institute<br>1815 North Fort Myer Drive<br>Arlington, VA 22209 |
| ASHRAE | American Society of Heating, Refrigerating and Air Conditioning Engineers<br>345 East 47th Street<br>New York, NY 10017 |
| ASTM | American Society for Testing and Materials<br>1916 Race Street<br>Philadelphia, PA 19103 |
| CDA | Copper Development Association<br>405 Lexington Avenue<br>New York, NY 10017 |
| CSA | Canadian Standards Association<br>178 Rexdale Blvd.<br>Rexdale 603<br>Ontario, Canada M9W IR3 |
| FIA | Factory Insurance Association<br>85 Woodland Street<br>Hartford, CT 06102 |
| ISA | Instrument Society of America<br>400 Stanwix Street<br>Pittsburg, PA 15222 |
| NEMA | National Electrical Manufacturers Association<br>2101 L Street N.W.<br>Washington, DC 20037 |

NFPA/T3.27.4

## SOURCE OF DOCUMENTS (continued)

NFPA
National Fluid Power Association, Inc.
3333 North Mayfair Road
Milwaukee, WI 53222

NPFC
Commanding Officer, Naval Publications and Forms Center
5801 Tabor Avenue
Philadelphia, PA 19120
Attn: NPFC 105

PNEUROP
Secretariat: BCAS
8 Leicester Street
London W. C. 2H 7 BL

SAE
Society of Automotive Engineers
400 Commonwealth Drive
Warrendale, PA 15095

TEM
Tubular Exchanger Manufacturers Association
331 Madison Avenue
New York, NY 10017

UL
Underwriters Laboratories, Inc.
333 Pfingsten Road
Chicago, IL 60062

---

Note: This bibliography provides a list of standards related to compressed air dryers which have been issued by NFPA and other standards-writing organizations. Listing in this bibliography is only for reference and does not imply Association endorsement.

NFPA/T3.27.4

| Identification Number | Title | Source | Date of Issue |
|---|---|---|---|
| AAR/EM36 | Procedure, Inspection and Maintenance for Cleaning Refrigerant Lines of Mechanical Refrigeration Systems | AAR | 1960 |
| AAR/EM37 | Pressure Test, Inspection | AAR | 1960 |
| AIR-STD-14/7F | Compressed Air Characteristic Supply Pressure and Hoses | NPFC | 1975 |
| AMCA/300 | Standard Test Code for Sound Rating Air Moving Devices | AMCA | 1967 |
| AMCA/302 | Guide of Sound Loudness Rating for Non-Ducted Air Moving Devices | AMCA | 1965 |
| AMCA/402 | Standard Air Density Ratios at Various Altitudes and Temperatures | AMCA | 1966 |
| AMCA/500 | Test Methods for Louvers, Dampers, and Shutters | AMCA | 1975 |
| ANSI/B31.5 | Standard Refrigeration Piping | ANSI | 1966 |
| ANSI/N101.1 | Standard for Efficiency Test of Air Cleaning Systems Containing Devices for Removal of particles | ANSI | 1972 |
| ARI/430 | Standard for Central Station Air Handling Units (Air Conditioning and Refrigeration) | ARI | 1974 |
| ARI/495 | Standard for Refrigerant Liquid Receivers | ARI | 1968 |
| ARI/515 | Sealed Refrigerant Compressors and Condensing Units, 20 Horsepower and Smaller | ARI | 1960 |

NFPA/T3.27.4

| Identification Number | Title | Source | Date of Issue |
|---|---|---|---|
| ARI/520 | Standard for Positive Displacement Refrigerant Compressor and Condensing Units | ARI | 1968 |
| ARI/730 | Application of Suction Line Filters and Filter Driers in Refrigeration and Air Conditioning Systems | ARI | 1974 |
| ARI/750 | Standard for Thermostatic Refrigerant Expansion Valves | ARI | 1970 |
| ASHRAE/12 | Refrigeration Terms and Definitions | ASHRAE | 1958 |
| ASHRAE/15 | Safety Code for Mechanical Refrigeration | ASHRAE | 1964 |
| ASHRAE/17 | Method of Rating and Testing Expansion Valves, Refrigerant | ASHRAE | 1966 |
| ASHRAE/22 | Method of Testing for Rating Water Cooled Refrigerant Condensers | ASHRAE | 1961 |
| ASHRAE/23 | Methods of Testing for Rating Refrigerant Compressors | ASHRAE | 1967 |
| ASHRAE/28 | Standard Method of Testing Flow Capacity of Refrigerant Capillary Tubes | ASHRAE | 1972 |
| ASHRAE/35 | Method of Testing Dessicants for Refrigerant Drying | ASHRAE | 1966 |
| ASHRAE/51 (AMCA 210-74) | Laboratory Methods of Testing Fans for Rating | ASHRAE | 1975 |
| ASHRAE/63 | Method of Testing Liquid Line Refrigerant Driers | ASHRAE | 1968 |

NFPA/T3.27.4

| Identification Number | Title | Source | Date of Issue |
|---|---|---|---|
| ASTM/B280 | Specifications for Seamless Copper Tube for Air Conditioning and Refrigeration Field Service | ASTM | 1971 |
| ASTM/C667 | Reflective Insulation Systems for Equipment and Pipe Operating at Temperatures Above Ambient Air | ASTM | 1971 |
| CDA/405-9 | Fittings Handbook including Standard Types for Plumbing and Heating, Air Conditioning and Refrigeration | CDA | 1965 |
| CSA/C22.2 No. 120 | Commercial Refrigerated Equipment | CSA | 1974 |
| FIA/XYZ36 | Protection of Paper Machines Equipped with High Velocity Hot Air Dryers | FIA | 1961 |
| ISA/S39.3 | Standard Control Valve Sizing Equations for Compressible Fluids | ISA | 1973 |
| MIL-A-51009A | Air Filter-Medium | NPFC | 1962 |
| MIL-D-83934A | Dryer, Air, Low Pressure | NPFC | 1972 |
| NEMA/MG1-1 | Standard for Definitions for Motors and Generators | NEMA | 1972 |
| NEMA/MG1 | Standard for Motors and Generators | NEMA | 1973 |
| NFPA/T3.27.1 | Glossary of Terms for Compressed Air Dryers | NFPA | 1972 |
| NFPA/T3.27.2 | Conditions for Rating Compressed Air Dryers | NFPA | 1975 |
| NFPA/T3.27.3 | Methods for Testing Compressed Air Dryers | NFPA | 1976 |

| Identification Number | Title | Source | Date of Issue |
|---|---|---|---|
| PNEUROP Committee 15 Report 678 | Compressed Air Dryers, Specification and Testing | PNEUROP | |
| SAE/J246 | Standard for Spherical Sleeve (Compression) Tube Fittings | SAE | 1971 |
| SAE/J513C | Standard for Refrigeration Flare Type Tube Fittings | SAE | 1967 |
| TEM/N2 | Standard Terminology for Tubular Heat Exchanger Components | TEM | 1968 |
| UL/109 | Tube Fittings for Flammable and Combustible Fluids, Refrigeration Service and Marine Use | UL | 1972 |
| UL/207 | Standard for Safety for Refrigerant Containing Components and Accessories | UL | 1973 |
| UL/303 | Safety Standard for Condensing and Compressor Units | UL | 1971 |
| UL/984 | Safety Standard for Sealed (Hermetic Type) Motor Compressors for Air Conditioning and Refrigeration Equipment | UL | 1971 |
| UL/1004 | Safety Standard for Electric Motors for Appliances and Equipment | UL | 1972 |

# NOTES

**NFPA Information Report**
**T3.28.11-1982**

## AN INDUSTRY STANDARD FOR FLUID POWER

NFPA Information Report

A Bibliography of

Fluid Logic Devices

published by
**NATIONAL FLUID POWER ASSOCIATION, INC.**
3333 N. Mayfair Road / Milwaukee, WI 53222 / 414-778-3344 / TLX 26898

NFPA/T3.28.11

## SOURCE OF DOCUMENTS

| | |
|---|---|
| ANSI | American National Standards Institute<br>1430 Broadway<br>New York, NY 10018 |
| CETOP | Eureopean Committee on Oil Hydraulic and Pneumatic Control<br>c/o AHEM<br>192 - 198 Vauxhall, Bridge Road<br>London SWIV IDX ENGLAND |
| NFPA | National Fluid Power Association, Inc.<br>3333 North Mayfair Road<br>Milwaukee, WI 53222 |
| NPFC | Commanding Officer, Naval Publications and Forms Center<br>5801 Tabor Avenue<br>Philadelphia, PA 19120<br>Attn: NPFC 105 |
| SAE | Society of Automotive Engineers<br>400 Commonwealth Drive<br>Warrendale, PA 15096 |

**Note:** This Bibliography provides a list of standards related to fluid logic devices which have been issued by NFPA and other standards-writing organizations. Listing in this Bibliography is only for reference and does not imply Association endorsement.

NFPA/T3.28.11

| Identification Number | Title | Source | Date of Issue |
|---|---|---|---|
| ANSI/B93.14 - 1971 (R1979) | Methods of Presenting Basic Performance Data for Fluidic Devices | NFPA (ANSI) | 1971 |
| ANSI/B93.38 - 1976 (R1981) | Method of Diagramming for Moving Parts Fluid Controls | NFPA (ANSI) | 1976 |
| CETOP/RP33P | Symbols and definitions for operations of logic and relative functions | CETOP | 1970 |
| CETOP/RP49R | Technological symbols for fluid logic and related devices with and without moving parts | CETOP | 1972 |
| MIL - L - 60196 | Logic Flip Flop | NPFC | 1968 |
| MIL - STD - 1306A | Fluerics Terminology and Symbols | NPFC | 1972 |
| MIL- STD-1361A | Fluidics Test Methods and Instrumentation | NPFC | 1973 |
| NFPA/T3.7.2 - 1968 (R1980) | Graphic Symbols for Fluidic Devices and Circuits | NFPA | 1968 |
| SAE ARP 993A - 1967 (R1969) | Fluidic Technology | SAE | 1967 |